Molecular Biology Intelligence Unit

The Genetic Code and the Origin of Life

Lluís Ribas de Pouplana, Ph.D.
The Scripps Research Institute
La Jolla, California, U.S.A.
and
ICREA and Barcelona Institute for Biomedical Research
Barcelona Science Park, Barcelona, Spain

Landes Bioscience / Eurekah.com
Georgetown, Texas
U.S.A.

Kluwer Academic / Plenum Publishers
New York, New York
U.S.A.

The Genetic Code and the Origin of Life

Molecular Biology Intelligence Unit

Landes Bioscience / Eurekah.com
Kluwer Academic / Plenum Publishers

Kluwer Academic / Plenum Publishers, 233 Spring Street, New York, New York, U.S.A. 10013
http://www.wkap.nl/

Please address all inquiries to the Publishers:
Eurekah.com / Landes Bioscience, 810 South Church Street, Georgetown, Texas, U.S.A. 78626
Phone: 512/ 863 7762; FAX: 512/ 863 0081
http://www.eurekah.com
http://www.landesbioscience.com

ISBN 978-1-4419-3413-0 e-ISBN 978-0-387-26887-3

The Genetic Code and the Origin of Life edited by Lluís Ribas de Pouplana. Landes / Kluwer dual imprint. Landes series: Molecular Biology Intelligence Unit.

Library of Congress Cataloging-in-Publication Data

The genetic code and the origin of life / [edited by] Lluís Ribas de Pouplana.
p. ; cm. -- (Molecular biology intelligence unit)
Includes bibliographical references and index.

1. Evolutionary genetics. 2. Genetic code. I. Ribas de Pouplana, Lluís. II. Series: Molecular biology intelligence unit (Unnumbered)
[DNLM: 1. Genetic Code. 2. Evolution, Molecular. QH 450.2 G3275 2004]
QH390.G445 2004
572.8'633--dc22

2004015694

To Berta, Bernat, and Irene.
The future of the origin.

CONTENTS

EDITOR

Lluís Ribas de Pouplana, Ph.D.
The Scripps Research Institute
La Jolla, California, U.S.A.
and
ICREA and Barcelona Institute for Biomedical Research
Barcelona Science Park
Barcelona, Spain
Chapter 8

CONTRIBUTORS

Ricardo Amils
Centro de Biología Molecular
UAM-CSIC
Universidad Autónoma de Madrid, Cantoblanco
Madrid, Spain
Chapter 7

Jeffrey L. Bada
Scripps Institution of Oceanography
University of California as San Diego
San Diego, California, U.S.A.
Chapter 1

Arturo Becerra
UNAM
Cd. Universitaria
México, D.F., México
Chapter 3

Oliver Botta
Scripps Institution of Oceanography
University of California as San Diego
San Diego, California, U.S.A.
Chapter 1

Carlos Briones
Centro de Astrobiología
Torrejón de Ardoz
Madrid, Spain
Chapter 7

James R. Brown
Bioinformatics Division
GlaxoSmithKline
Collegeville, Pennsylvania, U.S.A.
Chapter 2

Donald H. Burke
Indiana University
Bloomington, Indiana, U.S.A.
Chapter 4

Luis Delaye
UNAM
Cd. Universitaria
México, D.F., México
Chapter 3

Jocelyne DiRuggiero
Department of Cell Biology and Molecular Genetics
University of Maryland
College Park, Maryland, U.S.A.
Chapter 11

Jonathan Filée
Institut de Génétique et Microbiologie
Université Paris-Sud XI
Orsay, Cedex, France
Chapter 10

Patrick Forterre
Institut de Génétique et Microbiologie
Université Paris-Sud XI
Orsay, Cedex, France
Chapter 10

George E. Fox
Department of Biology and Biochemistry
University of Houston
Houston, Texas, U.S.A.
Chapter 6

Stephen J. Freeland
Department of Ecology
and Evolutionary Biology
Princeton University
Princeton, New Jersey, U.S.A.
Chapter 13

Rob D. Knight
Department of Molecular, Cellular
and Developmental Biology
University of Colorado
Boulder, Colorado, U.S.A.
Chapter 5, 13

Laura F. Landweber
Department of Ecology
and Evolutionary Biology
Princeton University
Princeton, New Jersey, U.S.A.
Chapter 13

Antonio Lazcano
UNAM
Cd. Universitaria
México, D.F., México
Chapter 3

David R. Liu
Department of Chemistry
Harvard University
Cambridge, Massachusetts, U.S.A.
Chapter 14

Thomas J. Magliery
Yale University
Molecular Biophysics and Biochemistry
New Haven, Connecticut, U.S.A.
Chapter 14

Hannu Myllykallio
Institut de Génétique et Microbiologie
Université Paris-Sud XI
Orsay, Cedex, France
Chapter 10

Ashwinikumar K. Naik
Department of Biology and Biochemistry
University of Houston
Houston, Texas, U.S.A.
Chapter 6

Frank T. Robb
Center of Marine Biotechnology
University of Maryland Biotechnology
Institute
Baltimore, Maryland, U.S.A.
Chapter 11

Manuel A.S. Santos
Department of Biology
University of Aveiro
Aveiro, Portugal
Chapter 12

Paul Schimmel
The Scripps Institute
La Jolla, California, U.S.A.
Chapter 8

Mathias Sprinzl
Laboratorium für Biochemie
Universität Bayreuth
Bayreuth, Germany
Chapter 9

Mick F. Tuite
Department of Biosciences
University of Kent
Kent, United Kingdom
Chapter 12

Michael Yarus
Department of Molecular, Cellular
and Developmental Biology
University of Colorado
Boulder, Colorado, U.S.A.
Chapter 5

FOREWORD

Early Thoughts on RNA and the Origin of Life

The full impact of the essential role of the nucleic acids in biological systems was forcefully demonstrated by the research community in the 1950s. Although Avery and his collaborators had identified DNA as the genetic material responsible for the transformation of bacteria in 1944, it was not until the early 1950s that the Hershey-Chase experiments provided a more direct demonstration of this role. Finally, the structural DNA double helix proposed by Watson and Crick in 1953 clearly created a structural framework for the role of DNA as both information carrier and as a molecule that could undergo the necessary replication needed for daughter cells.

Research continued by Kornberg and his colleagues in the mid-1950s emphasized the biochemistry and enzymology of DNA replication. At the same time, there was a growing interest in the role of RNA. The 1956 discovery by David Davies and myself showed that polyadenylic acid and polyuridylic acid could form a double-helical RNA molecule but that it differed somewhat from DNA. A large number of experiments were subsequently carried out with synthetic polyribonucleotides which illustrated that RNA could form even more complicated helical structures in which the specificity of hydrogen bonding was the key element in determining the molecular conformation. Finally, in 1960, I could show that it was possible to make a hybrid helix. The RNA molecule polyadenylic acid and the DNA molecule polydeoxythymidylic acid could form a double helix, even though it was known that the conformation of the DNA backbone and the RNA backbone were different from each other. This suggested a molecular model for the production of an RNA strand on a DNA strand. Work by Hurwitz, Stevens and Weiss on purification of RNA polymerase eventually led to identification of the system in which DNA-dependent RNA could be synthesized. This set the stage early in 1961 for experiments by several groups that suggested that a rapidly-turning-over RNA molecule called messenger RNA was active in directing the synthesis of proteins on ribosomes, a result that was confirmed by experiments of Nirenberg and associates which showed that the RNA molecule polyuridylic acid directed the synthesis of polyphenylalanine on ribosomes.

In the early 1960s I was pondering the impact of all of this newly acquired knowledge on problems of evolution and the origin of life. The most popular ideas about the origin of life in the 1940s was that espoused by the Russian scientist Oparin, who expressed the belief that life began in coaccervates of polypeptide chains that formed specialized environments leading to the production of enzymatic activity and eventually to living systems. This was a view that I felt was likely to be incorrect since it did not explain the fundamental role of the nucleic acids in providing the information needed for specifying biological systems.

When the chemist and spectroscopist Michael Kasha was a Visiting Professor at Harvard for the 1960-1961 term, we spent some time doing experiments together. It was toward the end of his stay that he approached me to ask if I would contribute a chapter to a book that he and Bernard Pullman were editing as a Festschrift for Albert Szent-Gyorgyi. This book, with the title *Horizons in Biochemistry* (Academic Press, New York, 1962), represented an opportunity for me to put down a number of thoughts that I had about the origin of living systems. I wrote an article with the title "On the Problems of Evolution and Biochemical Information Transfer" (pp. 103-126). It presented a brief overview of the way information was transmitted from DNA to RNA and eventually to directing the synthesis of proteins through the interaction of transfer RNA molecules with messenger RNA in ribosomes. In this essay, I stressed the fact that life could not have originated with protein molecules since it could not explain how nucleic acids came to control protein synthesis. I stressed that it was more likely that polynucleotides were the origin of living systems. I postulated a primitive environment in which polynucleotide chains are able to act as a template or as a somewhat inefficient catalyst for promoting the polymerization of the complementary nucleotide residues to build up an initial two-stranded molecule. Such an inefficient system could be followed by denaturation of the nucleic acid duplex and continuation of the process which would ultimately lead to an increasingly larger number of nucleic acid polymers. I then outlined various ways in which the polymerization of nucleic acids might be coupled with an inefficient polymerization of amino acids. Of key importance in this view was the development of primitive activating enzymes that would begin to relate a specific nucleic acid sequence to the assembly of specific amino acids. Thus, "life" was viewed as starting with a coupling of nucleic acid polymerization and amino acid polymerization although it was stressed that the prototype of this reaction may have been in a form quite different from that which we observe today.

In another section I asked the question, why are there two nucleic acids in contemporary biological systems? It seemed reasonable to believe that both RNA and DNA stemmed from a common precursor. However, it was apparent in contemporary biological systems that DNA seems to act as a major carrier of genetic information, while the RNA molecule is used to convert the genetic information into actual protein molecules. However, I noted that RNA molecules are also able to carry genetic information as in RNA-containing viruses. Thus, it seemed reasonable to speculate that the first polynucleotide molecule was initially an RNA polymer that was able to convey genetic information as well as organize amino acids into specific sequences to make proteins.

This article, published in 1962, was probably the first statement to suggest that RNA was the fundamental nucleic acid involved in the origin of living systems.

It is interesting to note that, in this same essay, I discussed the method by which the newly discovered messenger RNA was made. I suggested the possibility that messenger RNA may be made in vivo as complementary copies of one or both strands of DNA. If both strands are active, then the DNA would produce two RNA strands, and only one of these might be active as messenger RNA in protein synthesis. The other strand, I speculated, might be a component of a control or regulatory system. This is probably the first statement of an anti-sense function for RNA molecules. It also suggests the possibility that RNA could have other regulatory functions.

These statements were published over 40 years ago. Today we have a wealth of information that strengthens the role of RNA in the early evolution of life. The discovery of ribozymes by Cech and the more recent discovery of micro-RNAs that have a variety of functions in controlling the development of biological systems suggests that these may be trace evidence of what has been called the "RNA world", meaning an era in early evolution in which RNA played a dominant role in both replication and in carrying out a number of chemical modifications leading to the organization of present-day biological systems.

Given our present much more extensive knowledge base concerning the role of nucleic acids in biological systems, it leaves open the issue of when "life" actually began. Preceding the arrival of RNA, there must have been an enormous complexity of chemical reactions. The recent research by Eschenmoser and colleagues points out the possible participation of the four carbon threose nucleic acid polymers as precursors of present-day RNA molecules. It is likely that we will not be able to define a precise event that led to the origin of life. Rather, we are likely to view a growing level of molecular complexity that eventually yields a system that we would call "living" but for which it would be very hard to define a unique point at which we can say that life began. Of course, a key element is the extent to which all of these processes in early evolutionary history were error prone. These errors provided the substrate for Darwinian selection since, among the errors in the system, some will create efficiencies that form the basis for selection that eventually provides the direction for molecular evolution.

Alexander Rich
Department of Biology
Massachusetts Institute of Technology
Cambridge, Massachusetts, U.S.A.

PREFACE

It has been said that everybody thinks that he understands evolution but nobody really does. A key word in the previous phrase is 'everybody'. Unlike many other disciplines of biology that are rarely pondered upon by laymen, interest in evolution far precedes the application of scientific method to the problem. The question 'where do we come from?' is the cradle of all forms of pantheism and continues to be one of the central questions of modern biology. The complexity of the problem is such that pessimistic predictions abound on our chances of ever understanding the origin of life. And yet, we are making extraordinary progress.

The field advances mainly through a punctuated cycle of large-scale theoretical predictions and breakthroughs followed by the slow gathering of experimental evidence. Among the most spectacular successes of the field one must surely count the prediction and exploration of the role of RNA, which has provided a satisfactory background to the origin of modern species and a crucial insight into extant cell metabolism. On the other hand, we are still far from understanding the transition from a chemical world to this RNA world that gave rise to life on earth.

This book deals mainly with the processes that led from an established RNA world to the modern rule of the genetic code. The first three chapters of the book describe general concepts regarding the origin of life. These are followed by two reviews of theoretical and experimental studies of the RNA world that illustrate our current understanding of this crucial phase of life evolution. Finally, the biochemistry and the evolution of the central components of the modern genetic code are dissected individually. The goal of this structure is to provide the reader with relevant and detailed information on each of these aspects of the evolution of the code within a temporal perspective of the general evolution of life.

I need to acknowledge the generosity of the authors that have contributed to this volume. They have written excellent reviews and have endured my demanding messages with considerable patience. Many people have read and commented chapters of the book or its general organization. I need to thank especially Ricard Amils, Gustaf Arrenhius, Hugues Bedouelle, Kirk Beebe, Oliver Botta, Jim Brown, Stephen Cusack, Patrick Forterre, Magali Frugier, Rezha Gadhiri, Jerry Joyce, Roshan Kumar, Antonio Lazcano, Leslie Orgel, John Reader, Julius Rebek, Manuel Santos, Paul Schimmel, Bill Waas, Malcolm White, and Xianglei Yang.

Finally, my warmest thanks to Landes Bioscience, Ron Landes, and Cynthia Conomos, for their efforts during the long process of publication of this volume.

Lluís Ribas de Pouplana
Barcelona 2004

PREFACE

Chapter 1

The Early Earth

Oliver Botta and Jeffrey L. Bada

Introduction

The Earth is so far the only place in the Universe where life is known to exist. Is the Earth special, or are there other places both in our own solar system and beyond where life may have originated and either became extinct or still exists today? Hopefully, in the not to distant future we may find out. During the coming decades, spacecraft will search for evidence of life on Mars and Jupiter's moon Europa, which are considered to be the most promising places for the existence of extant or extinct extraterrestrial life within our solar system. Using remote sensing techniques, we will also begin to look for signs of life's chemistry on the extrasolar planets, which seem to be omnipresent companions of many main sequence stars. If the conditions that resulted in the origin of life on Earth are common throughout the Universe, it seems almost certain that life must exist elsewhere. However, to evaluate whether the Earth is a unique place, or simply an average rocky planet around an average star, we must access what the Earth was like before life began and how these conditions contributed to the processes thought to be involved in the origin of life.

Raw Materials

The 'Big Bang' produced all the hydrogen now in the Universe, as well as about one third of the helium. In the early Universe there were no elements heavier than these simple light elements. Thus any solar systems that might have formed during those early distant times must have been made up primarily of gas giant planets. This implies that life, at least with respect to carbon based life as we know it, would not have arisen during the Universe's early existence.

And yet, about 10 billion years (Gyr) after the "Big Bang", the nebula that produced our own solar system had a complete inventory of the chemical elements, ranging from the light to the very heavy. Rocky planets could form from these materials, and carbon, oxygen, nitrogen, iron, sulfur and other trace elements that make life possible were present in abundance. Where did these other elements come from?

The main reaction by which stars like our Sun obtain their energy is the fusion of four hydrogen atoms into a ^{4}He nucleus, a process called 'hydrogen burning'. As stars age, their hydrogen is eventually consumed and exhausted and the star then swells into a bloated object called a Red Giant. It is in Red Giant stars that the next stage of energy and element production begins by a process called "helium burning" in which carbon is the principal product. Carbon production involves first the fusion of two ^{4}He atoms to produce an unstable ^{8}Be nucleus, which in the interiors of hot helium burning stars exists long enough to fuse with another ^{4}He atom, yielding a ^{12}C nucleus. Fusion of another ^{4}He atom with a ^{12}C atom yields ^{16}O. All carbon and oxygen that are present in all the organic compounds on Earth and in the Universe were once part of Red Giant stars.

Essentially all of the remaining chemical elements, including the very heaviest ones, are synthesized during the death of a star. In fact, most of the heavy elements are created during

The Genetic Code and the Origin of Life, edited by Lluís Ribas de Pouplana.

great cosmic explosions, called 'supernovae', that are the inevitable fate of aged giant stars that are at least 5-10 times more massive than our Sun. While much of this process used to be merely informed speculation on the part of astronomers, it has now actually been seen taking place. A supernova in a satellite galaxy of our own Milky Way was observed in great detail in 1987 using state-of-the-art astrophysical methods, and these observations confirmed that heavy elements had been made in abundance during the explosion.[1]

As stars age and die, their elemental waste products are strewn throughout the Universe and become incorporated into accumulating interstellar clouds. Parts of these clouds eventually collapse, and contract to form denser clumps of material, which contain dense cores, the localized sites of star formation within the cloud. Further infall of material leads to the formation of protostars, and finally young stars. These star forming regions are characterized by the presence of one or several young stars that have a very high flux in the ultraviolet part of the electromagnetic spectrum. Some new stars in the well-known Orion nebula observed with the Hubble Space Telescope showed dark disks of material around them that are thought to be the initial stages of forming planetary systems.[2]

In the recent years, many extrasolar planets and collapsing dust clouds have now been found that estimates of how many planets —and how many Earth-like planets— there might be in the Milky Way and in the Universe at large are continuously revised. It now seems that a large fraction of stars give rise to planetary systems some which probably have Earth-like planets. It thus appears that there are many potential places in the Universe where life could have arisen.

But, with respect to the origin of life on Earth, we still need to know some important things. How long ago did the Earth form? What was the juvenile Earth and solar system like? What does the early rock record on the Earth tell us about this ancient period of Earth history?

Formation of the Solar System

The formation of solar systems such as ours is thought to take place in a surprisingly short period of time. Within 10 to 20 Myr after the initial collapse of the interstellar cloud, planetesimals started to form in the inner region of the accretionary disk of dust and gas that surrounded our young star. Accretionary disks around stars, similar to the one that surrounded our young Sun, have been observed recently with the Hubble Space Telescope.[3] Astronomical observations of stars in their early evolutionary stage have shown that the lifetime of the accretionary disks is $\sim 10^7$ years,[4,5] in agreement with the estimates from our own solar system.

Generally, the current scenario for the formation of the planets and smaller objects in our solar system suggests that within the residual accretionary disk small bodies and dust particles began to stochastically accumulate to form larger and larger planetesimals. The orbits of these planetesimals, of which thousands had formed, were not circular, but eccentric, leading to gravitational interactions and collisions. As these interactions continued, planetary embryos were formed by low-velocity collisions, and due to their increased gravitational pull, these began to be the dominant bodies within the disk. Computer simulations of these dynamical processes show that in about 50% of the cases numerous small rocky planets formed near the star while larger gaseous planets formed further out.[5] Although there were originally many more rocky planets in the inner part of the solar system, most of them were either pulled into the Sun or ejected out of the solar system, leaving only the four present day terrestrial planets, Mercury, Venus, Earth and Mars, in the inner solar system.

A significant number of extrasolar systems have been detected in the last several years (for a review see ref. 7). However, these systems appear to be very unlike our solar system, with large Jupiter-sized planets in very close orbits around the parent stars, implying that there may be completely different solar system formation mechanisms. The validity of the formation hypothesis for our own system, and therefore the formation of terrestrial planets that can harbor life, is still based on only our own case. Future planned space missions such as the NASA Terrestrial Planet Finder (TPF) or the ESA Eddington and Darwin spacecraft will be designed to detect Earth-like planets around other stars and help us to learn more about the abundance and formation of extrasolar planets, and perhaps their habitability

Formation of the Earth

We know today from the uranium/lead dating method of meteorites that the solar system, and by inference the Earth, is 4.55 ± 0.11 Gyr old. Our planet probably formed from planetesimals of material that has a composition similar to that of meteorites known as ordinary chondrites. This primitive class of meteorites contains less volatiles than other solar system objects such as carbonaceous chondrites and comets and are probably representative of the material that formed in region around 1 Astronomical Unit (AU), the distance of the Earth to the Sun.

Other types of meteorites have provided additional information about the formation of our planet. The meteorite types achondrites, stony-iron and iron meteorites are products of melting and differentiation that took place on their parent bodies and these meteorites can therefore provide constraints on the timescales over which planetesimal accretion and subsequent differentiation took place.[8] Recent results indicate that these processes occurred fast on meteorite parent bodies, but the inference to the large planets such as the Earth or Mars is not straightforward, mainly due to the controversy over whether the core formation in these planets was a single global melting event or if the accreting planetesimals were already differentiated. Other work suggests that core formation in the Earth took place < 80 Myr after differentiation of the planetesimals in which iron meteorites formed.[9,10] Based on Hf-W (hafnium-tungsten) systematics of SNC meteorites, which are widely believed to be samples of the Martian crust, it was concluded that core formation on Mars took place within the first ~30 Myr of solar system history.[11]

The Earth is the only terrestrial planet with a large moon. The current favored scenario for the formation of the moon involves a collision, around 50 Myr after the formation of the inner planets, between the Earth and a planetesimal about the size of Mars.[12,13] Some of the debris from that collision went into a close orbit, probably around 25,000 km, around the Earth and eventually aggregated to form the moon. Due to this close distance, and to the much higher spin rate of the Earth at that time, tidal forces were three hundred times stronger than today, greatly deforming the freshly formed crust, and if global oceans existed, causing oceanic tides much higher than today. These tides might have played a significant role in the formation of life on Earth, as we will discuss later. If this collision theory for the origin of the moon is indeed correct, it also provides indirect support to the hypothesis that there were probably many smaller planetary bodies that formed in the inner solar system other than the four that are present today.

Even after the moon formed, the Earth was still bombarded by bolides 10s of km in diameter at a frequency of about one collision every 1000 years or so. These impacts, as well as the decay of radioactive elements in the Earth's interior, caused the planet to stay in a molten state for a few million years or less after its initial accretion.[14] During this so-called "Hadean" period, the temperatures of the Earth were so high that the heavier elements sank towards the center of the planetary body forming the core, leaving the lighter elements to form the mantle and crust. As already mentioned, the timing of this differentiation is absolutely crucial for the origin of life, because it defined the oxidation state of the crust and consequently the composition of the atmosphere. The formation of the core itself is important for other reasons as well. On one hand, it depleted the crust and the mantle of the heavy elements and made carbon one of the most abundant elements in these regions. On the other hand, a metallic core is a prerequisite to the formation of a strong magnetic field around the planet. A magnetic field acts like a protection shield against the harsh particle radiation that is present everywhere in the solar system and the galaxy.

The oldest rocks that have been dated on the Earth are those found in the Acasta Gneiss complex in Canada, which were formed over 4 Gyr ago.[15] However, zircon crystals that are up to ~4.4 Gyr old have been extracted out of younger rocks from Australia.[16] The chemical and isotopic data obtained from one of the zircon crystals indicates the presence of large bodies of water, perhaps even oceans, as early as 4.3 to 4.4 Gyr, nearly a billion years earlier than the earliest evidence for cellular life on Earth.[16,17]

The Early Atmosphere, Ocean and Climate

During the Hadean period, the volatile compounds that were trapped inside the accreting planetesimals were released from the molten rock to form a secondary atmosphere. Any primary atmosphere (if one existed at all) must have been lost, as evidenced by the depletion of rare gases in Earth's atmosphere compared to cosmic abundances.[18] As a consequence of the simultaneous formation of Earth's core with accretion, the metallic iron was removed from the upper mantle, which would allow the volcanic gases to remain relatively reduced and produce a very early atmosphere that contained species such as CH_4, NH_3 and H_2. Since the temperature at the surface was high enough to prevent any water from condensing, the atmosphere would have consisted mainly of superheated steam along with these other gases.[19] However, this secondary atmosphere may have been lost several times during large impact events such as the one that formed the moon, and would have been replaced by further outgassing from the interior and resupply from later impactors.

It is believed that the impactors during the latter stages of the accretion process have originated from further out in the solar system and would have been comparable in composition to comets (see Fig. 1). Comets, whose volatile compounds are the most pristine materials surviving from the formation of the solar system, may have supplied a substantial fraction of the volatiles on the terrestrial planets, perhaps including organic compounds that may played a role in the origin of life on earth (see below). It has been suggested that the water present currently on the Earth was provided entirely from this source. However, recent measurements of the deuterium enrichment of water in comets Halley, Hyakutake and Hale-Bopp indicate that only a fraction of it was delivered by comets, whereas the largest fraction was trapped during the earlier accretionary phase.[20] The volatiles on comets are more oxidized than the ones in asteroids due to their similarity to interstellar ices and their higher water/rock ratio.

Without going into details on other effects such as ingassing of volatiles and loss of H_2 to space, it can be assumed that the atmosphere that developed on the Earth over the period 4.4 to 3.8 Gyr ago (perhaps several times if was erased by large impact events) was basically a mix of volatiles delivered by volatile rich impactors such as comets and outgassing from the interior of an already differentiated planet. This atmosphere was probably dominated by water steam until the surface temperatures dropped to ~100°C (depending on the pressure), at which point water condensed out to form early oceans.[16] The reduced species, which were mainly supplied by volcanic outgassing, are very sensitive to UV radiation that penetrated through the atmosphere due to the lack of a protective ozone layer. These molecules were probably destroyed by photodissociation, although there might have been steady state equilibrium between these two processes that allowed a significant amount of these reduced species to be present in the atmosphere. Overall, however, the atmosphere was dominated by oxidized species such as CO_2, CO and N_2. A similar atmosphere is present on Venus today, although it is much more dense than the atmosphere of the early Earth.

The climate on the early Earth at this stage depended mainly on two factors: the luminosity of the Sun and the radiative properties of the atmosphere. Standard theoretical solar evolution models predict that the Sun was about 30 % less luminous than today.[21] If the atmosphere of the early Earth was the same as it is now, the entire surface of the planet would have been frozen. However, as discussed extensively by Kasting,[18,19] a CO_2 rich atmosphere may have been present throughout the Hadean and Early Archean period and this would have resulted in a significant Greenhouse effect that would have prevented the oceans on the early Earth from freezing. Since there were probably no major continents existing during that period of time, silicate weathering (the long-term loss process on for CO_2 today) would have been low, and CO_2 would have been primarily contained in the atmosphere and ocean. Even with the assumption of a 70% present solar luminosity, a steady-state atmosphere containing ~ 10 bars of CO_2 could have resulted in to a mean surface temperature approaching 100°C

In summary, the current models for the early terrestrial atmosphere suggest that it consisted of a weakly reducing mixture of CO_2, N_2, CO, and H_2O with lesser amounts of H_2, SO_2, and

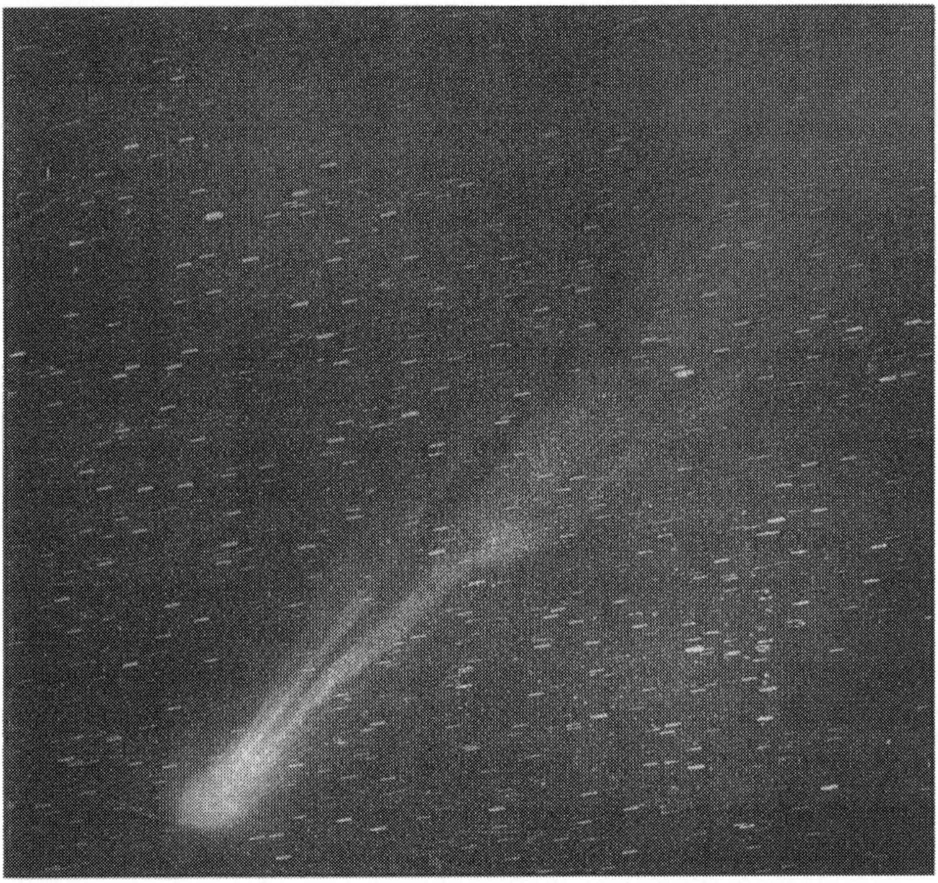

Figure 1. Comets, such as this one photographed in 1892 by E. R. Barnard (taken from R. S. Ball, "The Story of the Heavens", Cassell and Company LTD, London, Paris, New York, Melbourne, 1900), may have supplied the early Earth with some of the reagents (HCN, aldehydes/ketones, etc.) needed for the abiotic syntheses as well as some of the organic compounds needed for the origin of life.

H_2S. Reduced gases such as CH_4 and NH_3 are considered to be nearly absent or present only in localized regions near volcanoes or hydrothermal vents.

There is, however, the possibility that the CO_2 concentrations in the early atmosphere were not high enough to prevent the formation of an ice-covered ocean.[22] If this was the case, the thickness of the ice sheet has been estimated to be on the order of 300 m, which would have been thin enough to allow melting by an impactor of ~ 100 km in diameter. The frequency of impacts of such ice-melting bolides has been estimated to be one event every 10^5-10^7 years between about 3.6 and 4.5 Gyr ago, suggesting that there could have been periodic thaw-freeze cycles associated with the ice-melting impacts. The precursor compounds imported by the impactor or synthesized during the impact, such as HCN, would have been washed into the ocean during the thaw period. In addition, CH_4, H_2, CO and NH_3 derived from hydrothermal vents would have been stored in the unfrozen ocean below the ice layer which protected these gases from the ultraviolet radiation[23]. Following a large impact, the trapped gases would have been expelled into the atmosphere where they could have persisted for some time before they were destroyed by photochemical reactions.

Organic Compounds on the Early Earth

Today, organic compounds are so pervasive on the Earth's surface that it is hard to image the Earth devoid of organic material. However, during the period immediately after the Earth first formed some 4.5 Gyr ago, there would have been no organic compounds present on its surface. This was because soon after accretion, the decay of radioactive elements heated the interior of the young Earth to the melting point of rocks. Volcanic eruptions expelled molten rock and hot scorching gases out of the juvenile Earth's interior creating a global inferno. In addition, the early Earth was also being peppered by mountain-sized planetesimals, the debris left over after the accretion of the planets. Massive volcanic convulsions, coupled with the intense bombardment from space, generated surface temperatures so hot that the Earth at this point could very well have had an "ocean" of molten rock.

Although temperatures would have slowly decreased as the infall of objects from space and the intensity of volcanic eruptions declined, elevated temperatures likely persisted for a few hundred million years after the formation of the Earth.[16] During this period, temperatures would have probably been too hot for organic compounds to survive. Without organic compounds, life as we know could not exist. However, by roughly 4 Gyr ago, and perhaps even a lot earlier, the Earth's surface had cooled to the point that liquid water could exist and global oceans began to form.[16] It was during this period that organic compounds would have first started to accumulate on the Earth's surface, as long as there were natural pathways by which they could be synthesized, or sources from elsewhere that could supply them to the Earth.

The origin of life of life as we know it on Earth required the presence of liquid water and an inventory of prebiotic organic compounds from which could undergo further chemical processing so that life could emerge. Although it appears that liquid water in the form of large oceans was present on the surface of the early Earth,[16,17] the source of the required organic molecules on the primordial Earth is not clear.

Sources of Prebiotic Organic Compounds

Based on our current knowledge of the last common ancestor to all life today, the first organisms were probably some sort of heterotroph, which means that not only water and energy sources had to be available, but also a minimal set of organic compounds. In contrast, it would have been possible that the first organisms were autotrophic, meaning that they could convert CO_2 directly into the reduced organic molecules they need to live. But heterotrophic organisms are much simpler than autotrophic organisms, which require an elaborate array of protein biosynthesis reactions and enzymes in order to fix CO_2. Therefore, a heterotrophic origin of life has been widely accepted, mainly based on insights gained over the last fifty years.[23]

Nearly a century ago, the Russian biochemist Alexander Oparin[24] and the British biochemist J. B. S. Haldane[25] proposed a theory that organic compounds could have been synthesized on the early Earth when gases in the atmospheric were subjected to some type of energy. Inspired by these ideas, and with the assumption that the atmosphere of the early Earth was reducing as proposed by his mentor Harold Urey,[26] Stanley Miller was the first to experimentally demonstrate the possible synthesis of organic compounds under prebiotic conditions.[27] He constructed an apparatus in which he could simulate the interaction between an atmosphere and an ocean (Fig. 2). As an energy source, Miller chose a spark discharge, considered to be the second largest energy source, in the form of lightning and coronal discharges, on the early Earth after UV radiation. Miller filled the apparatus with various mixtures of methane, ammonia and hydrogen as well as water, which was then heated during the experiment. A spark discharge between the tungsten electrodes, which simulated lightning and corona discharges in the early atmosphere, was produced using a high frequency tesla coil with a voltage of 60,000 V. The reaction time was usually a week or so and the maximum pressure 1.5 bars. With this experimental setup, Miller was able to transform almost 50% of the original carbon (in the form of methane) into organic compounds. Although almost of the synthesized organic

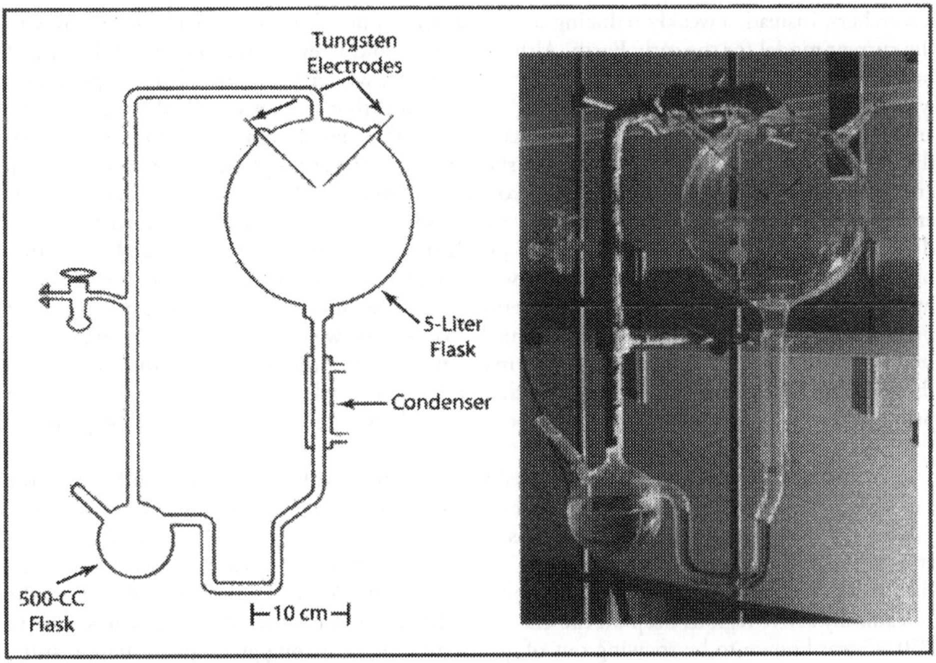

Figure 2. A diagram (right) of the original Miller-Urey apparatus showing the various components. A photograph of the actual apparatus is shown on the left (courtesy of Stanley L. Miller).

material was an insoluble tar-like solid, he was able to isolate amino acids and other simple organic compounds from the reaction mixture. Glycine, the simplest amino acid, was produced in 2% yield (based on the original amount of methane carbon), whereas alanine,[28] the simplest chiral amino acid, showed a yield of 1%. Miller was able to demonstrate that the alanine that was produced was a racemic mixture (equal amounts of D- and L-alanine). This provided convincing evidence that the amino acids were produced in the experiment and were not biological contaminants somehow introduced into the apparatus.

The other organic compounds that Miller was able to identify made it possible for him to propose a possible reaction pathway for the amino acids.[29] The proposed synthetic mechanism had actually been discovered in the mid-19th century by the German chemist A. Strecker.[30] It features the reaction of hydrogen cyanide, ammonia and carbonyl compounds (aldehydes or ketones) to form cyanohydrins, which would then undergo hydrolysis to form the α-amino acids. Depending on the concentration of ammonia in the reaction mixture, varying amounts of α-hydroxy acids are produced as well, which is what Miller found, with larger relative amounts of hydroxy acids being formed in a reaction mixture containing less ammonia.

It is important to note that neither purines nor pyrimidines, the nucleobases that are part of DNA and RNA, were investigated in the mixtures of the original Miller-Urey experiment. However, in experiments carried out soon after Miller's experiment by J. Oró and coworkers, the formation of adenine from ammonium cyanide solutions was demonstrated.[31,32] Later, it could be shown that the abiotic synthesis of purines and other heterocyclic compounds also works under the same conditions than the original Miller-Urey experiment, but in much smaller yields than the amino acids.[33] In addition, it has been found that guanine can be produced in a direct "one-pot" synthesis from the polymerization of ammonium cyanide.[34]

However, as should be clear from the earlier discussion, the atmospheric composition that formed the basis of the Miller-Urey experiment is not considered to be plausible by many researchers. Instead, a weakly reducing or neutral atmosphere is to be more in agreement with the current model for the early Earth. Although Miller and Urey originally rejected the idea of nonreducing conditions for the primitive atmosphere, Miller later carried out experiments with CO and CO_2 model atmospheres.[35,36] He found that not only were the yields of the amino acids reduced, but that glycine was basically the only amino acid synthesized under these conditions. He found a trend, which showed that as the atmosphere became less reducing and more neutral, the yields of synthesized organic compounds decreased drastically. The presence of methane and ammonia appeared to be especially important for the formation of a diverse mixture of amino acids. The main problem in the synthesis of amino acids and other biologically relevant organic compounds with nonreducing atmospheres is the formation of hydrogen cyanide (HCN), which is an intermediate in the Strecker pathway and an important precursor compound for the synthesis of nucleobases.[37] However, as mentioned earlier, localized high concentrations of reduced gases may have existed around volcanic eruptions and in these localized environments reagents such as HCN, aldehydes and ketones may have been produced, which after washing into the oceans could have become involved in the prebiotic synthesis of organic molecules.

Because of supposed problems associated with the direct Miller-Urey type syntheses on the early Earth, a completely different hypothesis for the "home-grown" synthesis of organic compounds has been proposed. The hypothesis is based on one of the great oceanographic discoveries of the 20th century. In February 1977, using the submersible research vessel *Alvin*, scientists from the Woods Hole Oceanographic Institution (WHOI) approached one of the volcanically active regions deep off the Galápagos Islands, where water at temperatures of up to 350 °C was known to be spewing out of geothermal vents. Strong gradients of temperature, acidity and chemical composition were found to be present at the vent/seawater interface. The big surprise came when the crew of the *Alvin* detected rich and complex communities of animals around the dark, chimney-like openings of the vents. From the analysis of samples brought back from these vents, it has shown that bacteria which oxidize the H_2S that spews from the vents form the low end of the food chain for these communities. These miniature deep-sea ecosystems are, however, not completely independent from solar radiation, since the chemosynthetic energy that the bacteria at these modern sites use to oxidize the sulfur-compounds comes from molecular oxygen (O_2), which was produced by photosynthesis at the ocean surface.

A group of researchers, known colloquially as 'ventists', believe that the remarkable properties of the hydrothermal vent environments, particularly their protection from the harsh conditions caused by large impact events, might have played an important role in the origin of life.[38] Since it is thought by some that the last common ancestoral organism of all extant life on Earth was a thermophile, several researchers have proposed the hypothesis that the organic compounds necessary for the origin of life were actually synthesized under vent conditions. For example, Shock and coworkers have calculated that thermodynamic-based equilibria favors the formation of compounds such as amino acids at hydrothermal vent temperatures,[39] especially in vents associated with off-axis systems.[40] However, at elevated temperatures associated with vent discharges, amino acids and other biomolecules have been found to rapidly decompose.[41,42] For example, amino acids are destroyed in time scales of minutes. The rate of hydrolysis for RNA at pH 7 extrapolated to elevated temperatures gives a half-life of 2 min at 250 °C for the hydrolysis of every phosphodiester bond; at 350 °C the half-life is 4 s. For DNA, the half-lives for depurination of each nucleotide at pH 7 are nearly the same as the hydrolysis rates for RNA.[42] It has been pointed out by Lazcano that if the origin of life was sufficiently long, all the complex organic compounds in the ocean, whether derived from hown-grown synthesis or from exogenous delivery, would be destroyed by passage through the hydrothermal vents.[43] It thus appears that hydrothermal vents are much more effective in regulating the concentration of critical organic molecules in the oceans rather than playing a significant role in their direct synthesis.

Prebiotic Organic Compounds from Space

Because of the difficulties discussed in the previous sections with the concept of "home-grown" synthesis of amino acids and nucleobases as a major source of these compounds for the origin of life, other hypotheses about the origin of these compounds were developed. In the early 1990s, Chyba and Sagan proposed that the exogenous delivery of organic matter by asteroids, comets and interplanetary dust particles (IDPs) could have played a significant role in seeding the early Earth with the compounds necessary for the origin of life.[44] They drew this conclusion from the knowledge about the organic composition of meteorites. It is important to note that, if this concept is valid, impacts on the early Earth not only created devastating conditions which made it difficult for life to originate, but at the same time perhaps delivered the raw material necessary for setting the stage for the origin of life. In an even wider view, this hypothesis could have profound implications on the abundance of life in the universe. The origin of the essential organic compounds needed for the origin of life is not constrained by the conditions on a particular planet. But rather organic compound synthesis is a ubiquitous process that takes place on primitive planetary bodies such as asteroids and comets. The possibility for the origin of life is thus considerably increased, provided the essential organic compounds are delivered intact to a planet that is suitable for further chemical evolution.

Carbonaceous chondrites, a class of stony meteorites, are considered to be the most primitive objects in the solar system in terms of their elemental composition, yet they feature a high abundance of carbon, more than 3 weight-% in some cases. The most extensively analyzed meteorites for organic compounds include the CMs Murchison (fell in 1969 in Victoria, Australia) and Murray (1950, Kentucky, USA) and the CI Orgueil (1864, France). The carbon phase is dominated by an insoluble fraction, with the rest being soluble compounds (Table 1). PAHs make up the majority (up to 80%) of the of the soluble organic matter, followed by the carboxylic acids, the fullerenes and amino acids, which are about an order of magnitude less abundant.[45] Other important compounds in context with the origin of life are the nucleobases. The purines adenine, guanine, xanthine and hypoxanthine have been detected as well as the pyrimidine uracil in concentrations of 200 to 500 parts per billion (ppb) in the CM chondrites Murchison and Murray and in the CI chondrite Orgueil.[46-48] In addition, a variety of other nitrogen-heterocyclic compounds including pyridines, quinolines and isoquinolines were also identified in the Murchison meteorite.[49]

More recently, it was found the CI type meteorites such as Orgueil contain a distinct amino acid composition in comparison to the CMs.[50] The simple amino acid mixture, consisting of just glycine and β-alanine, found in CI carbonaceous chondrites is interesting in the sense that it has been generally thought that a wide variety of amino acids were required for the origin of life. However, among the candidates for the first genetic material is peptide nucleic acid (PNA), a nucleic acid analogue in which the backbone does not contain sugar or phosphate moieties.[51,52] For the PNA backbone, achiral amino acids such as glycine and β-alanine, possibly delivered by CI type carbonaceous chondrites to the early Earth, may have been the only amino acids needed for the origin of life.

The Prebiotic Soup and the First Living Entities

The organic material on the early Earth before life existed, regardless of its source, would have likely consisted of an wide array of different types of compounds, including amino acids, nucleobases, fatty acids, aromatic and heteroaromatic hydrocarbons among others. How these abiotic organic constituents on the prebiotic Earth were assembled into the first living entities is highly contentious. In modern biological systems, these compounds are part of oligomeric or polymeric molecular structures needed for catalysis and replication, so simple abiotic compounds were only the starting points for the chemical evolution that followed. For example, amino acids are the monomeric building blocks for proteins and enzymes, the structural and catalytic units without which life as we know it can not exist. Also, DNA and RNA, the molecules that encode and transcribe the genetic information in all terrestrial organisms, are com-

Table 1. Abundances of soluble organic compounds found in meteorites. Amino acids concentrations have been determined for several CI and CM chondrites. All other data are for the CM chondrite Murchison (except those for the polycyclic aromatic hydrocarbons and the fullerenes, which are from Yamato-791198 and Allende, respectively). Taken from ref. 45.

Compound Class		Concentration (ppm)
Amino Acids		
	CM meteorites	17-60
	CI meteorites	~ 5
Aliphatic hydrocarbons		> 35
Aromatic hydrocarbons		3300
Fullerenes		> 100
Carboxylic acids		> 300
Hydroxycarboxylic acids		15
Dicarboxylic acids & Hydroxydicarboxylic acids		14
Purines & Pyrimidines		1.3
Basic N-heterocycles		7
Amines		8
Amides	linear	> 70
	cyclic	> 2
Alcohols		11
Aldehydes & Ketones		27
Sulphonic acids		70
Phosphonic acids		2

prised by mononucleotides, which contain nucleobases such as adenine, guanine, thymine, cytosine, and uracil, attached to a sugar-phosphate backbone. The most widely accepted scenario for the transition from abiotic to biotic chemistry is that the simple monomeric compounds present in the prebiotic soup somehow underwent polymerization, perhaps with the assistance of clays and minerals, and formed longer and longer chains or polymers which over time became increasingly more complex with respect to both their structures and properties.[53] Eventually, some of these polymers acquired the capacity to replicate, one of the fundamental and most important properties of living organisms.

What the first molecular self-replicating entity consisted of and how replication was accomplished is not known, but several suggestions have recently been made. The most likely candidates are nucleic acid analogs of DNA and RNA such as PNA. Stanley Miller and coworkers have found that the building blocks of PNA molecules, the nucleobase derivatives adenine- and guanine-N^9-acetic acid and uracil- and cytosine-N^1-acetic acid as well as and N-(2-aminoethyl)glycine (AEG), the molecule that makes up the PNA backbone, can be synthesized under likely prebiotic conditions and, under favorable conditions, could have been major constituents of the primitive milieu.[52] Still to be worked out are mechanisms for the polymerization of the monomers, but preliminary results indicate that AEG oligomerizes more efficiently at 100 °C than mixture of α-amino acids at higher temperatures. However, even with PNA-like genetic informational molecules stability may be a problem. This could be overcome because the stability is highly sequence-dependent and the breakdown may be partly alleviated by blocking or acetylating the N-terminus.[52] However, other possibilities need to be considered because there may be other backbones and bases that were more abundant and more efficient for early biotic replication.

Life as We Do Not Know It: Nonheterotrophic Hypotheses for the Origin of Life

There is a more radical hypothesis which has been proposed that discards the whole idea of a primordial soup. This 'metabolic-life" hypothesis, promoted primarily by Günter Wächtershäuser and coworkers in Germany, claims that life at the time of origin consisted of nothing more than a sequential series of chemical reactions that are catalyzed on mineral surfaces.[54] According to this theory, the first living systems on Earth were based a type of autotrophic metabolism of low-molecular weight constituents such as CO and CO_2 which are converted into biologically relevant compounds such as pyruvate at high temperature (100-250 °C)/high-pressure (0.2 - 200 MPa) vent-type conditions. The metabolism theory claims that life, at least in its beginnings, was nothing more than a continuous chain of mineral surface-associated self-sustaining chemical reactions with no requirement for genetic information. A primitive type of reductive citric acid cycle is often cited as a model. There is some experimental support for the hypothesis, although the conditions for the various individual reaction steps are very different,[55,56] and it remains to be established if the conditions used in these laboratory experiments are geophysically plausible and are therefore relevant to the origin of life. Of the various metabolic reaction schemes that have been proposed and investigated none have been demonstrated to be autocatalytic, nor are there any empirical indications that this indeed is even possible in a prebiotic context.

Whether a set of self-sustaining set of purely chemical reactions really constitutes a system that can be considered alive is debatable. Nevertheless, if self-sustaining reaction chains did arise on the early Earth, they could have played an important role in enriching the prebiotic soup in some molecules that were perhaps not readily synthesized by other abiotic reactions or derived from space. In this context, the metabolism theory can be viewed as simply a component of the prebiotic soup theory. But, regardless of its initial complexity, self-maintaining chemical-based metabolic life could not have evolved in the absence of a genetic replicating mechanism insuring the maintenance, stability, and diversification of its components. In the absence of any hereditary mechanisms, autotrophic reaction chains would have come and gone without leaving any direct descendants able to resurrect the process. Life as we know it consists of both chemistry and information. If metabolic life ever did exist on the early Earth, to convert it to life as we know it would have required the emergence of some type of information system under conditions that are favorable for the survival and maintenance of genetic informational molecules.[57]

Finally, in an even more radical theoretical hypothesis, it has been suggested that the origin of life did not occur on the surface of the Earth, but inside the crust. This "subterranean model" is based on the idea that there exists a biosphere that feeds off abiogenic hydrocarbons (petroleum) formed deep inside the Earth.[58] These supposed subterranean organisms are completely independent of photosynthesis. It is assumed that the habitability of the subsurface is enhanced in comparison to the surface, particularly during the early Archean. Therefore, it is concluded that there is a higher probability for the origin of life to occur in the deep subsurface and life then made its way to the surface when the surface became habitable after the end of the heavy bombardment. There are many fatal flaws in this hypothesis. For one, there is no evidence that abiotic hydrocarbons are formed deep within the Earth. In addition, according to the current view on the formation of the Earth which was discussed earlier, no primordial hydrocarbons would have survived the accretion process of the Earth (or any other planet) since the temperatures reached during accretion were high enough to decompose any reduced carbon compound in a short period of time.

Conclusions

The early history of the Earth, the first 100 Myr or so, was dominated by the hot accretion of the planet followed by a relatively rapid cooling. There is evidence for the presence of liquid water on the surface at 4.4 Gyr ago. The proto-atmosphere, if it existed at all, was probably reduced, but it was removed from the planet early on by large impact events. The first "real" atmosphere was produced by outgassing from the crust, and was dominated by oxidized gases such as CO_2, CO and N_2 with lesser amounts of H_2 and CH_4. Only trace amounts of oxygen were present. In such an atmosphere, organic compounds, and amino acids in particular, with spark discharge as an energy source ("Miller-Urey-type") are not produced. The influx of extraterrestrial organic compounds delivered by comets, asteroids, meteorites and IDPs may have been a major source of the compounds necessary for a heterotrophic origin of life on Earth. These compounds, perhaps in combination with other chemical products on the early Earth, may have been interacted on mineral surfaces in drying lagoons or other appropriate locations on a turbulent and chaotic surface to form more and more complex compounds. Eventually, through the process of chemical evolution, some of the compounds that developed the capability for catalytic activity and/or information storage (replication) became dominant. This scenario is a logical, but still a highly speculative pathway about the beginning of the first living entities, and there are still big gaps between the various stages that need to be filled. The following chapters address some of these remaining questions, such as the formation of the first nucleic acids and the role of replicating polypeptides in this scenario.

References

1. For a review, see: McCray R. Supernova 1987A revisited. Annu Rev Astron Astrophys 1993; 31:175-216.
2. For pictures and press releases see: http://oposite.stsci.edu/pubinfo/pictures.html
3. Schneider G, Smith BA, Becklin EE et al. NICMOS imaging of the HR 4796A circumstellar disk. Astrophys J 1999; 513:L127-L130.
4. Storm SE. The early evolution of stars. Rev Mex Astron Astrofis 1994; 29:23-29.
5. Natta A, Grinin VP, Mannings V. Properties and evolution of disks around premain-sequence stars of intermediate mass. In: Mannings V, Boss AP, Russell SS, eds. Protostars and Planets IV. Tucson, University of Arizona Press, 2000:559-587.
6. Marcy GW. Personal communication.
7. Marcy GW, Cochran WD, Mayor M. Extrasolar planets around main-sequence stars. In: Mannings V, Boss AP, Russell SS, eds. Protostars and Planets IV. Tucson, University of Arizona Press, 2000:1285-1311.
8. Wadhwa M, Russell SS. Timescales of accretion and differentiation in the early solar system: The meteoritic record. In: Mannings V, Boss AP, Russell SS, eds. Protostars and Planets IV. Tucson, University of Arizona Press, 2000:995-1018.
9. Lee DC, Halliday AN. Hafnium-tungsten chronometry and the timing of terrestrial core formation. Nature 1995; 378:771-774.
10. Halliday AN, Rehkämper M, Lee DC et al. Early evolution of the Earth and the Moon: New constraints for Hf-W isotope geochemistry. Earth Planet Sci Lett 1996; 142:75-89.
11. Lee DC, Halliday AN. Core formation on Mars and differentiated asteroids. Nature 1997; 388:854-857.
12. Hartmann WK, Philips RJ, Taylor GJ. Origin of the Moon. Houston, Lunar and Planetary Institute, 1986.
13. Lee DC, Halliday AN, Snyder GA et al. Age and origin of the Moon. Science 1997; 278:1098-1103.
14. Maher KA, Stevenson DJ. Impact frustration of the origin of life. Nature 1988;331:612-614.
15. Bowring SA, Williams IS. Prisocan (4.00-4.03 Ga) orthogneisses from northwestern Canada. Contrib Mineral Petrol 1999; 134:3-16.
16. Wilde SA, Valley JW, Peck WH et al. Evidence from detrital zircons for the existence of continental crust and oceans on the Earth 4.4 Gyr ago. Nature 2001; 409:175-178.
17. Mojzsis SJ, Arrhenius G, McKeegan KD et al. Evidence for life on Earth before 3,800 million years ago. Nature 1996; 384:55-59.
18. Kasting JF. Earth's early atmosphere. Science 1993; 259:920-926.

19. Kasting JF. Early evolution of the atmosphere and ocean. In: Greenberg JM, Mendoza-Gomez CX, Pirronello V, eds. The Chemistry of Life's Origin. The Netherlands: Kluwer Academic Publishers, 1993:149-176.
20. Meier R, Owen TC, Matthews HE et al. A determination of the HDO/H_2O ratio in comet C/ 1995 O1 (Hale-Bopp). Science 1998; 279:842-844.
21. Gilliland RL. Solar evolution. Global Planet Change 1989; 1:35-55.
22. Bada JL, Bigham C, Miller SL. Impact melting of frozen oceans on the early Earth: Implications for the origin of life. Proc Natl Acad Sci USA 1994; 91:1248-1250.
23. Lazcano A, Miller SL. The origin and early evolution of life: Prebiotic chemistry, the preRNA world, and time. Cell 1996; 85:793-798.
24. Oparin AI. Proiskhozhdenie zhiny. Moscow, 1924, translated as "The Origin of Life". In: Bernal JD, ed. The Origin of Life. Cleveland and New York: 1957:Appendix I, 199-234.
25. Haldane JBS. Rationalist annual (1929). Reprinted. In: Bernal JB, ed. The Origin of Life. Cleveland and New York, 1957:243-249.
26. Urey HC. The planets. New Haven: Yale University Press, 1952.
27. Miller SL. A production of amino acids under possible primitive Earth conditions. Science 1952; 117:528-529.
28. Miller SL. Production of some organic compounds under possible primitive Earth conditions. J Am Chem Soc 1955; 77:2351-2361.
29. Miller SL. The mechanism of synthesis of amino acids by electric discharges. Biochim Biophys Acta 1957; 23:480-489.
30. Strecker A. Über die künstliche Bildung der Milchsäure und einem neuen dem Glycocoll homologen Körper. Ann Chem 1850; 75:27.
31. Oró J. Synthesis of adenine from ammonium cyanide. Biochim Biophys Res Commun 1960; 2:407-412.
32. Oró J, Kimball AP. Synthesis of purines under possible primitive Earth conditions. I. Adenine from hydrogen cyanide. Arch Biochem Biophys 1961; 94:217-227.
33. Yuasa S, Flory D, Basile B et al. Abiotic synthesis of purines and other heterocyclic compounds by the action of electrical discharges. J Mol Evol 1984; 21:76-80.
34. Levy M, Miller SL. Oró J. Production of guanine from NH^4CN polymerizations. J Mol Evol 1999; 49:165-168.
35. Miller SL, Urey HC. Organic compound synthesis on the primitive earth. Science 1959; 130:245-251.
36. Schlesinger G, Miller SL. Prebiotic synthesis in atmospheres containing CH_4, CO, and CO_2. I. Amino Acids. J Mol Evol 1983; 19:376-382.
37. Ferris JP, Joshi PC, Edelson EH et al. HCN: A plausible source of purines, pyrimidines and amino acids on the primitive Earth. J Mol Evol 1978; 11:293-311.
38. Holm NG, Andersson EM. Abiotic synthesis of organic compounds under the conditions of submarine hydrothermal vents: A perspective. Planet Space Sci 1995; 43:153-159.
39. Shock EL. Geochemical constraints on the origin of organic compounds in hydrothermal systems. Orig Life Evol Biosphere 1990; 20:331-367.
40. Kelley DS, Karson JA, Blackman DK et al. An off-axis hydrothermal vent field near the Mid-Atlantic Ridge at 30° N. Nature 2001; 241:145-149.
41. Bernhardt G, Lüdemann HD, Jaenicke R et al. Biomolecules are unstable under "black smoker" conditions. Naturwissenschaften 1984; 71:583-586.
42. Miller SL, Bada JL. Submarine hot springs and the origin of life. Nature 1988; 334:609-611.
43. Lazcano A. The tempo and mode(s) of prebiotic evolution. In: Cosmovici CB, Bowyer S, Wertheimer D, eds. Astronomical and biochemical origins and the search for life in the universe. Editrice Compositori, 1997: 70-80.
44. Chyba C, Sagan C. Endogenous production, exogenous delivery and impact-shock synthesis of organic molecules: An inventory for the origins of life. Nature 1992; 355:125-132.
45. Botta O, Bada JL. Extraterrestrial organic compounds in meteorites. Surv Geophys 2002; 23:411-467.
46. Van der Velden W, Schwartz AW. Search for purines and pyrimidines in the Murchison meteorite. Geochim Cosmochim Acta 1977; 41:961-968.
47. Stoks PG, Schwartz AW. Uracil in carbonaceous meteorites. Nature 1979; 282:709-710.
48. Stoks PG, Schwartz AW. Nitrogen-heterocyclic compounds in meteorites: Significance and mechanisms of formation. Geochim Cosmochim Acta 1981; 45:563-569.
49. Stoks PG, Schwartz AW. Basic nitrogen-heterocyclic compounds in the Murchison meteorite. Geochim Cosmochim Acta 1982; 46:309-315.

50. Ehrenfreund P, Glavin DP, Botta O et al. Extraterrestrial amino acids in Orgueil and Ivuna: Tracing the parent body of CI type carbonaceous chondrites. Proc Natl Acad Sci USA 2001; 98:2138-2141.
51. Egholm M, Buchardt O, Nielsen PE et al. Peptide Nucleic Acid (PNA). Oligonucleotide Analogues with an Achiral Peptide Backbone. J Am Chem Soc 1992; 114:1895-1897.
52. Nelson KE, Levy M, Miller SL. Peptide nucleic acids rather than RNA may have been the first genetic molecule. Proc Natl Acad Sci USA 2000; 97:3868-3871.
53. Ferris JP, Hill Jr AR, Liu R et al. Synthesis of long prebiotic oligomers on mineral surfaces. Nature 1996; 381:59-61.
54. Wächtershäuser G. Life as we don't know it. Science 2000; 289:1307-1308.
55. Huber C, Wächtershäuser G. Activated acetic acid by carbon fixation on (Fe,Ni)S under primordial conditions. Science 1997; 245:245-247.
56. Cody GD, Boctor NZ, Filley TR et al. Primordial carbonylated iron-sulfur compounds and the synthesis of pyruvate. Science 2000; 289:1337-1340.
57. Bada JL, Lazcano A. Some Like It Hot, but Not the First Biomolecules. Science 2002; 296:1982-1983.
58. Gold T. The deep hot biosphere. New York: Springer-Verlag, 1999.

CHAPTER 2

Reconstructing the Universal Tree of Life

James R. Brown

Abstract

The universal tree of life depicts the evolutionary relationships of all living things by grouping them into one of three Domains of life; the Archaea (archaebacteria), Bacteria (eubacteria) and Eucarya (eukaryotes). The "canonical universal tree" topology is actually a composite of phylogenies based on single ribosomal RNA gene trees and duplicated, paralogous protein gene trees. The salient features of the canonical universal tree are: (1) all three Domains are mono/holophyletic; (2) Archaea and eukaryotes are sister groups with the Bacteria at the root; and (3) thermophilic bacteria are the earliest evolved bacterial lineage. Recent studies based on new genome sequence data suggest that the universal tree has been "uprooted" by extensive horizontal gene transfer (HGT). However, the scope of HGT is still unclear and reports of extensive *trans*-Domain HGT based on sequence homology, without supporting phylogenetic analysis, need careful reconsideration. Phylogenetic analysis of combined conserved proteins suggests that there is still underlying support for the concept of the universal tree.

Introduction

The universal tree of life is the depiction of the evolutionary relationships among all living organisms. The tacit supposition of the universal tree is that all living things are related genetically, however distant. Key support for this assumption comes from the subject of this book, the genetic code, which is ubiquitous with remarkably little variation. Furthermore, the basic processes of DNA replication, transcription and translation are preserved in all cells which adds support to the notion of common, if distant, origins.

While science has long attempted to classify living things, modern universal tree construction truly began with molecular evolutionary studies. Sixty years ago, Chatton[1] and Stanier and van Niel[2] proposed subdividing life into two fundamental groups, prokaryotes and eukaryotes (summarized in ref. 3). Later, the key features distinguishing prokaryotes from eukaryotes were better defined, namely, the lack of internal membranes (such as the nuclear membrane and endoplasmic recticulum), and replication by binary fission rather than mitosis.[4,5] However, neither detailed morphology nor extensive biochemical phenotyping provided sufficient phylogenetic signal for reconstructing evolutionary relationships among prokaryotic species let alone their relationships to eukaryotes.

In the late 1970s, Woese, Fox and coworkers initiated the field of molecular prokaryotic systematics by digesting in vivo labeled 16S ribosomal RNA (rRNA) using T1 ribonuclease to produce oligonucleotide "words" then analyzing the results data using dendograms. Their rRNA dendograms showed that some unusual methanogenic "bacteria" were significant offshoots from the main bacterial clade.[6] So deep was the split in the prokaryotes that Woese and Fox[7] named the methanogens and their relatives "archaebacteria", which relayed their distinctness from the true bacteria or "eubacteria" as well as met contemporary preconceptions that these

The Genetic Code and the Origin of Life, edited by Lluís Ribas de Pouplana.

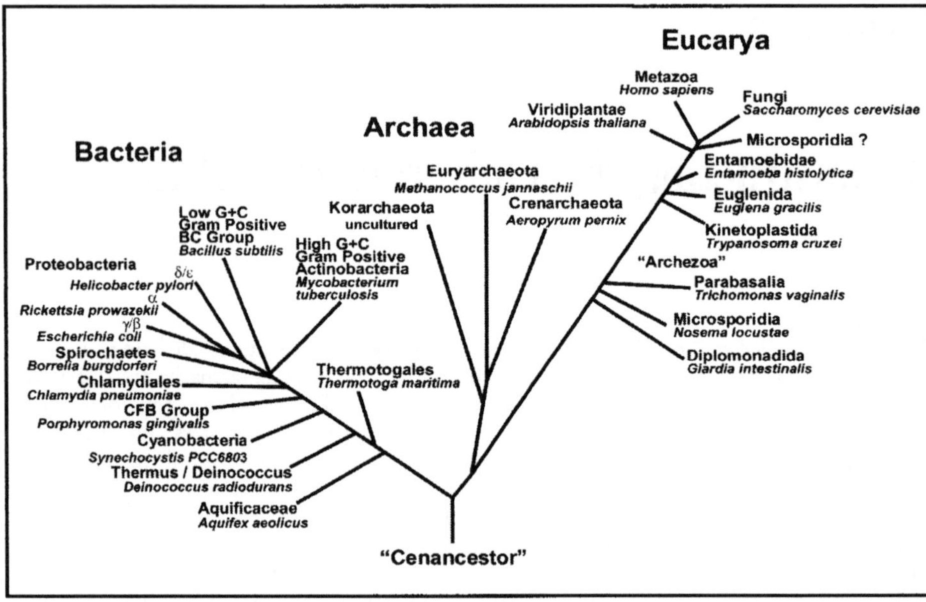

Figure 1. Schematic drawing of the universal tree showing the relative positions of evolutionary pivotal groups in the domains Bacteria, Archaea, and Eucarya. The phylum or other higher order name is given for key groups of organisms with a representative species named in italics below. The location of the root (the cenancestor) corresponds with that proposed by reciprocally rooted gene phylogenies (see text). The question mark beside Microsporidia denotes recent suggestions that it might branch higher in the eukaryotic portion of the tree.[120] (Branch lengths have no meaning in this tree). Figure adapted from ref. 13.

organisms might have thrived in the environmental conditions of a younger Earth. Thus, their findings challenged the fundamental subdivision of living organisms into prokaryotes and eukaryotes thereby upsetting the assumption that evolution progressed directly from simple (prokaryotes) to more complex entities (eukaryotes).

In 1990, Woese, Kandler and Wheelis[8] formally proposed the replacement of the bipartite prokaryote-eukaryote division with a new tripartite scheme based on three urkingdoms or Domains; the Bacteria (formally eubacteria), Archaea (formally archaebacteria) and Eucarya (eukaryotes, still the more often used name). The rationale behind this revision came from a growing body of biochemical, genomic and phylogenetic evidence which, when viewed collectively, suggested that the Archaea were unique from eukaryotes and the Bacteria. The discovery of the Archaea was a significant event, which added a new dimension to the construction of the universal tree since evolutionary relationships between the three major subdivisions had to be considered (Fig. 1).

Topology of the Universal Tree

The obvious challenge in universal tree reconstruction is determining which Domain evolved first and, therefore, is the root of the universal tree. Assuming that each Domain is monophyletic there are three possible answers (depicted respectively in Fig. 2) (1) Bacteria diverged first from a lineage producing Archaea and eukaryotes (AE tree) or (2) eukaryotes diverged from a fully prokaryotic clade, consisting of Archaea and Bacteria (AB tree) or (3) the Archaea diverged first such that Bacteria and eukaryotes (BE tree) are sister groups.

In terms of species diversity and carbon biomass, the Archaea are far from insignificant. Early interest in the Archaea was motivated by their remarkable success in flourishing in the harshest of environments, which earned them the title of "extremophiles". However, more

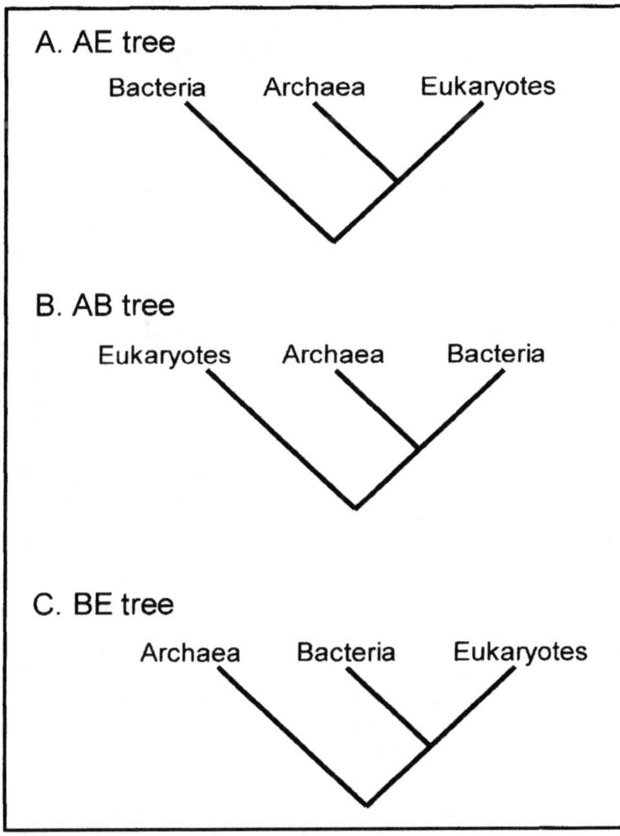

Figure 2. Three possibilities for the rooting of the universal tree. A) Bacteria diverged first from a lineage producing Archaea and eukaryotes (called here the AE tree); B) Eukaryotes diverged from a fully prokaryotic clade, consisting of Bacteria and Archaea (the AB tree) or; C) the Archaea diverged first such that eukaryotes and Bacteria are sister groups (the BE tree).

recent studies show that many archaeal species are "mesophiles", living in oceans, lakes, soil, and even animal guts.[9]

Prior to whole genome sequence data, considerable knowledge had accumulated on the comparative biochemistry, and cellular and molecular biology of the Archaea (for a review see refs. 10-13). Archaea seem to have a few unique biochemical and genetic traits as well as a variety of metabolic regimes, which deviate from known metabolic pathways of Bacteria and eukaryotes, and are not simply particular environmental adaptations. Recent genome comparisons found 351 archaea-specific "phylogenetic footprints" or combinations of genes uniquely shared by two or more archaeal species but not found in either bacteria or eukaryotes.[14] However, such inventories might over estimate the number of unique functional proteins since hyperthermophilic Archaea and Bacteria tend to have more split genes compared to their mesophilic counterparts.[15] Archaeal and bacterial species are definitely prokaryotes with generally similar ranges of cell sizes, genes linked in operons, large circular chromosomes often accompanied by one or more smaller circular DNA plasmids, and lacking nuclear membranes and organelles.

However, Archaea and eukaryotes share significant components of DNA replication, transcription, and translation, which are either not found in Bacteria or replaced by an evolution-

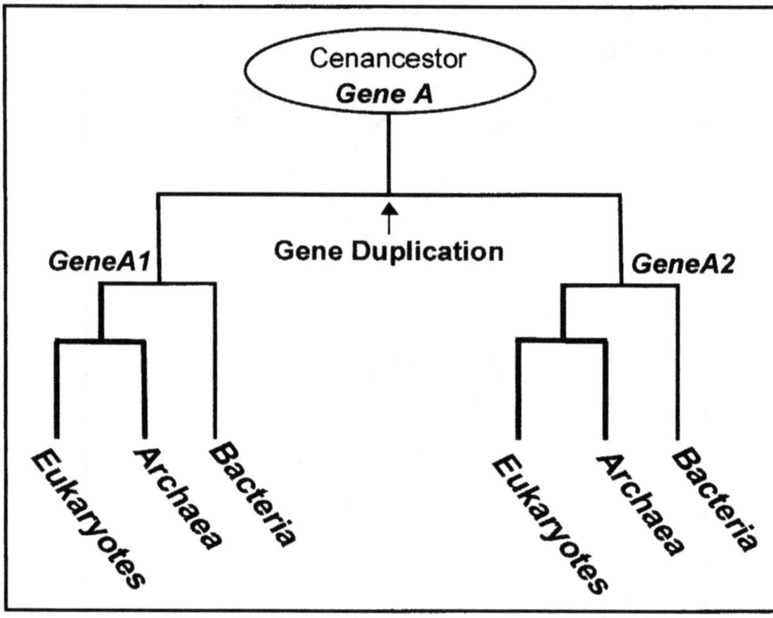

Figure 3. Conceptual rooting of the universal tree using paralogous genes. Gene A was duplicated in the cenancestor such that all extant organisms have paralogous copies, gene A1 and gene A2. The two genes are sufficiently similar to allow for the construction of reciprocally rooted trees thus rooting the tree of one paralog with that of the other. The topology depicted here, Archaea and eukaryotes as sister groups with the root in Bacteria, has been consistently supported by paralogous trees (see text).

ary unrelated (analogous) enzyme. Many DNA replication and repair proteins are homologous between Archaea and eukaryotes but completely absent in Bacteria.[16] While the archaebacterium, *Pyrococcus abyssi*, was recently shown to have a bacteria-like origin of DNA replication, most of its replication enzymes are eukaryote-like.[17,18] Archaeal DNA scaffolding proteins are remarkably similar to eukaryotic histones.[19] Eukaryotes and the Archaea have similar transcriptional proteins, such as multi-subunit DNA-dependent RNA polymerases,[20] as well as sharing translation initiation factors not found in the Bacteria.[21,22] Thus, based on cellular and genetic components, the Archaea seem to occupy a middle ground between the Bacteria and eukaryotes, a conclusion which serves little in resolving the rooting problem. Only in molecular phylogenetics lies such hope.

The lack of an outgroup to all living things meant that the rooting of the universal tree could only be resolved by using paralogous genes to construct reciprocally rooted trees (Fig. 3). Iwabe and coworkers[23] aligned amino acids from five conserved regions shared by the elongation factors (EF) Tu/1α and EF-G/2 genes of the archaebacterium, *Methanococcus vannielii*, and several species of Bacteria and eukaryotes. According to protein sequence similarity and neighbor-joining trees, both EF-1α and EF-2 genes of Archaea were more similar to their respective eukaryotic, rather than bacterial, homologs. Gogarten and coworkers[24] developed composite trees based on duplicated ATPase genes where the V-type A and V-type B occurs in Archaea and eukaryotes and the F_0F_1-type β and F_0F_1-type α occurs in Bacteria. In agreement with the elongation factor rooting, reciprocally rooted ATPase subunits trees also showed that the Archaea, represented by a sole species *Sulfolobus acidocaldarius*, were closer to eukaryotes than to Bacteria.

Subsequent paralogous protein rootings based on aminoacyl-tRNA synthetases[25,26] and carbamoylphosphate synthetase[27] confirmed the rooting in the Bacteria and linking Archaea and eukaryotes as sister groups. If one argues that enzymes involved in DNA replication, transcription and translation, so-called "information" genes, are core to living things then the evolutionary scenario suggested by paralogous gene trees seems particularly reasonable. Thus emerged the "canonical" universal tree with the Archaea and eukaryotes being sister groups, the rooting in the Bacteria, and all three Domains as monophyletic groups.

Uprooting the Universal Tree

Despite the convincing results from paralogous gene trees, the rooting of the universal tree has not been without controversy. Phylogenetic analyses using alternative methods and expanded data sets raised questions about the rooting of the universal tree and the monophyly of the Archaea.[28-30] Philippe and coworkers[31,32] have maintained that phylogenies of distantly related species are strongly affected by saturation for multiple mutations at nearly every amino acid position in a protein. Unequal mutation rates between different species can lead to long branch attraction effects. However, a greater issue is the degree to which horizontal gene transfers between the Domains of life have affected the actual viability of constructing a definitive universal tree.

The increasing size of sequence databases adds to the species richness of universal trees. Perhaps not surprisingly, nature provides plenty of exceptions to the canonical universal tree paradigm. In most cases, the key hypothesis invoked has been horizontal gene transfer or HGT. Simply stated, HGT is the exchange of genes between organisms which are not directly related by evolutionary descent. Many examples of HGT between closely related species are known, such as the transfer of bacterial antibiotic resistance genes.[33] The extent and nature of more ancient HGT events, (i.e., *trans*-Domain HGT between species of one Domain to species of another Domain), is an important and open evolutionary question[34-36] which is further considered for the remainder of this chapter.

Among the first documented *trans*-Domain HGT events involved ATPase subunits which were actually key in rooting the universal tree. Archaeal V-type ATPases were reported for two bacterial species, *Thermus thermophilus*[37] and *Enterococcus hiraea*,[38] while a bacterial F_1- ATPase β subunit gene was found in the Archaea, *Methanosacrina barkeri*.[39] Consequently, Forterre and coworkers[40] suggested that the ATPase subunit gene family had not been fully determined, and that other paralogous family members might be discovered Hilario and Gogarten[41] believed that the observed distribution of ATPase subunits was the result of a few, rare HGTs. In support of the latter view, broader surveys have failed to detect archaeal V-type ATPases in other bacterial species.[42]

The HGT debate was amplified by a growing number of examples where single gene trees, although not uniquely rooted, had irreconcilable topologies to that of the canonical universal tree.[43] In 1995 Golding and Gupta[44] examined the phylogenetic trees for 24 universally conserved proteins and found only nine with the AE tree topology. Although subsequent phylogenetic analyses by Gupta and Golding[45] and Roger and Brown[46] slightly modified the number of protein trees with AE topologies, a significant number of proteins still conflicted with the canonical universal tree. Feng, Cho and R.F. Doolittle[47] found that in the 34 universal protein trees they constructed, AE, AB and BE clusters occurred in the phylogenies for 8, 11, and 15 proteins, respectively. A broader survey involving phylogenetic analysis of 66 proteins found that AE, AB, and BE topologies occurred for 34, 21, and 11 protein trees, respectively, with the remaining trees having indeterminate relationships among the Domains.[13] New genome sequence data have further reduced the AE list with additional examples of horizontal gene transfer between eukaryotes and bacteria, such as isoleucyl-tRNA synthetases.[48]

Genomes and HGT

Genomes are being sequenced at a remarkable pace, the progress of which can be followed at number of websites including those of the NCBI Genome (http://www.ncbi.nlm.nih.gov/PMGifs/Genomes/bact.html) and TIGR Microbial (http://www.tigr.org/tdb/mdb/mdb.html) Databases. This new abundance of sequence data has resulted in a more, not less, confusing picture of the universal tree. Comparative analysis of archaea, bacterial and eukaryotic genomes suggest that relatively few genes are entirely conserved across all genomes. Important biochemical pathways appear to be incomplete in some organisms. In some instances, a protein has been discovered to take over the catalytic role of an unrelated protein, so-called nonorthologous gene replacement.[49]

Phylogenetic analyses of conserved proteins suggest that *trans*-Domain HGT has been extensive. Lake and colleagues suggest that based on their propensity for HGT, genes could be divided into two categories, informational and operational genes.[50] Informational genes, which include the central components of DNA replication, transcription and translation, are less likely to be transferred between genomes than operational genes involved with cell metabolism. The fact that informational gene products, at least qualitatively, have more complex interactions might restrict their opportunities for genetic exchange and fixation.[51] Additional support for this view is the conservation of genomic context for translation-associated genes in bacteria.[52]

Despite their critical role in protein synthesis and ancient origins (without them interpretation of the genetic code would be impossible), aminoacyl-tRNA synthetases have been extensively shuttled between genomes (for a review see refs. 53-55). Phylogenetic trees suggest that class I isoleucyl-tRNA synthetases may have been transferred from an early eukaryote to bacteria as a specific adaptation to resist a natural antibiotic compound.[48] Orthologous genes to eukaryotic glutaminyl-tRNA synthetase occur in many proteobacteria and *D. radiodurans* but not in other Bacteria or the Archaea.[56] Archaea and some bacteria, Spirochaetes, share novel type of lysyl-tRNA synthetases[57] and phenylalanyl-tRNA synthetases.[55,58,59]

Metabolic genes can have surprising species distributions such as the mevalonate pathway for isoprenoid biosynthesis. The mevalonate pathway has been well studied in humans because 3-hydroxy-3-methylglutaryl coenzyme A [HMG-CoA] reductase is the target for the statin class of cholesterol-lowering drugs. The mevalonate pathway was long believed to be specific to eukaryotes since most bacteria utilize an evolutionary unrelated metabolic route for isoprenoid biosynthesis, the pyruvate/GAP pathway. However, recent genome surveys and phylogenetic analyses have found not only HMGCoA reductase but also four other enzymes in the mevalonate pathway in Gram-positive coccal bacteria.[60-62] The genes are also found in the Archaea and the bacterial spirochaete, *Borrelia burgdorferi*. However, the mevalonate pathway is absent from the completely sequenced genome of a closely related Spirochaete, *Treponema pallidum*, and the Archaea have likely substituted an analogous protein for at least one enzyme in the pathway.[63] In those Bacteria with the mevalonate pathway, the genes encoding component enzymes are tightly linked suggesting that all genes might have been transferred simultaneously. Genes contributing products to a common metabolic pathways might be more readily fixed in the recipient genome than isolated, individual genes, which, in turn, would favor the organization of pathway genes into tightly linked operons.[64,65]

Cautionary Notes on the HGT Hypothesis

Recent science news reports have painted the picture that significant fractions of the scientific community engaged in genomics and universal tree studies have taken "a sky is falling" attitude towards the possibility of reconstructing cellular evolution in light of widespread HGT.[66,67] In summary, their view is that while phylogenetic approaches are still useful for mapping the evolution of individual proteins, HGT has significantly confounded the reconstruction of the universal tree, hence, any discerned patterns in early genome evolution are suspect.[68] However, there is a need to critically evaluate methods for detecting HGT, which in some cases, can lead to overestimates of its occurrence.[36,69]

Reports of HGT without supporting phylogenetic analyses should be carefully scrutinized. Comparative studies based on BLAST[70] analyses have concluded that HGT has extensively occurred between Archaea and Bacteria. Koonin and coworkers[71] found that 44 % of the gene products of the archaebacterium, *Methanococcus jannaschii* were more similar to bacterial over

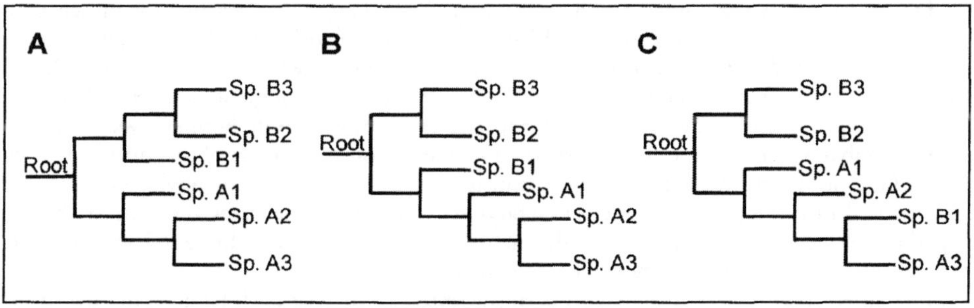

Figure 4. Detection of horizontal gene transfer (HGT) from phylogeny. Hypothetical protein trees for three bacterial species (B1-B3) and three archaeal species (A1-A3). A) The true rooting of the tree postulates a split between the Archaea and Bacteria, which results in two monophyletic clusters. B) The lowest branching bacterial species, B1, has a more rapid rate of amino acid substitution than other bacterial species which results in phylogenetic programs as well as homology searching software implicating the Archaea as the closest relatives. At first glance, the tree would suggest HGT between B1 and Archaea. However, the clustering of species is actually the result of the new position of the root, which was shifted by the "attraction" of the B1 branch to the outgroup, the Archaea. C) Strong phylogenetic evidence for HGT is the "imbedding" of a distantly related in-group species within the outgroup and away from the root. In this example, bacterial species B1 clusters with a more derived archaeal species, A3, which strongly suggests HGT occurred from the Archaea (A3) to Bacteria (B1).

eukaryotic proteins while only 13% were more like eukaryotic proteins. Nelson and coworkers[72] reported that 24% of proteins from *Thermotoga maritima*, a thermophilic bacterium with a deep rRNA tree lineage, were most similar to archaeal proteins.

However, deep branching species of one Domain are susceptible to arbitrary clustering with species from the other Domains, such as bacterial thermophiles with the Archaea and eukaryotes.[36,73] Differences in evolutionary rates can lead to an incorrect rooting which will result in mistaken occurrences of HGT between the deep branching species and the outgroup (Fig. 4A and 4B). Conversely, protein trees where an in-group species is solidly embedded within an outgroup clade provide strong evidence for HGT (Fig. 4C). Consequently, phylogenetic analysis suggests that *T. maritima* received far fewer genes from the Archaea than first estimated by homology searches.[72,73] Phylogenetic analyses of putative archaeal-like proteins from *Deinococcus radiodurans*, a bacterium which branches nearly as deeply as *Thermotoga in rRNA trees*, suggests that HGT involving either Archaea or eukaryotes occurred for fewer than 1% of its total genome complement.[74]

Some remarkable claims of direct HGTs from bacteria to vertebrates were made in the historic publication of the first draft of the human genome sequence by the International Human Genome Sequencing Consortium (IHGSC) in 2001.[75] In the paper, they stated that as many as 113 vertebrate genes, some only found in humans, were the result of direct HGT from bacteria. This conclusion was based on BLASTP score analyses where the expect value (E-values) of human gene matching a bacterial gene was 9 orders of magnitude greater than the value to the closest related nonvertebrate eukaryote gene. The possibility of direct bacteria to vertebrate HGT has several important evolutionary and medical ramifications. First, any gene transferred and fixed in the genome of a multicellular organism, like vertebrates, would need to be introduced into the germ cell line. Second, bacterial genes could only be functionally expressed in vertebrate genomes if they could readily adapt to the eukaryotic gene regulon. Finally, there are serious public health concerns if the human gene pool could become permanently contaminated from bacterial genes as a consequence of infection or the ingestion of genetically modified foods. However, three independent studies concluded that there was no evidence for HGT from bacteria to vertebrates.[76-78]

In our study,[78] we examined all 28 cases where the IHGSC[75] had verified the presence of the gene in the human genome by PCR. BLAST[70] searches of additional databases, in particular nonvertebrate EST databases (i.e., the National Center for Biotechnology Information "EST others" database), revealed many homologs in nonvertebrates (i.e., fungi, nematodes and insects) which were previously undetected. In other instances, a nonvertebrate homolog was found in public databases but at a threshold above the E-value cut-off of 9 orders of magnitude used in the IHGSC study. However, alignment of multiple sequences followed by phylogenetic analyses, resulted in monophyletic clades of eukaryotes with both vertebrates and nonvertebrates together. Of the 28 genes examined, only one instance of possible vertebrate to bacteria HGT was found. There was no evidence of bacteria to vertebrate HGT.

Hypothetical HGT events have also been suggested by analysis of differences in nucleotide composition (G+C content) between donor and recipient coding regions.[79] However, intragenomic base composition can be highly variable between chromosomal regions which could lead to over estimates in the number of transferred genes.[80,81] Arguably, genes might be more likely to be transferred in clusters, such as operons, particularly if the genes encode several proteins in a common biochemical pathway.[64] Thus, patterns of gene position or context across genomes might be useful indicators of HGT. However, even simple operons can vary greatly among closely related species or be identical among highly unrelated ones. An example is the organization of the two genes coding the alpha and beta subunits of phenylalanyl-tRNA synthetase which are cotranscribed in most species of Bacteria and Archaea but have become dispersed in the genomes of others through what appears to be multiple, independent events.[56]

In summary, reports of HGT need to be critically evaluated. Proper scientific inquiry should begin with the assumption of the null hypothesis, which, in the case of comparative genomic studies, is that HGT has not occurred and that all genes evolved by direct inheritance. Only after adopting such a stance, can we begin to grasp the true role of HGT in genome evolution.

Possible HGT Patterns and Processes

In addition to the detection of *trans*-Domain HGT, there are issues about the magnitude, directionality and timing of this phenomena are discussed below in the context of the three possible topologies of the universal tree.

First, trees which depict Archaea and eukaryotes as sister groups (the AE tree in Fig. 2) largely result from the phylogenetic analyses of proteins involved in DNA replication, transcription and translation.[13] Archaea seem to utilize a wider range of eukaryote-type proteins for these processes than Bacteria. Paralogous gene trees also position Archaea and eukaryotes as sister groups although it has been suggested that such results are idiosyncratic due to more rapid rates of evolutionary change in Bacteria.[82]

Among the three possible universal tree scenarios, only trees with the AE clustering depict, even if occasionally, all three Domains to be monophyletic simultaneously.[13] If extensive polyphyly (species from different Domains in the same clade) is evidence for HGT then, by default, monophyly indicates evolution in the absence of HGT. Given the large universe of genes, Domain monophyly appears to be a rare occurrence. However, the existence of some monophyletic gene trees should suggest that their topology reflects the underlying evolutionary trajectory of the species involved without the complication of HGT. If true, then the overall scenario of cellular evolution, heavily diluted by HGT events, remains the canonical universal tree with a rooting in the Bacteria with Archaea and eukaryotes as sister groups. However, the persistence of monophyly in universal trees is highly dependent upon the diversity of species sampled. Notably, genome sequences from simple, single-cell eukaryotes will likely reveal instances of *trans*-Domain HGT previously unnoticed in higher eukaryotes.[83]

Second, there are phylogenies where Archaea and Bacteria are closest relatives (the AB tree in Fig. 2). However, in those trees, one or both Domains are always para/polyphyletic groups. Such tree topologies are evidence for HGT between Archaea and Bacteria, the patterns for which can be often complex. The genes and species implicated in Archaea-Bacteria HGT are highly varied. Glutamine synthetases,[84] glutamate dehydrogenase[85] and HSP70[86] of Archaea are closely related to orthologs from Gram-positive bacteria. Hyperthermophilic archaeal and bacterial species share a reverse gyrase which is likely a common adaptation to life at extremely high temperatures.[87] Catalase-peroxidase genes appear to have been exchanged between Archaea and pathogenic proteobacteria.[88] Two component signal transduction systems in the Archaea as well as fungi and slime molds were likely acquired from the Bacteria.[89] However, as discussed above, similarities between Bacteria and Archaea are not always conclusive evidence for HGT events. Species forming low branches in the two Domains can be attracted or cluster together because of rooting artifacts. In addition, gene distributions shared by Bacteria and Archaea but not eukaryotes might be caused by gene loss or replacement in eukaryotes rather than HGT between Archaea and Bacteria.

The third universal tree topology, Bacteria and eukaryotes as closest relatives or the BE tree (Fig. 2), might result from specific bi-directional gene transfers. Some bacterial species appear to have acquired genes from eukaryotes such as the glutaminyl-tRNA synthetase gene.[53,90] On the other hand, eukaryotes have likely integrated a large number of bacterial genes as a consequence of endosymbiosis related to mitochondria and plastid biogenesis. The endosymbiosis theory of organelle origins[91] is a widely accepted fact. However, the deeper consequences of endosymbiosis to eukaryotic genome evolution are just being revealed by genome sequencing projects. Genome comparisons and phylogenetic analyses involving *Arabidopsis thaliana* and *Synechocystis* sp., suggest that plants obtained from 1.6% (~400 genes) to 9.2% (~2200 genes) of their gene complement from cyanobacterium, the bacterial progenitor of plastids.[92] Phylogenies for many conserved proteins, such as the glycolytic pathway enzymes suggest bacterial origins for many eukaryotic genes (for a review see ref. 13). The occurrence of mitochondria-targeted genes in simple protists which both lack mitochondria (amitochondrial) and appear as early evolved eukaryotic lineages, suggests endosymbiotic transfer of genes to the nuclear genome occurred early in the evolution of eukaryotes.[93-97] In some instances, the organelle gene has either contributed a new function or replaced the original orthologous gene in the genome of the host. However, other phylogenetic trees, namely of aminoacyl-tRNA synthetases, suggest that patterns of integration of bacterial genes in the eukaryotic genome via endosymbiosis might be more complex.[83,98]

Universal Trees Based on Multiple Datasets

Construction of universal trees based on the distribution of genes is a logical use of genomic sequence data in evolutionary biology. The underlying principal of this approach is that species with the largest proportion of common genes should be more recently diverged than species with fewer shared genes. There are several important methodological considerations such as distinguishing orthologous genes from paralogous ones, accurate prediction of genes, and normalization of gene inventories across genomes. Although employing somewhat different approaches, studies which constructed universal trees from gene distributions generally found tree topologies remarkably similar to that of the canonical universal tree and rRNA tree.[15,99,100] However, it has been argued that while genome inventories might tell us about the similarities in the contents of genomes from different species, the nuisances of HGT involving universally conserved genes are lost.[101]

Potentially, gene order could also be used to reconstruct phylogenies of bacteria and archaea since many recognizable operon organizations occur across these two Domains. However, gene order is poorly conserved between species and is unlikely to be a useful phylogenetic marker[102,103] although overall neighborhoods of genes on the chromosome might be preserved because of functional and regulatory consequences.[59]

On the other hand, the combination or concatenation of multiple protein datasets derived from genome sequences might be useful for the phylogenetic reconstruction of universal trees. Phylogenies based on concatenated protein datasets are potentially more robust and representative of the evolutionary relationships among species since the number of phylogenetically informative sites and sampled gene loci are greatly increased. The main principle behind combining data is that it allows for the amplification of phylogenetic signal, and increased resolving power, in cases where signal is masked by homoplasy (similarities in amino acids for reasons other than inheritance) among the individual gene data sets. Such protein datasets have helped resolve evolutionary relationships among photosynthetic bacteria[104] and eukaryotic protists.[105]

By definition, a universally conserved protein occurs in every organism. The increasing number of completely sequenced genomes will invariably lead to the shrinking of this inventory since the odds will increase for finding exceptional cases. For example, the 70 kilo-Dalton heat shock protein (HSP70), once thought to be highly conserved from the perspective of both amino acid substitutions and species distribution, is absent from several species of Archaea.[106] In many cases, the biochemical function is still required but an evolutionary unrelated enzyme serves as the catalyst. Arguably, only those proteins found in all completely sequenced genomes are conserved enough to provide a continuous picture of all lineages back to the last universal common ancestor. Fortunately, the contemporary collection of completely sequence genomes represents fairly diverse groups of Bacteria, Archaea and eukaryotes. Therefore, for purposes of universal tree reconstruction, the list of completely conserved proteins across the three Domains is unlikely to be further reduced with new genomes.

Recently, we constructed universal trees based on the combined alignments of proteins conserved across 45 species from all three Domains.[107] Proteins were selected on fairly strict criteria of being conserved across all species and being orthologous (i.e., paralogs or duplicated proteins within a species were eliminated from the entire analysis). For eukaryotes, where two copies of a gene might exist, one targeted to the mitochondria and the other to the cytoplasm, only the latter was used since the cytoplasmic version best tracks the evolution of the eukaryotic nucleus. The determined number of conserved proteins, 23, was far fewer than previous genomic studies (Table 1). For example, the Clusters of Orthologous Groups of proteins (COGs) database (http://www.ncbi.nlm.nih.gov/COG/xindex.html) reports for 34 complete genomes, a total of 78 completely conserved proteins.[108] However, we included several additional genomes, a few which were incomplete at the time of the study. In addition, if the collection of organisms is diverse, then the likelihood increases that particular lineages, by chance, have lost a particular pathway or replaced components with analogous proteins. Our list, shown in Table 1, represents the most highly conserved or widely found proteins known to date. The edited multiple sequence alignment of the concatenated dataset of 23 proteins was 6591 amino acids in length, which was far larger than any single protein dataset, and is the largest applied to universal tree reconstruction.

Similar to universal rRNA trees, all combined protein dataset phylogenetic trees strongly supported the monophyly of the three Domains (Fig. 5). On average, archaeal and eukaryotic species were slightly more similar to each other than either was to Bacteria. However, it cannot be confirmed that Archaea and Eucarya share a last common ancestor since the tree is unrooted. Within each Domain, the branching order of most nodes are well supported by bootstrap replications (> 70%). Although fewer genomes of Archaea and eukaryotes have been completely sequenced, branching orders of those species were consistent with contemporary views of organism evolution.

In the Bacteria, the major subdivisions of Bacillus/Clostridium (low G+C Gram positives), Spirochaetes, and Proteobacteria were strongly supported as being monophyletic, as postulated by the universal rRNA trees. However, a major departure was the placement of Spirochaetes (represented by the species *Treponema pallidum* and *Borrelia burgdorferi*) as the first bacterial branch rather than thermophiles (*Aquifex aeolicus* and *Thermotoga maritima*). While the basal position of Spirochaetes is incompatible with hypotheses regarding the thermophilic origins of life, there are suggested instances of HGT between Spirochaetes and Archaea, such as class I lysyl-tRNA synthetases.[54] In the combined protein alignment phylogenetic method, the inclusion of such proteins would tend to move the Spirochaete branch to a more basal position in the bacterial clade.

Table 1. Proteins included in concatenated alignments, the number of residues, and the support for domain monophyly in individual protein trees[107]

	Cellular Function	Protein Name	Number of Amino Acids[a]	Support for Domain Monophyly[b]		
				Archaea	Bacteria	Eucary
1	translation	alanyl-tRNA synthetase	502	100	–	100
2		aspartyl-tRNA synthetase[c]	249	–	100	100
3		glutamyl-tRNA synthetase[c]	188	50 (–)	100	100
4		histidyl-tRNA synthetase	166	–	–	100 (93)
5		isoleucyl-tRNA synthetase	552	–	–	–
6		leucyl-tRNA synthetase[c]	358	–	100	100
7		methionyl-tRNA synthetase	306	–	–	99
8		phenylalanyl-tRNA synthetase b subunit	177	–	–	100
9		threonyl-tRNA synthetase	305	–	– (34)	100
10		valyl-tRNA synthetase	538	–	–	100
11		initiation factor 2[c]	337	–	100	100
12		elongation factor G[c]	536	64(87)	100	100
13		elongation factor Tu[c]	340	– (42)	100	100
14		ribosomal protein L2[c]	192	46(–)	100	100
15		ribosomal protein S5[c]	131	46(19)	100	100(99)
16		ribosomal protein S8[c]	118	–	100	100
17		ribosomal protein S11[c]	110	–	100	100
18		aminopeptidase P	95	–	–	–
19	transcription	DNA-directed RNA polymerase b chain[c]	537	99(78)	100	100
20	DNA replication	DNA topoisomerase I[c]	236	–	100	100
21		DNA polymerase III subunit[c]	194	46(49)	100	100(95)
22	metabolism	signal recognition particle protein[c]	298	71(39)	100	100
23		rRNA dimethylase	126	–	–	100(98)
		full alignment length[d]	6591			
		truncated alignment length[e]	3824			

[a] Length of alignments after removing ambigously aligned regions. [b] Occurrence of monophyletic nodes in 100 bootstrap replicated datasets of protein distance/neighbor-joining and maximum parsimony methods (in paratheses where maximum parsimony values differ from those of the neighbor-joining consense tree). Dash indicates that the nodes were not monophyletic. [c] Proteins included in both the full and truncated alignments. [d] Length of multiple sequence alignment, which included all proteins, used to produce phylogeny in Figure 5. [e] Length of multiple sequence alignment, which excluded proteins where the Bacteria were not monophyletic, used to produce phylogeny in Figure 6. Table adapted from ref. 107.

Examination of the individual gene trees revealed topologies where the Domains, primarily the Bacteria, were not monophyletic thus implicating possible instances of HGT (Table 1). Interestingly, none of the 23 individual protein trees suggested that hyperthermophilic bacteria, the species *Thermotoga maritima* and *Aquifex aeolicus*, exchanged genes with either eukary-

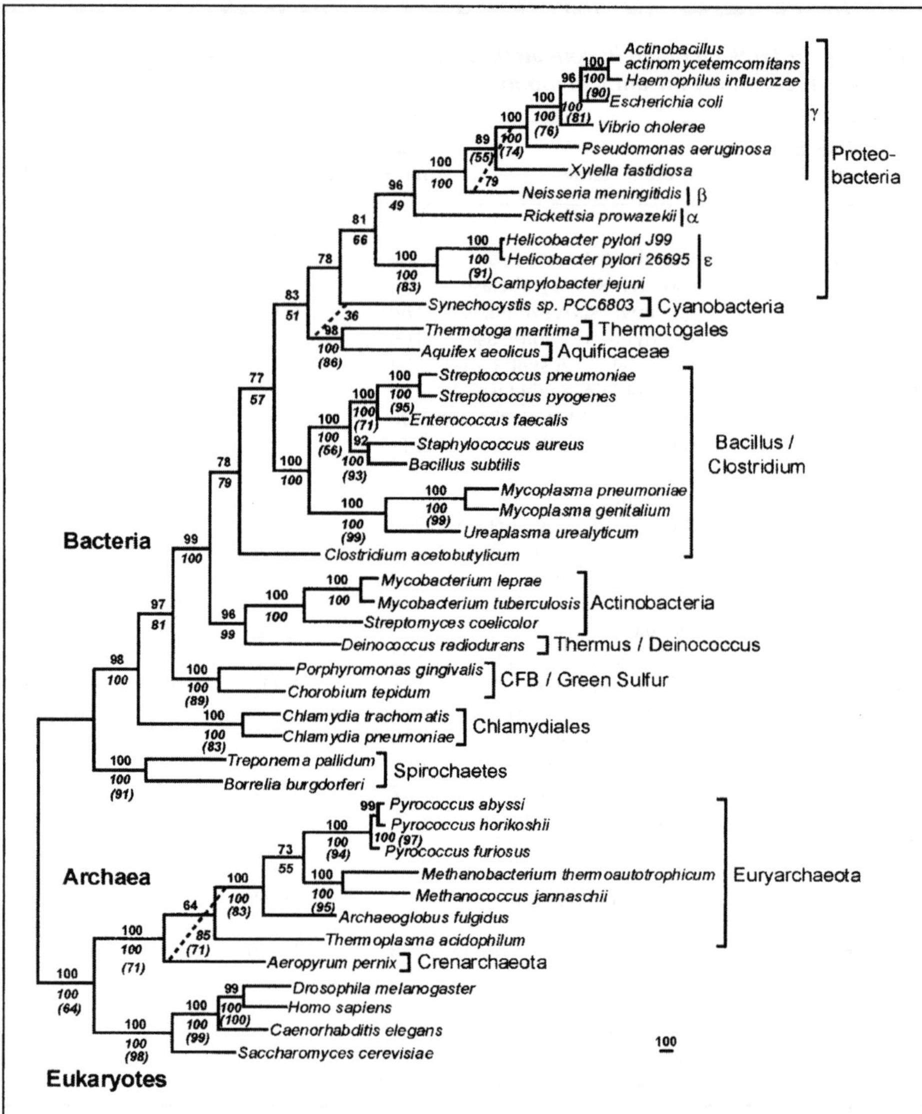

Figure 5. Universal tree based on 23 combined protein datasets.[107] Minimal length maximum parsimony universal tree based on 23 combined protein datasets is shown. Spirochaetes are placed as the lowest branching Bacteria. Numbers along the branches show the percent occurrence of nodes in 50% or greater of 1000 bootstrap replicates of maximum parsimony[122] (plain text) and neighbor joining[123] (italicized text) analyses or 1000 quartet puzzling steps of maximum likelihood[124] analysis (in parentheses). Dashed lines show occasional differences in branching orders in neighbor-joining trees. Scale bar represents 100 amino acid residue substitutions. CFB stands for the Cytophaga-Flexibacter-Bacteroides group of bacteria. For a full explanation of methods of construction see ref. 107. Figure adapted from ref. 107.

otes or the Archaea. When nine putatively horizontally transferred proteins were removed from the combined protein dataset, the truncated combined protein alignment was reduced to 3824 amino acids (Table 1).

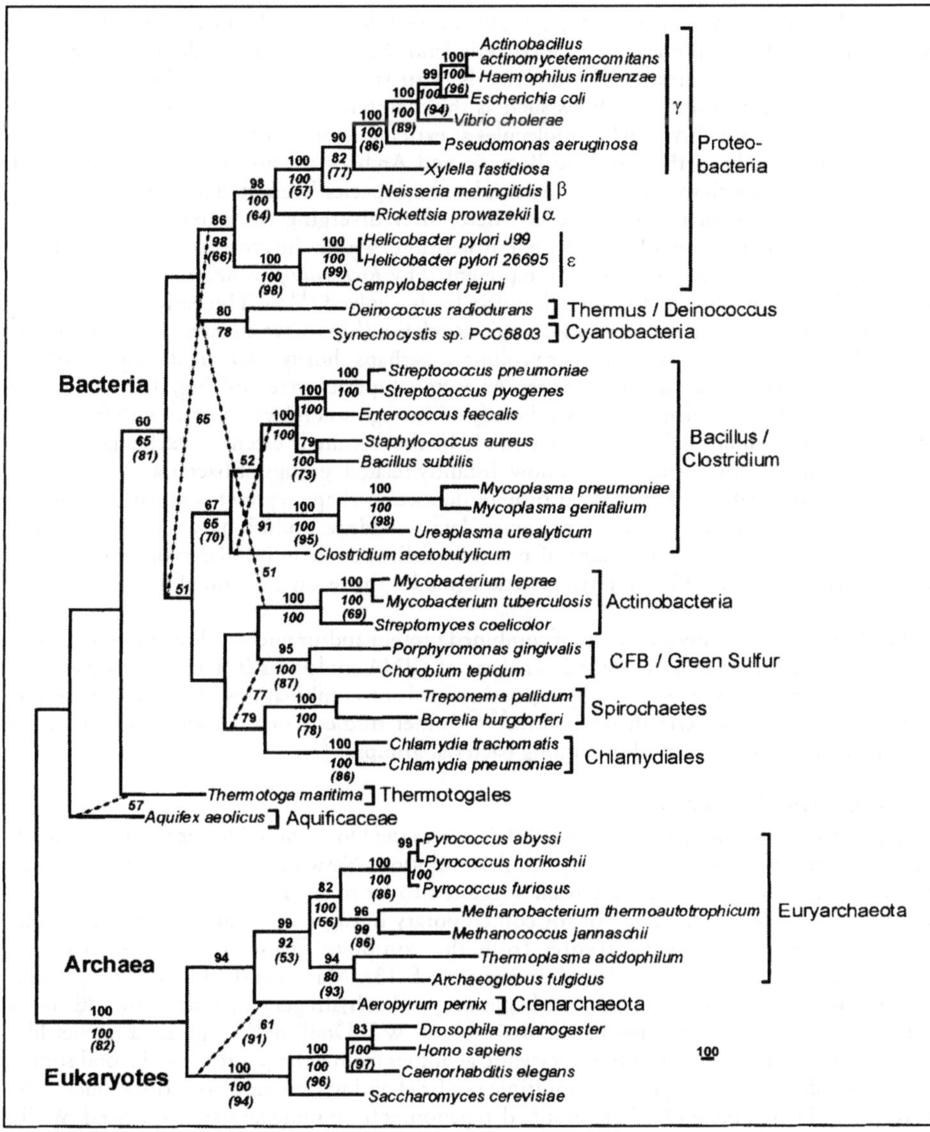

Figure 6. Universal tree based on 14 combined protein datasets. Minimal length maximum parsimony universal tree based on 14 proteins, with 9 horizontal gene transfer proteins removed, is shown. The tree shows Thermophiles as the basal group in Bacteria. Methods and labels are the same as Figure 5 and ref. 107. Figure adapted from ref. 107.

In contrast to the combined alignment of 23 proteins, phylogenetic trees based on the alignment of 14 nonHGT proteins agreed with universal rRNA trees in the placement of hyperthermophilic species, *A. aeolicus* and *T. maritima*, as the lowest branching bacterial lineages while Spirochaetes were a derived group (Fig. 6). However, high G+C and low G+C Gram-positives were not collectively monophyletic as previously reported for rRNA and other molecular markers.[109] The clustering of Chlamydiales, CFB and Spirochaetes together is also novel relative to rRNA trees.[110] The agreement between the dataset that excluded horizontal transferred genes (truncated protein tree) and the rRNA tree, in the placement of extreme thermophiles as the basal lineage in the Bacteria lends further support to the theory that life evolved at high temperatures.[110-112] However, there are still many unresolved issues surrounding the "hot" origin of life hypothesis such as the maintenance of extracellular biochemical reactions[18] and the stability of RNA molecules at extreme temperatures.[113]

Genes found only in thermophilic Bacteria and Archaea are just as likely to be shared syplesiomorphies, which were later lost in other bacterial species. Truncated protein trees showed a fundamental division in the Bacteria where, after diverging from hyperthermophiles, Proteobacteria split from all other bacteria. Furthermore, within the Proteobacteria, the earliest diverged group is the alpha-subdivision, represented by *Rickettsia prowazekii*, from which the endosymbiont progenitor of the mitochondria likely evolved.[114,115] The early emergence of alpha-Proteobacteria suggests that endosymbiotic relationships between eukaryotes and bacteria could have occurred early in cellular evolution, perhaps shortly after the divergence of the Domains Bacteria, Archaea and eukaryotes. As bacterial species were evolving, they could have shared genes with early eukaryotes either directly or through secondary transfers with free-living relatives of endosymbionts. The net result would be the seemingly extensive exchange of genes between eukaryotes and many diverse, now distantly related, groups of bacteria.

Phylogenetic analysis of combined protein datasets perhaps represents an important approach in the utilization of genome sequence data to address evolutionary questions. While HGT has likely played an important, if not fully defined, role in cellular evolution perhaps genomes have retained sufficient phylogenetic signal for the reconstruction of meaningful universal trees.

In addition, phylogenetic analysis of combined protein and/or nucleotide alignments might be a useful alternative to phylogenetic analysis of rRNA molecules in bacterial systematics. While some analyses suggest the phylogenetic signal for combinations of certain conserved proteins within the Bacteria might be low,[55,116] other studies based on wider collections of proteins support new relationships among bacterial groups.[102]

Concluding Remarks

The apparent occurrence of extensive HGT across the Domains of life has prompted much speculation on its significance to early cellular evolution. Networks of genetic interactions at the base of the universal tree have been suggested to be so intense as to render useless the concept of a single cellular ancestor for contemporary lineages.[41,117] Other radical positions discuss the emergence of eukaryotes from the complete fusion of genomes from an archaebacterium and bacterium (for a review see ref. 13). Martin and Müller[118] proposed a more stepwise progression to eukaryotes beginning with a hydrogen-dependent host, likely an archaebacterium, and a respiring bacterial symbiont. W.F. Doolittle[119] suggests a ratchet-like addition of bacterial content to the eukaryotic genomes from either a prokaryotic food source or gene transfers as a consequence of multiple but brief endosymbiotic associations. Such controversies will either be resolved or amplified as genomes from more taxa are sequenced. While HGT has certainly unsettled the universal tree of life, it is premature to say that the tree has been permanently uprooted.[121]

Acknowledgments

Preliminary sequence data reported in ref. 107 and used to construct the trees in Figures 5 and 6 were obtained from various public databases. *Chlorobium tepidum, Enterococcus faecalis, Porphyromonas gingivalis*, and *Streptococcus pneumoniae* sequence data were obtained from The Institute for Genomic Research (TIGR) through the website at http://www.tigr.org and were funded by the U.S. Department of Energy (DOE), the National Institute of Allergy and Infectious Disease (NIAID) of NIH, TIGR, and the Merck Genome Research Institute. Preliminary sequence data for *Actinobacillus actinomycetemcomitans* were obtained from the *Actinobacillus* Genome Sequencing Project, University of Oklahoma ACGT and B.A. Roe, F. Z. Najar, S. Clifton, Tom Ducey, Lisa Lewis and D.W. Dyer through the website http://

www.genome.ou.edu/act.html which was supported by USPHS/NIH grant from the National Institute of Dental Research. Preliminary sequence data for *Pyrococcus furiosus* were obtained from the Utah Genome Center, Dept. of Human Genetics, University of Utah through the website http://www.genome.utah.edu/sequence.html which was supported by DOE. Preliminary sequence data for *Streptomyces coelicolor* were obtained from The Sanger Center through the website at http://www.sanger.ac.uk/Projects/S_coelicolor/.

References

1. Chatton E. Titres et Travauz Scientifiques. Setes Sattano Italy 1937.
2. Stanier RY, van Niel CB. The main outlines of bacterial classification. J Bacteriol 1941; 42:437-463.
3. Doolittle WF, Brown JR. Tempo, mode, the progenote, and the universal root. Proc Natl Acad Sci USA 1994; 91:6721-6728.
4. Stanier RY, van Niel CB. The concept of a bacterium. Arch Microbiol 1962; 42:17-35.
5. Stanier RY. Some aspects of the biology of cells and their possible evolutionary significance. Symp Soc Gen Microbiol 1970; 20:1-38.
6. Fox GE, Magrum LJ, Balch WE et al. Classification of methanogenic bacteria by 16S ribosomal RNA characterization. Proc Natl Acad Sci USA 1977; 74: 4537-4541.
7. Woese CR, Fox GE. Phylogenetic structure of the prokaryotic domain: The primary kingdoms. Proc Natl Acad Sci USA 1977; 51: 221-271.
8. Woese CR, Kandler O, Wheelis ML. Towards a natural system of organisms: Proposal for the domains Archaea, Bacteria and Eucarya. Proc Natl Acad Sci USA 1990; 87:4576-4579.
9. Stein JL, Simon MI. Archaeal ubiquity. Proc Natl Acad Sci USA 1996; 93: 6228-6230.
10. Danson MJ. Central metabolism of the Archaea. In: Kates M, Kushner DJ, Matheson AT, eds. The Biochemistry of Archaea (Archaebacteria). Amsterdam: Elsevier, 1993:1-24.
11. Kates M, Kushner DJ, Matheson AT. The Biochemistry of Archaea (Archaebacteria). Amsterdam: Elsevier, 1993.
12. Keeling PJ, Charlebois RL, Doolittle WF. Archaebacterial genomes: eubacterial form and eukaryotic content. Current Opinions in Genetics and Development 1994; 4:816-822.
13. Brown JR, Doolittle WF. Archaea and the prokaryote to eukaryotes transition. Microbiology and Molecular Biology Reviews 1997; 61:456-502.
14. Graham DE, Overbeek R, Olsen GJ et al. An archaeal genomic signature. Proc Natl Acad Sci USA 2000; 97: 3304-3308.
15. Snel B, Bork P, Huynen MA. Genome phylogeny based on gene content. Nature Genetics 1999; 21:108-110.
16. Edgell DR, Doolittle WF. Archaea and the origin(s) of DNA replication proteins. Cell 1997; 89:995-998.
17. Myllykallio H, Lopez P, López-Garcia P et al. Bacterial mode of replication with eukaryotic-like machinery in a hyperthermophilic archaeon. Science 2000; 288:2212-2215.
18. Kelman Z. The replication origin of archaea is finally revealed. Trends in Biochem Sci 2000; 25:521-523.
19. Reeve JN, Sandman K, Daniels CJ. Archaeal histones, nucleosomes and transcription initiation. Cell 1997; 89:999-1002.
20. Langer D, Hain J, Thuriaux P et al. Transcription in Archaea: similarity to that in Eucarya. Proc Natl Acad Sci USA 1995; 92:5768-5772.
21. Olsen GJ, Woese CR. Archaeal genomics – an overview. Cell 1997; 89:991-994.
22. Kyrpides NC, Woese CR. Universally conserved translation initiation factors. Proc Natl Acad Sci USA 1998; 95:224-228.
23. Iwabe N, Kuma K-I, Hasegawa M et al. Evolutionary relationship of Archaea, Bacteria, and eukaryotes inferred from phylogenetic trees of duplicated genes. Proc Natl Acad Sci USA 1989; 86:9355-9359.
24. Gogarten JP, Kibak H, Dittrich P et al. Evolution of the vacuolar H+-ATPase: Implications for the origin of eukaryotes. Proc Natl Acad Sci USA 1989; 86: 6661-6665.
25. Brown JR, Doolittle WF. Root of the universal tree of life based on ancient aminoacyl-tRNA synthetase gene duplications. Proc Natl Acad Sci USA 1995; 92:2441-2445.
26. Brown JR, Robb FT, Weiss R et al. Evidence for the early divergence of tryptophanyl- and tyrosyl-tRNA synthetases. J Mol Evol 1997; 45:9-16.
27. Lawson FS, Charlebois RL, Dillon J-AR. Phylogenetic analysis of carbamoylphosphate synthetase genes: evolution involving multiple gene duplications, gene fusions, and insertions and deletions of surrounding sequences. Mol Biol Evol 1996; 13:970-977.

28. Lake JA. Origin of the eukaryotic nucleus determined by rate-invariant analysis of rRNA sequences. Nature 1988; 331:184-186.
29. Rivera MC, Lake JA. Evidence that eukaryotes and eocyte prokaryotes are immediate relatives. Science 1992; 257:74-76.
30. Baldauf SL, Palmer JD, Doolittle WF. The root of the universal tree and the origin of eukaryotes based on elongation factor phylogeny. Proc Natl Acad Sci USA 1996; 93:7749-7754.
31. Lopez P, Forterre P, Philippe H. The root of the tree of life in the light of the covarion model. J Mol Evol 1999; 49:496-508.
32. Philippe H, Forterre P. The rooting of the universal tree of life is not reliable. J Mol Evol 1999; 49: 509-523.
33. Mazel D, Davies J. Antibiotic resistance in microbes. Cell Mol Sci 1999; 56:742-754.
34. de la Cruz F, Davies J. Horizontal gene transfer and the origin of species: lessons from bacteria. Trends Microbiol 2000; 8:128-133.
35. Eisen JA. Horizontal gene transfer among microbial genomes: New insights from complete genome analysis. Curr Opin Genet Dev 2000; 10:606-611.
36. Kyrpides NC, Olsen GJ. Archaeal and bacterial hyperthermophiles: Horizontal gene exchange or common ancestry? Trends Genet 1999; 15:298-299.
37. Tsutsumi S, Denda K, Yokoyama K et al. Molecular cloning of genes encoding major subunits of a eubacterial V-type ATPase from Thermus thermophilus. Biochim Biophys Acta 1991; 1098:13-20.
38. Kakinuma Y, Igarishi K, Konishi K et al. Primary structure of the alpha-subunit of vacuolar-type Na+-ATPase in Enterococcus hirae, amplification of a 1000 bp fragment by polymerase chain reaction. FEBS Lett 1991; 292:64-68.
39. Sumi M, Sato MH, Denda K et al. A DNA fragment homologous to F1-ATPase beta-subunit amplified from genomic DNA of Methanosarcina barkeri: Indication of an archaebacterial F-type ATPase. FEBS Lett 1992; 314:207-210.
40. Forterre P, Benachenhou-Lahfa N, Confalonieri F et al. The nature of the last universal ancestor and the root of the tree of life, still open questions. Biosystems 1993; 28:15-32.
41. Hilario E, Gogarten JP. Horizontal transfer of ATPase genes — the tree of life becomes the net of life. BioSystems 1993; 31:111-119.
42. Gogarten JP, Hilario E, Olendzenski L. Gene duplications and horizontal transfer during early evolution. In: Roberts DM, Alderson G, Sharp P et al, eds. Evolution of Microbial Life. Society for General Microbiology Symposia 54. Cambridge: Cambridge University Press, 1996:267-292.
43. Smith MW, Feng D-F, Doolittle RF. Evolution by acquisition: the case for horizontal gene transfers. Trends Biochem Sci 1992; 17:489-493.
44. Golding GB, Gupta RS. Protein-based phylogenies support a chimeric origin for the eukaryotic genome. Mol Biol Evol 1995; 12:1-6.
45. Gupta RS, Golding GB. The origin of the eukaryotic cell. Trends in Biochem Sci 1996; 21:166-171.
46. Roger AJ, Brown JR. A chimeric origin for eukaryotes reexamined. Trends Biochem Sci 1996; 21:370-371.
47. Feng D-F, Cho G, Doolittle WF. Determining divergence times with a protein clock: Update and reevaluation. Proc Natl Acad Sci USA 1997; 94:13028-13033.
48. Brown JR, Zhang J, Hodgson JE. A bacterial antibiotic resistance gene with eukaryotic origins. Curr Biol 1998; 8:R365-R367.
49. Koonin EV, Mushegian AR, Bork P. Nonorthologous gene displacement. Trends Genet 1996; 12:334-336.
50. Rivera MC, Jain R, Moore JE et al. Genomic evidence for two functionally distinct gene classes. Proc Natl Acad Sci USA 1998; 95:6239-6244.
51. Jain R, Rivera MC, Lake JA. Horizontal gene transfer among genomes: the complexity hypothesis. Proc Natl Acad Sci USA 1999; 96:3801-3806.
52. Lathe WC, Snel B, Bork P. Gene context conservation of a higher order than operons. Trends Biochem Sci 2000; 25:474-479.
53. Brown JR. Aminoacyl-tRNA synthetases: Evolution of a troubled family. In: Wiegel J, Adams M, eds. Thermophiles – the keys to molecular evolution and the origin of life. London: Taylor & Francis Group Ltd, 1998:217-230.
54. Wolf YI, Aravind L, Grishin NV et al. Evolution of aminoacyl-tRNA synthetases – analysis of unique domain architectures and phylogenetic trees reveals a complex history of horizontal gene transfer events. Genome Res 1999; 9:689-710.
55. Woese CR, Olsen GJ, Ibba M et al. Aminoacyl-tRNA synthetases, the genetic code, and the evolutionary process. Microbiol Mol Biol Rev 2000; 64:202-236.
56. Brown JR, Doolittle WF. Gene descent, duplication, and horizontal transfer in the evolution of glutamyl- and glutaminyl-tRNA synthetases. J Mol Evol 1999; 49:485-95.

57. Ibba M, Morgan S, Curnow AW et al. A euryarchaeal lysyl-tRNA synthetase: resemblance to class I synthetases. Science 1997; 278:1119-1122.
58. Teichmann SA, Mitchison G. Is there a phylogenetic signal in prokaryote proteins? J Mol Evol 1999; 49:98-107.
59. Brown JR. Genomic and phylogenetic perspectives on the evolution of prokaryotes. Systematic Biology 2001; 50:497-512.
60. Boucher Y, Doolittle WF. The role of lateral gene transfer in the evolution of isoprenoid biosynthesis pathways. Mol Microbiol 2000; 37:703-716.
61. Doolittle WF, Logsdon Jr JM. Archaeal genomics: Do archaea have a mixed heritage? Curr Biol 1998; 8:R209-R211.
62. Wilding EI, Brown JR, Bryant A et al. Identification, evolution and essentiality of the mevalonate pathway for isopentenyl diphosphate biosynthesis in Gram-positive cocci. J Bacteriol 2000; 182:4319-4327.
63. Smit A, Mushegian A. Biosynthesis of isoprenoids via mevalonate in Archaea: the lost pathway. Genome Res 2000; 10:1468-1484.
64. Lawrence JG. Selfish operons and speciation by gene transfer. Trends Microbiol 1997; 5:355-359.
65. Lawrence JG, Roth JR. Selfish operons – horizontal transfer may drive the evolution of gene clusters. Genetics 1996; 143:1843-1860.
66. Pennisi E. Genome data shake the tree of life. Science 1998; 280:672-674.
67. Pennisi E. Is it time to uproot the tree of life? Science 1999; 284:1305-1307.
68. Doolittle WF. Phylogentic classification and the universal tree. Science 1999; 284:2124-2128.
69. Kurland C. Something for everyone: Horizontal gene transfer in evolution. EMBO Reports 2000; 1:92-95.
70. Altschul SF, Madden TL, Schäffer AA et al. Gapped BLAST and PSI-BLAST: a new generation of protein database search programs. Nucleic Acids Res 1997; 25:3389-3402.
71. Koonin EV, Mushegian AR, Galperin MY et al. Comparison of archaeal and bacterial genomes: computer analysis of protein sequences predicts novel functions and suggests a chimeric origin for the archaea. Mol Microbiol 1997; 25:619-637.
72. Nelson KE, Clayton RA, Gill SR et al. Evidence for lateral gene transfer between Archaea and bacteria from genome sequence of Thermotoga maritima. Nature 1999; 399:323-329.
73. Logsdon Jr JM, Faguy DM. Evolutionary genomics: Thermotoga heats up lateral gene transfer. Curr Biol 1999; 9:R747-R751.
74. Olendzenski L, Liu L, Zhaxybayeva O et al. Horizontal transfer of archaeal genes into the deinococcaceae: detection by molecular and computer-based approaches. J Mol Evol 2000; 51:587-599.
75. International Human Genome Sequencing Consortium. Initial sequencing and analysis of the human genome. Nature 2001; 409:860-921.
76. Roelofs J, van Haastert PJM. Genomics: Genes lost during evolution. Nature 2001; 411:1013-1014.
77. Salzburg SL, White O, Peterson J et al. Microbial genes in the human genome: lateral transfer or gene loss? Science 2001; 292:1903-1906.
78. Stanhope MJ, Lupas AN, Italia MJ et al. Phylogenetic analyses do not support horizontal gene transfers from bacteria to vertebrates. Nature 2001; 411:940-944.
79. Ochman H, Lawrence JG, Groisman EA. Lateral gene transfer and the nature of bacterial innovation. Nature 2000; 405:299-304.
80. Lafay B, Lloyd AT, McLean MJ et al. Proteome composition and codon usage in spirochaetes: species-specific and DNA strand-specific mutational biases. Nucleic Acids Res 1999; 27:1642-1649.
81. Guindon S, Perrière G. Intragenomic base content variation is a potential source of biases when searching for horizontally transferred genes. Mol Bio Evol 2001; 18:1838-1840.
82. Brinkman H, Philippe H. Archaea sister group of bacteria? Indications from tree reconstruction artifacts in ancient phylogenies. Mol Biol Evol 1999; 16:817-825.
83. Chihade J, Brown JR, Schimmel P et al. Detection of an intermediate stage of mitochondria genesis. Proc Natl Acad Sci USA 2000; 97:12153-12157.
84. Brown JR, Masuchi Y, Robb FT et al. Evolutionary relationships of bacterial and archaeal glutamine synthetase genes. J Mol Evol 1994; 38:566-576.
85. Benachenhou-Lahfa N, Forterre P, Labedan B. Evolution of glutamate dehydrogenase genes: Evidence for paralogous protein families and unusual branching patterns of the archaebacteria in the universal tree of life. J Mol Evol 1993; 36:335-346.
86. Gupta RS, Golding GB. Evolution of HSP70 gene and its implications regarding relationships between archaebacteria, eubacteria and eukaryotes. J Mol Evol 1993; 37:573-582.
87. Forterre P, Bouthier de la Tour C, Philippe H et al. Reverse gyrase from thermophiles: probable transfer of a thermoadaptation trait from Archaea to Bacteria. Trends in Genet 2000; 16:152-154.

88. Faguy DM, Doolittle WF. Horizontal transfer of catalase-peroxidase genes between Archaea and pathogenic bacteria. Trends in Genet 2000; 16:196-197.
89. Koretke KK, Lupas AN, Warren PV et al. Evolution of two-component signal transduction. Mol Biol Evol 2000; 17:1956-1970.
90. Lamour V, Quevillon S, Diriong S et al. Evolution of the Glx-tRNA synthetase family: The glutaminyl enzyme as a case for horizontal gene transfer. Proc Natl Acad Sci USA 1994; 91:8670-8674.
91. Margulis L. Origin of eukaryotic cells. New Haven: Yale University Press, 1970.
92. Rujun T, William M. How many genes in Arabidopsis come from cyanobacteria? An estimate from 386 protein phylogenies. Trends in Genet 2001; 17:113-120.
93. Clark CG, Roger AJ. Direct evidence for secondary loss of mitochondria in Entamoeba histolytica. Proc Natl Acad Sci USA 1995; 92:6518-6521.
94. Germot A, Philippe H, Le Guyader H. Presence of a mitochondrial-type 70-kDa heat shock protein in Trichomonas vaginalis suggests a very early mitochondrial endosymbiosis in eukaryotes. Proc Natl Acad Sci USA 1996; 93:14614-14617.
95. Hashimoto T, Sánchez LB, Shirakura T et al. Secondary absence of mitochondria in Giardia lamblia and Trichomonas vaginalis revealed by valyl-tRNA synthetase phylogeny. Proc Natl Acad Sci USA 1998; 95:6860-6865.
96. Henze KA, Badr A, Wettern M et al. A nuclear gene of eubacterial origin in Euglena gracilis reflects cryptic endosymbioses during protist evolution. Proc Natl Acad Sci USA 1995; 92:9122-9126.
97. Keeling PJ, Doolittle WF. Evidence that eukaryotic triosephosphate isomerase is of alpha-proteobacterial origin. Proc Natl Acad Sci USA 1997; 94:1270-1275.
98. Brown JR, Italia MJ, Douady C et al. Horizontal gene transfer and the universal tree of life. In: Syvanen M, Kado CI, eds. Horizontal Gene Transfer. 2nd eds. Academic Press, 2002: In press.
99. Huynen M, Snel B, Bork P. Lateral gene transfer, genome surveys and the phylogeny of prokaryotes. Technical Comments. Science 1999; 286:1443a.
100. Lin J, Gerstein M. Whole-genome trees based on the occurrence of folds and orthologs: implications for comparing genomes on different levels. Genome Res 2000; 10:808-818.
101. Doolittle WF. Lateral gene transfer, genome surveys and the phylogeny of prokaryotes. Technical Comments. Science 1999; 286:1443a.
102. Wolf YI, Rogozin IB, Grishin NV et al. Genome trees constrcuted using five different approaches suggest new major bacterial clades. BMC Evolutionary Biology 2001 1:8.
103. Wolf YI, Rogozin IB, Kondrashov AS et al. Genome alignment, evolution of prokaryotic genome organization,and prediction of gene function using genomic content. Genome Res 2001; 11:356-372.
104. Xiong J, Inoue K, Bauer CE. Tracking molecular evolution of photosynthesis by characterization of a major photosynthesis gene cluster from Heliobacillus mobilis. Proc Natl Acad Sci USA 1998; 95:14851-14856
105. Baldauf SL, Roger AJ, Wenk-Siefert I et al. A kingdom-level phylogeny of eukaryotes based on combined protein data. Science 2000; 290:972-977.
106. Gribaldo S, Lumia V, Creti R et al. Discontinous occurrence of the hsp70 (dnaK) gene among Archaea and sequence features of HSP70 suggest a novel outlook on phylogenies inferred from this protein. J Bacteriol 1999; 181:434-443.
107. Brown JR, Douady CJ, Italia MJ et al. Universal trees based on large combined protein sequence datasets. Nature Genetics 2001; 28: 281-285.
108. Tatusov RL, Koonin EV, Lipman DJ. A genomic perspective on protein families. Science 1997; 278:631-637.
109. Shah HN, Gharbia SE, Collins MD. The Gram stain: a declining synapomorphy in an emerging evolutionary tree. Rev in Med Microbiol 1997; 8:103-100.
110. Olsen GJ, Woese CR, Overbeek R. The winds of (evolutionary) change: breathing new life into microbiology. J Bacteriol 1994; 176:1-6.
111. Woese CR. Bacterial evolution. Microbiol Rev 1987; 51:221-271.
112. Pace NR. Origin of life—Facing up to the physical setting. Cell 1991; 65:531-533.
113. Galtier N, Tourasse N, Gouy M. A nonhyperthermophilic common ancestor to extant life forms. Science 1999; 283:220-221.
114. Kurland C, Andersson SGE. Origin and evolution of the mitochondrial proteome. Microbiol Mol Biol Rev 2000; 64:786-820.
115. Andersson SG, Zomorodipour A, Andersson JO et al. The genome sequence of Rickettsia prowzekii and the origin of mitochondria. Nature 1998; 396: 133-140.
116. Hansmann S, Martin W. Phylogeny of 33 ribosomal and six other proteins encoded in an ancient gene cluster that is conserved across prokaryotic genomes: influence of excluding poorly alignable sites from analysis. Int J Syst Evol Microbiol 2000; 50:1655-1663.

117. Woese CR. The universal ancestor. Proc Natl Acad Sci USA 1998; 51:221-271.
118. Martin W, Müller M. The hydrogen hypothesis for the first eukaryote. Nature 1998; 392:37-41.
119. Doolittle WF. You are what you eat: A gene transfer ratchet could account for bacterial genes in eukaryotic nuclear genomes. Trends in Genet 1998; 14:307-311.
120. Keeling PJ, McFadden GI. Origins of microsporidia. Trends in Microbiol 1998; 6:19-23.
121. Brown JR. Ancient horizontal gene transfer. Nature Rev Gen 2003; 4:121-132.
122. Swofford DL. PAUP*. Phylogenetic Analysis Using Parsimony (*and Other Methods). Version 4.0b5. Sunderland: Sinauer Associates, 1999.
123. Felsenstein J. PHYLIP (Phylogeny Inference Package) version 3.6. Distributed by the author: http://evolution.genetics.washington.edu/phylip.html, Seattle: University of Washington, Seattle. 2000.
124. Strimmer K, von Haeseler A. Quartet puzzling: A quartet maximum likelihood method for reconstructing tree topologies. Mol Biol Evol 1996; 13:964-969.

Chapter 3

The Nature of the Last Common Ancestor

Luis Delaye, Arturo Becerra and Antonio Lazcano

Introduction

Until the late 1970s cellular evolution was assumed to be a continuous, unbroken chain of progressive transformations that begun with the emergence of life itself and continued until the endosymbiotic origin of eukaryotes marked the major biological discontinuity. This scheme was challenged when the comparison of small subunit ribosomal RNA (16/18S rRNA) sequences led to the construction of a trifurcated, unrooted tree in which all known organisms can be grouped in one of three major monophyletic cell lineages, i.e., the domains Bacteria (eubacteria), Archaea (archaeabacteria), and Eucarya (eukaryotes).[1] Information from one single molecular marker does not necessarily yield a precise reconstruction of evolutionary processes, but as shown by numerous phylogenies constructed from other genes such as those encoding polymerases, elongation factors, F-type ATPase subunits, heat-shock and ribosomal proteins, the identification of the three major lineages is not an artifact based solely upon the reductionist extrapolation of information derived from the rRNA tree, but a true reflection of an ancient trifurcation.

Cladistic analysis of rRNA sequences is acknowledged as a prime force in systematics, and from its very inception had a major impact in our understanding of cellular evolution. As shown by the unrooted rRNA trees, no single domain predates the other two and all three derive from a common ancestor. Recognition of the significant differences that exist between the transcriptional and translational machineries of the Bacteria, Archaea and Eucarya, which were assumed to be the result of independent evolutionary refinements, led to the conclusion that the primary branches were the descendants of a progenote, a hypothetical biological entity in which phenotype and genotype still had an imprecise, rudimentary linkage relationship.[2]

From an evolutionary point of view it is reasonable to assume that at some point in time the ancestors of all forms of life must have been less complex than even the simpler extant cells However, the conclusion that the last common ancestor (LCA) was a progenote was disputed over ten years ago when the analysis of homologous traits found among some of its descendants suggested that it was not a direct, immediate descendant of the RNA world, a protocell or any other prelife progenitor system. Under the assumption that horizontal gene transfer (HGT) had not been a major driving force in the distribution of homologous traits in the three domains, it was concluded that the LCA was a complex organism, much alike extant bacteria.[3,4] A decade ago the inventory of such shared features was small, but it was surmised that the sketchy picture developed with the limited databases would be confirmed when completely sequenced cell genomes from the three primary domains. This has not been the case: the availability of an increasingly large number of completely sequenced cellular genomes has sparked new debates, rekindling the discussion on the nature of the ancestral entity.[5] This is shown, for instance, in the diversity of names that have been coined to describe it: progenote,[2] cenancestor,[6] LUCA, last universal cellular ancestor,[7] and LCC, last common community,[8] among others. These terms are not truly synonymous, and they reflect the current controversies on the nature

The Genetic Code and the Origin of Life, edited by Lluís Ribas de Pouplana.

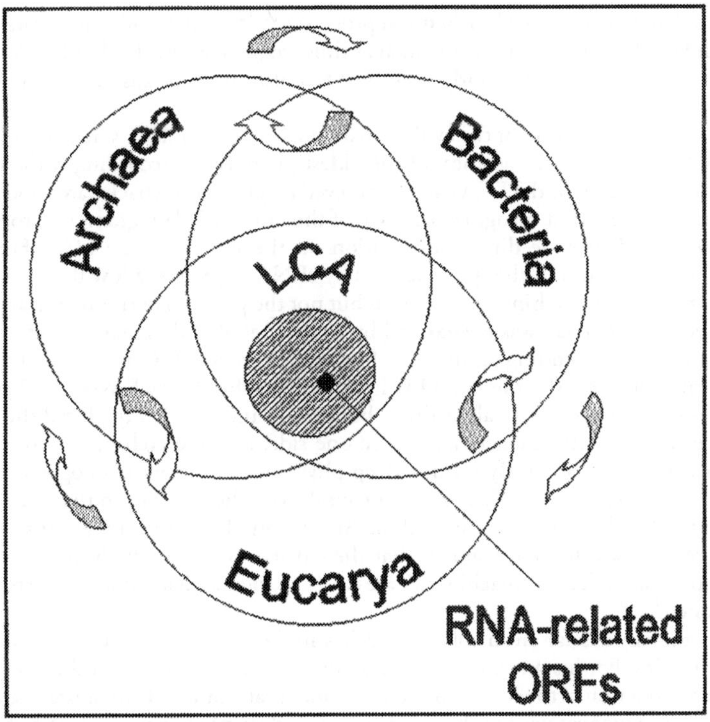

Figure 1. Venn diagram scheme indicating the most parsimonious characterization of the gene complement of the last common ancestor (LCA). The inner subset corresponds to highly conserved RNA metabolism-related sequences (see text), and the arrows indicate horizontal gene transfer (HGT) events, which is some cases involved endosymbiotic events.

of the universal ancestor and the evolutionary processes that shaped it. In this chapter we survey some the difficulties encountered in the description of the last common ancestor, and summarize ongoing discussions on its nature, reviewing briefly how this information can be used to infer earlier steps in biological evolution.

Universal Phylogenies and the Search for the Cenancestor

The traits shared by all known living beings are far to numerous and complex to assume that they evolved independently. Minor differences in the basic molecular processes of the three main cell lines can be distinguished, but all known organisms share the same genetic code and the same essential features of genome replication, gene expression, basic anabolic reactions, and membrane-associated ATPase mediated energy production. The molecular details of these universal processes not only provide direct evidence of the monophyletic origin of all extant forms of life, but also imply that the sets of genes encoding the components of these complex traits were frozen a long time ago, i.e., major changes in them are strongly selected against.

The variations that are observed in extant species can be easily explained as the outcome of divergent processes from an ancestral life form, *fons et origo* of all contemporary organisms. Of course, no geological remains will bear testimony of its existence, as the search for a fossil of the universal ancestor is bound to prove fruitless;[9] from a cladistic viewpoint, the LCA is merely an inferred inventory of features shared among extant organisms (Fig. 1), all of which are located at the tip of the branches of molecular phylogenies. However, if the term "universal distribution" is restricted to its most obvious sense, i.e., that of traits found in all completely sequenced genomes, then quite surprisingly the resulting repertoire is formed by relatively few features and by incompletely represented biochemical processes.[10-12] Analysis of some of the most likely *a priori* candidates for strict universality, such as those sequences involved in DNA replication, have turned out to be not only poorly preserved but also, in some cases, of polyphyletic origin.[13,14]

In principle, determination of the evolutionary polarity of character states in universal phylogenies should lead to the recognition of the oldest phenotype. Accordingly, the most parsimonious characterization of the LCA can be achieved by proceeding backwards and summarizing the features of the oldest recognizable node of the universal cladogram, i.e., rooting of the universal tree would provide direct information on the nature of the LCA. However, the plesiomorphic traits found in the space defined by rRNA sequences allow the construction of topologies that specify branching relationships but not the position of the ancestral phenotype. This phylogenetic *cul-de-sac* was overcomed by Iwabe[15] et al and Gogarten[16] et al, who analyzed paralogous genes encoding (a) the two elongation factors (EF-G and EF-Tu) that assist in protein biosynthesis; and (b) the α and β hydrophilic subunits of F-type ATP synthetases. Using different tree-constructing algorithms, both teams independently placed the root of the universal trees between the eubacteria, on the one side, and the archaea and the eukaryotic nucleocytoplasm on the other. By rooting deep phylogenies, ancient paralogous duplications provide the means to place the LCA in the universal tree. The conclusion that Bacteria are the oldest recognizable cellular phenotype, and the Archaea and Eucarya sister groups, is consistent with sequence analyses that have shown that the eukaryotic genes involved in the transcription/transcriptional molecular machineries are closer to their archaeal counterparts than to the eubacterial ones.[17-20]

However, the issue is far from solved, and has in fact been further complicated with the advent of genomics. For instance, Philippe and Forterre7 have argued that the bacterial root is a long-branch attraction artifact due to the mutational saturation of the more than 3.5 x 109 years-old marker sequences used in the construction of deep phylogenies. As part of an attempt to overcome this limitation, they have used a covarion-based phylogeny-building methodology that allows for rate variation of conserved sites under varying constraints, which led to cladograms with an eukaryotic root.

This conclusion has been enthusiastically embraced by Penny and Poole,[21] who in a number of publications have argued that the eucaryal fragmented genome (as indicated by the existence of separate chromosomes) and intranuclear RNA processing are evidence of the primitiveness of nucleated cell genomes, i.e., that the LUCA was a eukaryote. This hypothesis has been presented, albeit with somewhat different emphasis, by others.[5,22] However, there are several reasons that lead us to disagree with the proposal made by Penny and Poole (1999).[21] These include not only the presence of a widely distributed set of conserved set of DNA repair enzymes that are present in the three domains,[23] which may be interpreted as evidence of a cenancestral DNA genome, but also the following:

a. Although it is likely that the segmented genomes found among certain RNA viruses represent an evolutionary strategy to overcome the Eigen error threshold,[24] the average length of eukaryotic chromosomes is in general well above the size of each viral RNA genomic segment. Moreover, multiple chromosomes and other traits of eukaryotic genome architecture are not by themselves indicative of the antiquity of the eucaryal nucleocytoplasm; as summarized by Bendich and Drlica,[25] yeast telomerase-defficient cells are endowed with circular chromosomes, and other architectural features typical of eukaryotic genomes, such as polyploidy, linear chromosomes, and very large amounts of DNA have also been described in different prokaryotic species;
b. Intranuclear RNA processing is characterized self-splicing reactions of the immature RNA phosphodiester backbone. However, there is no conclusive evidence that intron self-splicing and ribozyme-mediated RNA processing are truly primordial activities: ribozymes with ligase activity and self-cleaving RNAs ribozymes are extremely abundant, and distinct mechanisms by which editing can occur have been described.[26] These observations demonstrate the polyphyly of ribozyme-mediated processes, and imply that not all of them are truly

vestigial activities, i.e., not all eukaryotic RNA processing is a relic of a preDNA/protein world but may be in fact a later development; and

c. Cholesterol and related sterols are hallmarks of nucleated cells. This is true even of anaerobic, amitochondrial ancient species such as Giardia lamblia, where cholesterol is furnished by its host. Although eucaryal genome architecture and sterol biosynthesis are independent features, the highly flexible eukaryotic internal membrane system which underlies the endoplasmic reticulum and the nuclear membrane, which defines the environment where RNA processing takes place, would not be possible in the absence of cholesterol. Since the anaerobic biosynthesis of cholesterol is not feasible, this suggests that, in contrast to prokaryotes, eukaryotes could have not appeared until free oxygen accumulated in the Precambrian environment. This strongly diminishes the likelihood of a eucaryal-like LCA.

Progenote Swarms or Prokaryote-Like Cenancestors

Analysis of an increasingly large number of genes and genomes has revealed major discrepancies with the topology of rRNA trees. As summarized by Brown (this volume) very often these differences have been interpreted as evidence of horizontal gene transfer (HGT) events between different species, questioning the feasibility of the reconstruction and proper understanding of early biological history.[27] Depending on their different advocates, a wide spectrum of mix-and-match recombination processes have been described, ranging from the lateral transfer of few genes, to cell fusion events involving organisms from different domains. There is clear evidence that genomes have a mosaic-like nature whose components come from a wide variety of sources.[28] However, not all sequences have the same likelihood of undergoing horizontal transfer events. Proteomic analysis of functional groups of sequences suggest that while housekeeping genes are more prone to HGT, genes involved in transcription, translation, and related process are less likely to be transferred.[29] On the one hand, these observations help to understand the peculiarities of metabolic gene phylogenies[30] and, on the other, the fact that even rRNA can undergo HGT events[31,32] supports contentions of a web-like pattern of early biological history.[27]

Reticulate phylogenies greatly complicate the inference of cenancestral traits. Driven in part by the impact of lateral gene acquisition, as revealed by the discrepancies of different gene phylogenies with the rRNA tree, and in part by the surprising complexity of the universal ancestor as suggested by direct backtrack characterizations of the oldest node of universal cladograms, Woese[33] proposed that the LCA was not a single organism, but rather a highly diverse population of metabolically complementary, cellular progenotes endowed with multiple, small linear chromosome-like genomes that benefited from massive multidirectional horizontal transfer events. According to this model, the essential features of translation and the development of metabolic pathways took place before the earliest branching event, but what led to the three domains was not a single ancestral lineage, but a rapidly differentiating community of genetic entities. This communal ancestor occupied as a whole the node located at the bottom of the universal tree, in which the decrease of sequence exchange and increasing genetic isolation would eventually lead to the observed tripartite division of the biosphere.

We have an alternative opinion. The genetic entities that formed the communal ancestor proposed by Woese[33] may have been extremely diverse, but an indication of their ultimate monophyletic origin from a sole progenitor is provided by universally distributed features such as the genetic code and the gene expression machinery. Did this hypothetical communal progenote ancestor diverged sharply into the three domains soon after the appearance of the code and the establishment of translation? Not necessarily. The origin of the mutant sequences ancestral to those found in all extant species, and the divergence of the Bacteria, Archaea, and Eucarya were not synchronous events, i.e., the separation of the primary domains took place later, perhaps even much later, than the appearance of the genetic components of their last common ancestor. Moreover, by definition, the node located at the bottom of the cladogram is the root of a phylogenetic tree, and corresponds to the common ancestor of the group under study. But names may be misleading. What we have been calling the root of the universal tree is in fact the tip of its trunk: inventories of LCA genes include sequences that originated in different precenancestral epochs.[11,34-36]

Universal gene-based phylogenies ultimately reach a single universal entity, but the bacterial-like LCA, which we favor, was not alone. Company must have been kept by its siblings, a population of entities similar to it that existed throughout the same period. They may have not survived, but some of their genes did if they became integrated via lateral transfer into the LCA genome. The cenancestor is one of the last evolutionary outcomes of a tree trunk of unknown length, during which the history of a long but not necessarily slow[37] series of ancestral events including lateral gene transfer, gene losses, and paralogous duplications probably played a significant role in the accretion of complex genomes.[3,38,39]

It is currently difficult to propose a unifying hypothesis. However, the scheme outlined here is supported by gene content trees, which exhibit an excellent broad-level agreement with rRNA-based phylogenies.[40-42] Such trees are not cladograms but phenograms, i.e., they are merely hierarchical representations of similarities and differences in gene content, where the presence or absence of a sequence is counted as a character. Since different lineages evolve at different rates, such overall similarity may be an equivocal indicator of genealogical relationships. Nevertheless, these trees are consistent with rRNA phylogenies, and do not support the hypothesis of massive HGT between distant species. Comparisons of combined orthologous protein data sets that exclude sequences that may have undergone lateral transfer are equally consistent with rRNA-based trees.[12] The robustness exhibited by these different methodologies indicate that although lateral gene transfer has played major role in cellular evolution, massive lateral transfer events between distant groups has not taken place. This suggests not only that the early history of life has not been completely obliterated by lateral transfer of genes,[43] but also that the role of reticulate evolution in defining the LCA as a progenote swarm may have been overstated.

The Nature of the Cenancestral Genome: DNA or RNA

Since all extant cells are endowed with DNA genomes, the most parsimonious conclusion is that this genetic polymer was already present in the cenancestral population. Woese[44,45] has suggested otherwise, arguing for a progenote-like universal ancestor endowed with a rapidly evolving genome formed by disaggregated, small-sized RNA molecules. This possibility was supported at least in part by the findings of Mushegian and Koonin,[46] who suggested that the absence of eucaryal or archaeal homologs of key components of DNA replication and nucleotide biosynthesis in the minimal gene set which resulted from the comparison of the Haemophilus influenzae and Mycoplasma genitalium genomes indicated that the cenancestor had used RNA as genetic polymer. Such conclusion is weakened by the limited data set analyzed, which consisted of only two parasitic bacterial genomes that have undergone extensive polyphyletic gene losses.[47] In a subsequent publication, however, Koonin and his collaborators analyzed a large set of primases, replicative polymerases, and other proteins involved in DNA replication, and have suggested an alternative scheme with a hybrid RNA/DNA cenancestral genetic system whose complex replication cycle involving reverse transcription.[48]

There are indeed manifold indications that RNA genomes existed during early stages of cellular evolution[49] but, as argued below, it is likely that double-stranded DNA genomes had become firmly established prior to the divergence of the three primary domains. The major arguments supporting this possibility are:

a. In sharp contrast with other energetically favorable biochemical reactions (such as phosphodiester backbone hydrolysis or the transfer of amino groups), the direct removal of the oxygen from the 2'-C ribonucleotide pentose ring to form the corresponding deoxy-equivalents is a thermodynamically much less-favored reaction, considerably reducing the likelihood of multiple, independent origins of biological ribonucleotide reduction;
b. demonstration of the monophyletic origin of ribonucleotide reductases (RNR) is greatly complicated by their highly divergent primary sequences and the different mechanisms by which they generate the substrate 3'-radical species required for the removal of the 2'-OH

group. However, sequence analysis and biochemical characterization of archaebacterial RNRs have shown their similarities with their eubacterial and eukaryotic counterparts, confirming their ultimate monophyletic origin;[50-52] and

c. sequence similarities shared by many ancient, large proteins found in all three domains suggest that considerable fidelity existed in the operative genetic system of their common ancestor, but such fidelity is unlikely to be found in RNA-based genetic systems.[3]

While accepting a DNA component in the LCA genome, Leipe et al[48] have underlined the highly divergent character of the main components of the (eu)bacterial replication machinery when compared with their archaeal/eukaryotic counterpart. Although it is possible to recognize the evolutionary relatedness of various orthologous DNA informational proteins (i.e., ATP-dependent clamp loader proteins, topoisomerases, gyrases, and 5'-3' exonucleases) across the entire phylogenetic spectrum,[14,13,48] comparative proteome analysis has shown that (eu)bacterial replicative polymerases and primases lack homologues in the two other primary kingdoms. As argued by Leipe et al[48] these observations can be explained by assuming a dual, independent origin of the DNA replication machineries of the Bacteria, on the one hand, and of the Archaea/Eucaryal on the other. Further convolutions have been added to the plot by Forterre,[53] who argued that the evolutionary separation between the replication components resulted from the nonorthologous displacement by rapidly evolving viral or plasmid-encoded gene products soon after the divergence of the three primary domains, as well by Villarreal and DeFilippis,[54] who in a similar vein have suggested a viral origin of nucleated cell DNA polymerases.

Evolution of enzymes in biological systems often involves the acquisition of new catalytic or binding properties by an existing protein scaffold. This has not been the case for the major types of polymerases, as shown by the identification of several nonhomologous classes of polymerases: primases, DNA polymerases, DNA-dependent RNA polymerases, replicases, and poly(A) polymerase, among others.[55] The polyphyletic origin of different polymerases and the large sequence space explored by DNA polymerases probably reflect the energetically favorable character of the enzyme-mediated synthesis of phosphodiester bonds in the presence of a template.

All DNA polymerases whose tertiary structure has been determined share a common overall architectural feature comparable to a right hand shape. Detailed analysis of the three-dimensional structures of the pol I, pol α, and reverse transcriptase families have shown that their palm subdomain, which catalyzes the formation of the phosphodiester bond, is homologous in all of them, while the fingers and thumb subdomains are different in all four of the families for which structures are known.[55] Homologous palm subdomains have also been identified in the viral T7 DNA- and RNA polymerases,[56] indicating that it can catalyze the template-dependent polymerization of ribo- and of deoxyribonucleotides (Fig. 2). More recently, the construction of a database of aligned crystal structures of DNA pol families A and B has allowed the precise identification of the conserved motifs described by Poch et al[57] in the catalytic palm subdomain of DNA polymerase families A(I) and B(II), and leading to its identification in the eukaryotic DNA polymerase δ and ζ subunits.[58]

As summarized by Forterre,[53] a nucleic acid replication enzymatic machinery requires, at the very least, a replicase, a primase, and a helicase, which are currently described as nonorthologues between the bacterial and the archaea/eukaryotic branches. Given the central role that is assigned to nucleic acid replication in mainstream definitions of life,[59] the lack of conservation and polyphyly of several of its key enzymatic components is somewhat surprising. However, the ample phylogenetic distribution of the catalytic palm subdomain and the relative template- and substrate specificities of polymerases[60,61] and helicases, suggest an explanation for the evolution of the DNA replication machinery simpler that those advocated by Leipe et al,[48] Forterre,[53] and Villarreal and DeFilippis.[54]

Our scheme assumes that the conserved palm subdomain described above is one of the oldest recognizable components of an ancestral cellular polymerase that may have acted both as a replicase and a transcriptase during the RNA/protein world stage. Once the advent of

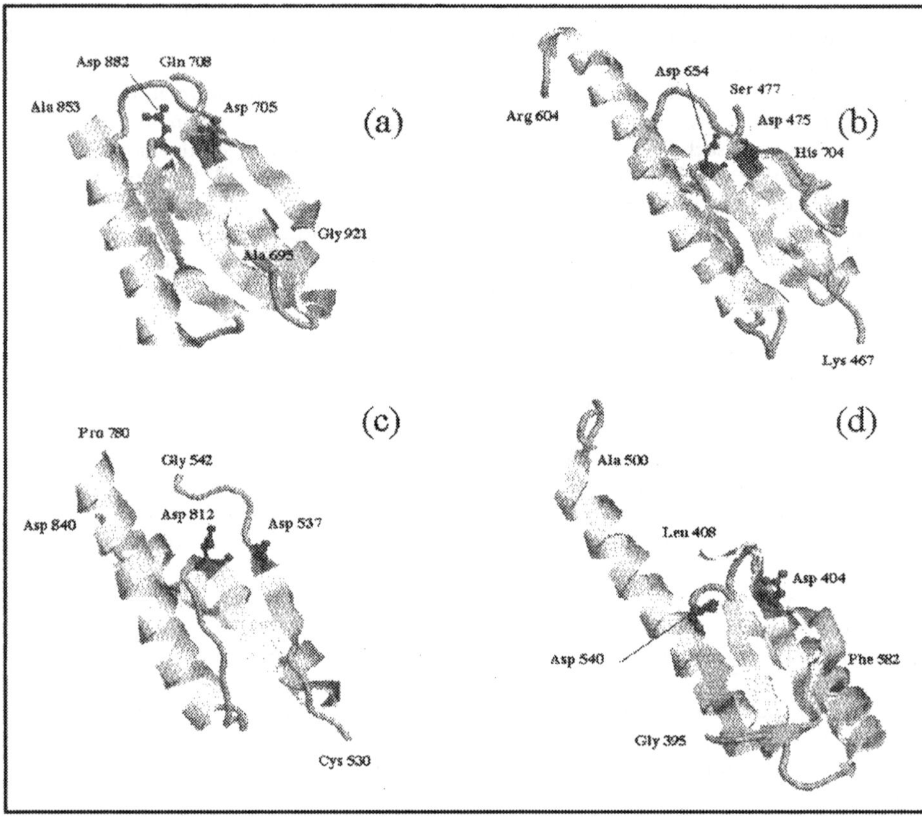

Figure 2. Conserved catalytic palm subdomain of the family I E. coli DNA polymerase I (a), the bacteriophage T7 DNA pol (b) and DNA-dependent RNA polymerase (c), and the family II Desulfurococcus DNA pol II (d). (Adapted from Brautigam and Steitz;[90] Cheetham et al;[91] Jeruzalmi and Steitz;[56] and Zhao et al.[92])

double-stranded DNA took place, relatively few mutations would have been required for the evolution of this RNA replicase into a DNA polymerase prior to the divergence of the three domains. Our hypothesis implies that this progenitor DNA polymerase was originally involved in the replication of the LCA genome, until its (eu)bacterial descendant (represented today by repair DNA pol II) underwent a nonorthologous displacement by the ancestor of the Escherichia coli replicative DNA pol III (DNA pol C) and its homologs. The structural homology of RNA- and DNA-helicase domains[62,63] suggest, on the other hand, the possibility of a nonspecific helicase inherited from the RNA/protein world that may have operated in unwinding double-stranded DNA until the evolution of the extant DNA helicases.

By analogy with the yeast and animal mitochondrial RNA polymerases, which play a dual role in transcription and in the initial priming required for DNA replication,[64] we propose that the original RNA polymerase described above catalyzed the formation of the RNA primer required for DNA replication. This hypothesis implies that extant bacterial and archaeal/eucaryotic primases are later independent evolutionary developments that displaced the cenancestral RNA polymerase from its primase function. As suggested above, this ancestral polymerase may have acted as a transcriptase during the RNA/protein stage, but the distribution of the highly conserved sequences of the oligomeric DNA-dependent RNA polymerase indicate that by the time the cenancestor diverged, a modern type of transcription had evolved. How this complex oligomeric transcriptase came into being can only be surmised at the time being.

Some Like It Very, Very Hot

The rooting of universal cladistic trees determines the directionality of evolutionary change and allows the recognition of ancestral from derived characters, i.e., primitive characters should appear in older, basal branches than do their derived counterparts. Determination of the rooting point of a tree normally imparts polarity to most or all characters.[65] It is, however, important to distinguish between ancient and primitive organisms. Organisms located near the root of universal rRNA-based trees are cladistically ancient, but they are not endowed with a primitive molecular genetic apparatus, nor seem to be more primitive in their metabolic abilities than their aerobic counterparts.

The situation is slightly different regarding the phylogenetic distribution of hyperthermophily, which appears to be a truly ancestral, primitive trait. Examination of the prokaryotic branches of unrooted rRNA trees had already suggested that the ancestors of both eubacteria and archaebacteria were extreme thermophiles, i.e., organisms that grow optimally at temperatures in the range 90° C and above.[66] Rooted universal phylogenies confirmed that hyperthermophiles are not randomly distributed in the universal tree, but are clearly located towards the lowest portion of molecular rRNA-based cladograms.[67] It is sometimes overlooked that the bacterial rooting of universal trees implies that hyperthermophilic bacteria such as Thermotoga and Aquifex are closer to the LCA than the oldest hyperthermophilic archaea, including the korearchaeota, which branch below the euryarchaeota/crenarchaeota split.[68] Some hyperthermophile sequences are displaced from their basal positions if molecular markers other than elongation factors or ATPase subunits are compared,[69] but the antiquity of hyperthermophiles appears to be well established,[45,67,70-72] and has received additional support from trees based on combined protein data sets from which sequences alignments that are candidates for HGT have been excluded.[12]

Backward extrapolation of the basal position of hyperthermophiles led not only to the hypothesis of a heat-loving LCA, but also of a high-temperature origin of life,[70] which according to some took place in extreme environments such as those found today in deep-sea vents[73] or in other sites in which mineral surfaces may have fueled the appearance of primordial chemoautolithotrophic biological systems.[74] However, all these views have been contested in one way or another, and are still open issues.[75] For instance, it is difficult to take for granted the possibility of hyperthermophilic universal ancestor endowed with a fragmented RNA genome proposed by Woese[33] with the extreme thermal fragility of RNA molecules.

The recognition that the deepest branches in rooted universal phylogenies are occupied by hyperthermophiles does not provides by itself conclusive proof of a heat-loving LCA. Analysis of the correlation of the optimal growth temperature of prokaryotes and the G+C nucleotide content of 40 rRNA sequences through a complex Markov model, has led Galtier et al[76] to conclude that the universal ancestor was a mesophile. If this is indeed the case, then the distribution of hyperthermophiles in rRNA-based phylogenies could be explained by: (a) lateral transfer of thermoadaptative traits;[77] (b) heat as a relic from early Archean high-temperature regimes that may have resulted from a severe impact regime;[78,79] (c) assuming that hyperthermophiles displaced older mesophiles when they adapted to lower temperatures, rather than being the sole survivors of an impact event.[80] It should be kept in mind, however, that since the time dimension is absent from the low G+C rRNA value inferred by Galtier et al,[76] it is possible that it corresponds not the cenancestor itself, but to one of its evolutionary predecessors, located along the trunk of the universal tree.

Thus, although no mesophilic organisms older than heat-loving bacteria have been discovered, it is possible that hyperthermophily is a secondary adaptation that evolved in early geological times.[78,81,82] Hyperthermophiles not only share the same basic features of the molecular machinery of all other forms of life; they also require a number of specific biochemical adaptations. Such adaptations may include histone-like proteins, RNA modifying enzymes, and reverse gyrase, a peculiar ATP-dependent enzyme that twists DNA into a positive supercoiled conformation.[81] Clues to the origin of hyperthermophily may be hidden in this list, and its evolutionary analysis may contribute to the understanding of the rather surprising phylogenetic distribution of the immediate mesophilic descendants of heat-loving prokaryotes,[67] which shows that at least five independent abandonment events of hyperthermophilic traits took place in widely separated branches of universal trees, one of which corresponds to the eukaryotic nucleocytoplasm.

Trimming the rRNA-Based Universal Trees

The conclusion that the LCA was a prokaryote-like organism similar to extant (eu)bacteria does not says much about its mode of energy acquisition and carbon sources. As summarized by Stetter,[67] the basal position of universal trees are occupied by heterotrophic and autotrophic hyperthermophiles, many of which live in sulphur-rich, extreme environments, with the deepest branches occupied by chemolithoautotrophs that have aerobic and anaerobic respiration. Direct extrapolation of these and other extremophile traits into the LCA has not been taken by granted by all. On the other hand, the irregular distribution of metabolic pathways and the large pool of sequences shared by extant species leads to a totipotent, phototroph LCA, unrealistically endowed with more biochemical attributes than some modern prokaryotes.[33,83] However, if multiple copies of every major gene family are assumed to have been already present in the LCA genome,[43] then the observed complex distribution patterns of bioenergetic and biosynthetic genes can be explained as the outcome of polyphyletic gene losses as the cenancestor descendants adapted to a wide variety of environments under different selection pressures.[38,39]

Although the timescale separating the LCA from the possible emergence of life is not a major problem given the rapid pace of prokaryotic evolution,[37] characterization of the cenancestral metabolic abilities can be hindered by several major problems. These include the horizontal acquisition of metabolic pathways, a possibility enhanced by likelihood of LGT of housekeeping genes,[29] and the fact that many open reading frames derived from complete genome sequencing projects remain unidentified (30 to 50% depending on the organism). It is possible that some of these ORFs correspond to rapidly evolving sequences encoding missing enzymes of incompletely reconstructed metabolic pathways.[84,85]

The inadequate biodiversity sampling that has shaped our current databases, which represent an extremely biased set of sequenced gene and genomes, also complicates our efforts. Quite understandably, medical and veterinarian interests have shaped the nature of extant genome databases from which many species are absent, perhaps even excluding members of every major biological group. Although clearly incomplete, the adequacy of fully sequenced genome databases for the reconstruction of ancestral states is probably greater than it is generally realized. There are, of course, many taxa we do not know about that are yet to be described. However, in spite of this strong limitation and of the extraordinary diversity of habitats and lifestyles, organisms share a surprising amount of enzymatic activities, metabolic routes, and basic biological functions, as reflected in genome replication, gene expression, and metabolic pathways. As the number of completely sequenced genomes has increased, the identification of new genes and functions common to all living beings has not expanded at the same rate (Fig. 3). The possibility that some of the enzymes of archaic pathways may have survived in unusual organisms suggest that considerable prudence should be exerted when attempting to describe the physiology of ancestral organisms. However, the sharp decline in the discovery of new, universally distributed sequences, which would correspond to an almost complete inventory of genes common to all living beings, should signal the approach to an almost complete universal set of genes (Fig. 3).

A more complicated issue is raised by the possibility that extant enzymes participated in alternative pathways which no longer exist or remain to be discovered,[86,87] a possibility that has begun to be explored by computer searches for alternative reaction pathways.[88] The discovery that carbamate kinase, which participates in fermentative ATP production, catalyzes the formation of carbamoyl phosphate in the archaea Pyrococcus furiosus and P. abyssi[89] shows that considerable attention should be given to the possibility that significant variations of the basic pathways may have existed in the past.

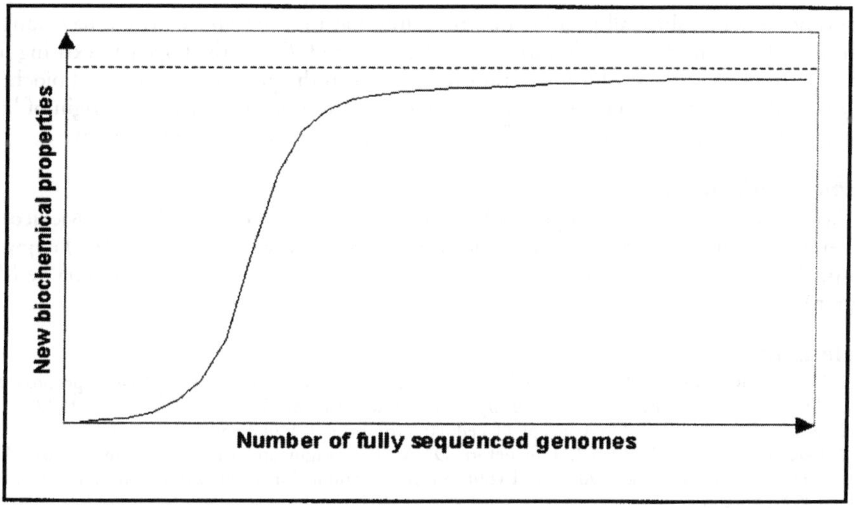

Figure 3. The discovery curve of universally distributed biochemical properties as inferred from proteome analysis. By analogy with the so called collector curve employed in ecology, it quantifies the analytic effort, assessed as the number of completely sequenced genomes analyzed, against the identification of additional biochemical features common to all cellular genomes.

Conclusions and Outlook

Understanding the biological attributes of the LCA and the evolutionary processes that shaped it has been defined as one of the major problems in evolutionary biology. This is not an overstatement, since it will assist in the comprehension of one of the major divergence events in the history of life, as well of paramount significance in understanding the different degrees of freedom that have been explored in the sequence- and three-dimensional spaces by the molecular components of central biological processes. Of course, current descriptions of the LCA are limited by the scant information available. It is hard, of course, to understand the evolutionary forces that acted on our distant ancestors, whose environments and detailed biological characteristics are forever beyond our ken.

Nevertheless, understanding the characteristics of the LCA may assist us in describing the entities that may have preceded it. Although we strongly favor an (eu)bacterial-like cenancestor, it is clear that biological evolution prior to the divergence of the three domains was not a continuous, unbroken chain of progressive transformation steadily proceeding towards the LCA. No evolutionary intermediate stages or ancient simplified version of the basic biological processes have been discovered in extant organisms. Did Woese's[33] differentiating communal progenote-like genetic entities existed during this period?

Molecular cladistics and comparative genomics may provide clues to the genetic organization and biochemical complexity of the earlier entities from which the cenancestor evolved may be derived from the analysis of conserved ORFs. Genes involved in RNA metabolism, i.e., ORFs whose products synthesize, degrade, or interact with RNA, are among the most highly conserved sequences common to all known genomes, and provide insights into an early stage in cell evolution during which RNA played a much more conspicuous biological role.[11,34,36] However, it is difficult to see how the applicability of molecular cladistics and comparative genomics can be extended beyond a threshold that corresponds to a period of cellular evolution in which protein biosynthesis was already in operation. Older stages are not yet amenable to molecular phylogenetic analysis. Although there have been considerable advances in the understanding of chemical processes that may have taken place before the emergence of the first living systems, life's beginnings are still shrouded in mystery. A cladistic approach to this problem is not feasible, since all possible intermediates that may have once existed have long since vanished. The temptation to do otherwise is best resisted. Given the huge gap existing in current descriptions of the evolutionary transition between the prebiotic synthesis of biochemical compounds and the cenancestor, it may be naive to attempt to describe the origin of life and the nature of the first living systems from the available rooted phylogenetic trees.

Acknowledgments

Work reported here was supported by project PAPIIT IN213598 (UNAM, Mexico). This paper was completed during a sabbatical leave of absence in which one of us (AL) enjoyed the hospitality of Dr. Ricardo Amils and his associates in the Universidad Autónoma de Madrid (Spain).

References

1. Woese CR, Kandler O, Wheelis ML et al. Towards a natural system of organisms, proposal for the domains Archaea, Bacteria, and Eucarya. Proc Natl Acad Sci USA 1990; 87:4576-4579.
2. Woese CR, Fox GE. The concept of cellular evolution. J Mol Evol 1977; 10:1-6.
3. Lazcano A, Fox GE, Oró J. Life before DNA: The origin and early evolution of early Archean cells. In: R.P. Mortlock, ed. The Evolution of Metabolic Function. Boca Raton, FL: CRC Press, 1992:237-295.
4. Lazcano A. Cellular evolution during the Early Archean: What happened between progenote and the cenancestor? Microbiologia SEM 1995; 11:185-198.
5. Doolittle WF. The nature of the universal ancestor and the evolution of the proteome. Curr Opin Struct Biol 2000; 10:355-358.
6. Fitch WM, Upper K. The phylogeny of tRNA sequences provides evidence of ambiguity reduction in the origin of the genetic code. Cold Spring Harbor Symp Quant Biol 1987; 52:759-767.
7. Philippe H, Forterre P. The rooting of the universal tree of life is not reliable. J Mol Evol 1999; 49:509-523.
8. Line MA. The enigma of the origin of life and its timing. Microbiology 2002; 148:21-27.
9. Gee H. In search of deep time. New York: The Free Press, 1999.
10. Tatusov RL, Koonin EV, Lipman DJ. A genomic perspective on protein families. Science 1997; 278:631-637.
11. Tekaia F, Dujon B, Lazcano A. Comparative genomics: Products of the most conserved protein-encoding genes synthesize, degrade, or interact with RNA. Abstracts of the 9th ISSOL Meeting San Diego, California, USA: July 11-16, 1999:Abstract c46:53.
12. Brown JR, Douady CJ, Italia MJ et al. Universal trees based on large combined protein sequence datasets. Nat Genet 2001; 28:281-285.
13. Edgell RD, Doolittle WF. Archaea and the origin(s) of DNA replication proteins. Cell 1997; 89:995-998.
14. Olsen G, Woese CR. Archaeal genomics: an overview. Cell 1997; 89:991-994.
15. Iwabe N, Kuma K, Hasegawa M et al. Evolutionary relationship of archaebacteria, eubacteria, and eukaryotes inferred from phylogenetic trees of duplicated genes. Proc Natl Acad Sci USA 1989; 86:9355-9359.
16. Gogarten JP, Kibak H, Dittrich P et al. Evolution of the vacuolar H+-ATPase, implications for the origin of eukayotes. Proc Natl Acad Sci USA 1989; 86:6661-6665.
17. Ouzonis C, Sander C. TFIIB, an evolutionary link between the transcription machineries of archaebacteria and eukaryotes. Cell 1992; 71:189-190.
18. Kaine BP, Mehr IJ, Woese CR. The sequence, and its evolutionary implications, of a Thermococcus celer protein associated with transcription, Proc Natl Acad Sci USA 1994; 91:3854-3856.
19. Brown JR, Doolittle WF. Archaea and the prokaryote to eukaryote transition. Microbiol Mol Biol Rev 1997; 61:456-502.
20. Koonin EV, Mushegian AR, Galperin MY et al. Comparisonof archaeal and bacterial genomes: computer analysis of protein sequences predicts novel functions and suggests a chimeric origin for the archaea. Mol Microbiol 1997; 25:619-637.
21. Penny D, Poole A. The nature of the universal common ancestor. Curr Opin Genet Dev 1999; 9:672-677.
22. Hartman H, Fedorov A. The origin of the eukaryotic cell: a genomic investigation. Proc Natl Acad Sci USA 2002; 99:1420-1425.
23. Eisen JA, Hanawalt PC. A phylogenomic study of DNA repair genes, proteins, and processes. Mutation Res 1999; 435:171-213.

24. Reanney DC. Genetic error and genome design. Cold Spring Harbor Symp Quant Biol 1987; 52:751-757.
25. Bendich AJ, Drlica K. Prokaryotic and eukaryotic chromosomes: What's the difference. BioEssays 2000; 22:481-486.
26. Gesteland RF, Atkins JF. eds. The RNA World: The Nature of Modern RNA Suggests a Prebiotic RNA World Cold Spring Harbor, New York:,Cold Spring Harbor Laboratory Press, 1993.
27. Doolittle WF. Phylogenetic classification and the universal tree. Science 1999; 284:2124-2128.
28. Ochman H, Lawrence JGM, Groisman EA. Lateral gene transfer and the nature of bacterial innovation. Nature 2000; 405:299-304.
29. Rivera MC, Jain R, Moore JE et al. Genomic evidence for two functionally distinct gene clasees. Proc Natl Acad Sci USA 1998; 95:6239-6244.
30. Alifano P, Fani R, Liò P et al. Histidine biosynthetic pathway and genes: structure, regulation, and evolution. Microbiol Rev 1996; 60:44-69.
31. Perez-Luz S, Rodríguez-Valera F, Lan R et al. Variation of the ribosomal operon 16S-23S gene spacer region in representatives of Salmonella enterica subspecies. J Bacteriol 1998; 180:2144-2151.
32. Yap WH, Zhang Z, Wang Y. Distinct types of rRNA operons exist in the genome of the actinomycete Thermonospora chromogena and evidence for horizontal transfer of an entire rRNA operon. J Bacteriol 1999; 181:5201-5209.
33. Woese CR. The universal ancestor. Proc Natl Acad Sci USA 1998; 95:6854-6859.
34. Delaye L, Lazcano A. RNA-binding peptides as molecular fossils. In: Chela-Flores J, Lemerchand G, Oró J, eds. Origins from the Big-Bang to Biology: Proceedings of the First Ibero-American School of Astrobiology (Dordrecht): Kluwer Academic Publishers, 2000:285-288.
35. Lazcano Araujo A. El último ancestro común. In: Martínez Romero E, Martínez Romero Y, eds. Microbios en Línea. UNAM, México, 2001:421-429.
36. Anantharaman V, Koonin EV, Aravind L. Comparative genomics and evolution of proteins involved in RNA metabolism. Nucleic Acid Res 2002; 30:1427-1464.
37. Lazcano A, Miller SL. How long did it take for life to begin and evolve to cyanobacteria? Jour Mol Evol 1994; 39:546-554.
38. Castresana J. Comparative genomics and bioenergetics. Biochem Biophys Acta 2001; 1506:147-162.
39. Snel B, Bork P, Huynen MA. Genomes in flux: the evolution of archaeal and proteobacterial gene content. Genome Res 2002; 12:17-25.
40. Fitz-Gibbon ST, House CH. Whole genome-based phylogenetic analysis of free-living organisms. Nucleic Acids Res 1999; 27:4218-4222.
41. Snel B, Bork P, Huynen MA. Genome phylogeny based on gene content. Nat Genet 1999; 21:108-110.
42. Tekaia F, Lazcano A, Dujon B. The genomic tree as revealed from whole proteome comparisons. Genome Res 1999b; 9:550-557.
43. Glansdorff N. About the last common ancestor, the universal life-tree and lateral gene transfer: A reappraisal. Mol Microbiol 2000; 38:177-185.
44. Woese CR. The primary lines of descent and the universal ancestor. In: Bendall DS, ed. Evolution from Molecules to Men. Cambridge: Cambridge University Press, 1983:209-233.
45. Woese CR. Bacterial evolution. Microbiol Reviews 1987; 51:221-271.
46. Mushegian AR, Koonin EV. A minimal gene set for cellular life derived by comparison of complete bacterial genomes. Proc Natl Acad Sci USA 1996; 93:10268-10273.
47. Becerra A, Islas S, Leguina JI et al. Polyphyletic gene losses can bias backtrack characterizations of the cenancestor. J Mol Evol1997; 45:115-118.
48. Leipe DD, Aravind L, Koonin EV. Did DNA replication evolve twice independently? Nucleic Acid Res 1999; 27:3389-3401.
49. Lazcano A, Guerrero R, Margulis L et al. The evolutionary transition from RNA to DNA in early cells. J Mol Evol 1988a; 27:283-290.
50. Tauer A, Benner SA. The B12-dependent ribonucleotide reductase from the archaebacterium Thermoplasma acidophila: An evolutionary solution to the ribonucleotide reductase conundrum. Proc Natl Acad Sci USA 1996; 94:53-58.
51. Riera J, Robb FT, Weiss R et al. Ribonucleotide reductase in the archaeon Pyrococcus furiosus: a critical enzyme in the evolution of DNA genomes. Proc Natl Acad Sci USA 1997; 94:475-478.
52. Freeland SJ, Knight RD, Landweber LF. Do proteins predate DNA? Science 1999 286:690-692.
53. Forterre P. Displacement of cellular proteins by functional analogues from plasmids or viruses could explain puzzling phylogenies of many DNA informational proteins. Mol Microbiol1999; 33:457-465.
54. Villarreal LP, DeFilippis VR. A hypothesis for DNA viruses as the origin of eukaryotic replication proteins. J Virol 2000; 74:7079-7084.

55. Steitz TA. DNA polymerases: structural diversity and common mechanisms. J Biol Chem1999; 274:17395-17398.
56. Jeruzalmi D, Steitz TA. Structure of T7 RNA polymerase complexed to the transcriptional inhibitor T7 lysozyme. EMBO J 1998 17:4101-4113.
57. Poch O, Sauvaget I, Delarue M et al. Identification of four conserved motifs among the RNA-dependent polymerase encoding elements. EMBO J 1989; 8:3867-3874.
58. Delaye L, Vázquez H, Lazcano A. The cenancestor and its contemporary biological relics: the case of nucleic acid polymerases. In: Chela-Flores J, Owen T, Raulin F, eds. First steps in the origin of life in the Universe. Dordrecht: Kluwer Academic Publisher, 2001:223-230.
59. Koshland DE. The seven pillars of life. Science 2002; 295:2215-2216.
60. Lazcano A, Fastag J, Gariglio P et al. On the early evolution of RNA polymerase. J Mol Evol 1988b; 27:365-37.
61. Siegel RW, Bellon L, Beigelman L et al. Use of DNA, RNA, and chimeric templates by a viral RNA-dependent RNA polymerase: evolutionary implications for the transition from the RNA to the DNA world. J Virol 1999; 73:6424-6429.
62. Theis K, Chen PJ, Skorvaga M et al. Crystal structure of UvrB, a DNA helicase adapted for nucleotide excision repair. EMBO J 1999; 18:6899-6907.
63. Caruthers JM, Johnson ER, McKay DB. Crystal structure of yeast initiation factor 4A, a DEAD-box RNA helicase. Proc Natl Acad Sci USA 2000; 97:13080-13085.
64. Schinkel AH, Tabak HF. Mitochondrial RNA polymerase: dual role in transcription and replication. Trends Genet 1989; 5:149-154.
65. Scotland RW. Character coding. In: Florey PL, Humphries CJ, Kitching IL et al, eds. Cladistics: A practical course in systematics. Oxford: Clarendon Press, 1992:14-43.
66. Achenbach-Richter L, Gupta R, Stetter KO et al. Were the original eubacteria thermophiles? System Appl Microbiol 1987; 9:34-39.
67. Stetter KO. The lesson of archaebacteria. In: Bengtson S, ed. Early Life on Earth, Nobel Symposium No. 84. New York: Columbia University Press, 1994:114-122.
68. Barns SM, Delwiche CF, Palmer JD et al. Perspectives on archaeal diversity, thermophily and monophily from environmental rRNA sequences. Proc Natl Acad Sci USA 1996; 92:2441-2445.
69. Forterre P. A hot topic: The origin of hyperthermophiles. Cell 1996; 85:789-792.
70. Pace N. Origin of life —facing up to the physical setting. Cell 1991; 65:531-533.
71. Di Giulio M. The universal ancestor lived in a thermophilic or hyperthermophilic environment. J Theoret Biol 2000a; 203:203-213.
72. Di Giulio M. The stage of the genetic code structuring took place at a high temperature. Gene 2000b; 261:189-195.
73. Holm NG, ed. Marine Hydrothermal Systems and the Origin of Life. Dordrecht: Kluwer Academic Publ, 1992.
74. Wächtershäuser G. The case for the chemoautotrophic origins of life in an iron-sulfur world. Origins Life Evol Biosph 1990: 20:173-182.
75. Wiegel J, Adams MWW, eds. Thermophiles: The keys to molecular evolution and the origin of life. London: Taylor and Francis, 1998.
76. Galtier N, Tourasse N, Gouy M. A nonhyperthermophilic common ancestor to extant life forms. Science 1999; 283:220-221.
77. Forterre P, Bouthier de la Tour C, Philippe H et al. Reverse gyrase from hyperthermophiles: Probable transfer of a thermoadaptation trait from Archaea to Bacteria. Trends Genet 2000; 16:152-154.
78. Sleep NH, Zahnle KJ, Kastings JF et al. Annihilation of ecosystems by large asteroid impacts on the early Earth. Nature 1989; 342:139-142.
79. Gogarten-Boekels M, Hilario E, Gogarten JP. The effects of heavy meteoritic bombardments of the early evolution —the emergence of the three domains of life. Origins Life Evol Biosph 1995; 25:251-264.
80. Miller SL, Lazcano A. The origin of life —did it occur at high temperatures? J Mol Evol 1995; 41:689-692.
81. Confalonieri F, Elie C, Nadal M et al. Reverse gyrase, a helicase-like domain and a type I topoisomerase in the same polypeptide. Proc Natl Acad Sci USA 1993; 90:4753-4758.
82. Lazcano A. Biogenesis, some like it very hot. Science 1993; 260:1154-1155.
83. Olsen G, Woese CR. Lessons from an archeal genome: what are we learning from Methanococcus jannaschii? Trends Genet 1996; 12:377-379.
84. Bono H, Ogata H, Goto S et al. Reconstruction of amino acid biosynthesis pathways from the complete genome sequence. Genome Res 1998; 8:203-210.
85. Velasco AM, Leguina JI, Lazcano A. Molecular evolution of the lysine biosynthetic pathways. J Mol Evol 2002; in press.

86. Zubay G. To what extent do biochemical pathways mimic prebiotic pathways? Chemtracts-Biochem. Mol Biol 1993; 4:317-323.
87. Becerra A, Lazcano A. The role of gene duplication in the evolution of purine nucleotide salvage pathways. Origins Life Evol Biosph 1998; 28:539-553.
88. Goto S, Bono H, Ogata H et al. Organizing and computing metabolic pathway data in terms of binary relations. In: Altman RB, Dunker K, Hunter L et al, eds. Pacific Symposium on Biocomputing. Singapore: World Scientific, 1996:175-186.
89. Alcántara C, Cervera J, Rubio V. Carbamate kinase can replace in vivo carbamoyl phosphate synthetase. Implications for the evolution of carbamoyl phosphate biosynthesis. FEBS Lett 2000; 484:261-264.
90. Brautigam CA, Steitz TA. Structural principles for the inhibition of the 3',5' exonuclease activity of Escherichia coli DNA polymerase I by phosphorothioates. J Mol Biol 1998; 277:363-377.
91. Cheetham GM, Jeruzalmi D, Steitz TA. Structural basis for initiation of transcription from an RNA polymerase-promoter complex. Nature 1999; 399:80-83.
92. Zhao Y, Jeruzalmi D, Moarefi I et al. Crystal structure of an archaeabacterial DNA polymerase. Struct Fold Des 1999; 7:1189-1199.

Chapter 4

Ribozyme-Catalyzed Genetics

Donald H. Burke

Summary

RNA World research in recent years has sought to establish whether ribozymes have the catalytic versatility and potency to transmit genetic information and to sustain a credible metabolism. At a minimum, organisms from just before the Protein Revolution would have had to catalyze nucleotide polymerization and invent the machinery for protein synthesis. There are now RNA enzymes (ribozymes) that catalyze the individual steps in each of these reactions. Some of the current challenges include increasing the vigor with which the individual reactions are catalyzed, strengthening the affinity and specificity of substrate recognition, integrating ribozymes into metabolic paths and coordinated networks of linked reactions, and deriving a ribozyme-catalyzed metabolic context to sustain the core reactions.

RNA World theories of the earliest evolution of life enjoy increasing acceptance and experimental support. The simplest statement of the RNA World theory is that our evolutionary history includes at least one organism that depended on RNA molecules both as the primary repository for genetic information and as the principle set of catalysts for cellular functions. In modern organisms, these two roles are predominantly filled by DNA and proteins, respectively. RNA World organisms are variously referred to as "ribo-organisms" or "ribocytes." The first ribocyte to make use of genetically encoded translation is called the "breakthrough organism." Far removed both from life's origins and from recognizably modern biochemistry, the descendents of the breakthrough organism are thought to have accumulated a broad diversity of proteins enzymes that took over nearly all of the functions of the cell. The transition from ribocytes to modern forms may have left traces of the ancestral state in the form of nucleotide cofactors, ribosomes built largely from RNA, and the requirement for ribonucleotides as biosynthetic precursors for deoxyribonucleotides. Speculation on how this transition may have taken place, and the relevant experimental evidence, are discussed in other chapters of this volume.

This review evaluates the degree to which ribozymes identified to date are adequate to the task of sustaining genetic information flow. Emphasis is on the reactions that underlie transmission of genetic information. The first section defines the aspects of RNA World theories that are within the purview of this endeavor. The next two sections describe the evolution, activities, and experimental challenges of ribozymes that catalyze each of the discrete reactions of replication and proteins synthesis. The fourth section addresses progress towards generating an RNA-catalyzed metabolism to support the flow of genetic information, and the fifth section offers concluding remarks. Additional reviews in related areas have been published elsewhere.[1,5-9]

Two RNA World Views

While significant challenges remain, the notion that ribozyme catalysis can sustain genetic information flow has survived the initial wave of experimental examination. By extension, the notion of RNA-based life has similarly survived. RNA World theories tend to come in two

The Genetic Code and the Origin of Life, edited by Lluís Ribas de Pouplana.

flavors, emphasizing either the abiotic chemical processes that led to the first living things (origins per se) or the early evolution of cellular organisms. The two views place different requirements on the chemical capacities of nucleic acids —which can be revealed by experimentation— without being mutually exclusive. Each is critically defined by a key, hypothesized transition:

Chemical view: Abiotic nucleotide chemistry → RNA-catalyzed biochemistry.

Biochemical view: RNA-based life → Protein/DNA-based life

The fundamental quest of the chemical approach is to connect abiotic chemistry with the first living entities, and is sometimes dubbed the "RNA early" view. Chemical reaction pathways appropriate to an early, lifeless Earth (or Mars or Europa or other world) are sought that yield self-replicating molecules —usually in the form of RNA or RNA-like polymers— or their constituent parts. Work in this area includes the prebiotic synthesis of the components of RNA (nucleobases, D-ribose sugars, nucleotides, and polyphosphorylated nucleotide monomers), the chemical properties of non-standard nucleobases and nucleotide analogs, non-enzymatic copying of monomers into random or templated polymers, and their encapsulation into proto-cells. Abiotic routes leading directly to RNA have thus far proven elusive. Alternate scenarios involving other genetic and biochemical infrastructures have been proposed as contexts for the initial appearance of RNA, and are collectively dubbed the "Pre-RNA-World".[1-4]

The fundamental quest of the early biochemical evolution approach is to trace extant biology and biochemistry backwards to an earlier, presumably simpler life form without proteins or DNA. It makes no specific claims as to the source(s) of the RNA or the mechanism by which life originated, nor does it specify the complexity of the biochemical context within which RNA arose. Work in this area addresses the chemical and biological versatility of ribozymes, the adaptive landscapes and evolutionary pathways along which new functions are derived from existing functional RNAs, the assembly of multiple ribozymes into complex metabolic pathways in vitro, the influence of engineered RNAs on the physiology of modern or artificial cells, and the roles of natural, non-protein-coding RNAs in existing cells. These questions are biochemical and biological as much as they are chemical in scope, and they seek to define the nature of life based on an RNA-catalyzed metabolism.

The ultimate test of RNA World theories lies less in establishing how an RNA World could have come into existence, and more in demonstrating whether RNA-based organisms can stay alive. One must define the requirements of life and ask whether RNA molecules have the chemical properties needed to fulfill those requirements. For the protein-catalyzed biochemistry of modern organisms to have come into existence within an RNA-catalyzed biochemistry, the ribozymes of that ancestral state must have catalyzed the transmission of genetic information through the various reactions of replication and protein synthesis. Additional ribozymes may also have been required catalyzed some elements of biosynthesis and energy extraction. Ribozymes that accelerate many of the requisite classes of reactions several orders of magnitude above background rates have been identified through in vitro selections, laying the foundation for future work to establish the validity of RNA-based life.

RNA-Catalyzed Genetics I: Nucleotide Polymerization

The Chemical Basics of Polymerization

The essence of replication is the sequential, templated addition of nucleotides onto a growing chain. A phosphate ester is formed at the expense of a phosphoanhydride when the terminal hydroxyl of the growing chain attacks the alpha phosphate of an incoming mononucleotide triphosphate to release pyrophosphate (Fig. 1). The large body of work that has gone into understanding natural ribozymes[8] creates a framework within which to understand polymerase ribozymes. Most natural ribozymes catalyze nucleophilic reactions at phosphate centers to yield the ligation of two strands, phosphoester hydrolysis, or phosphoester exchange. Three powerful catalytic strategies used by natural ribozymes include specific positioning of substrates,

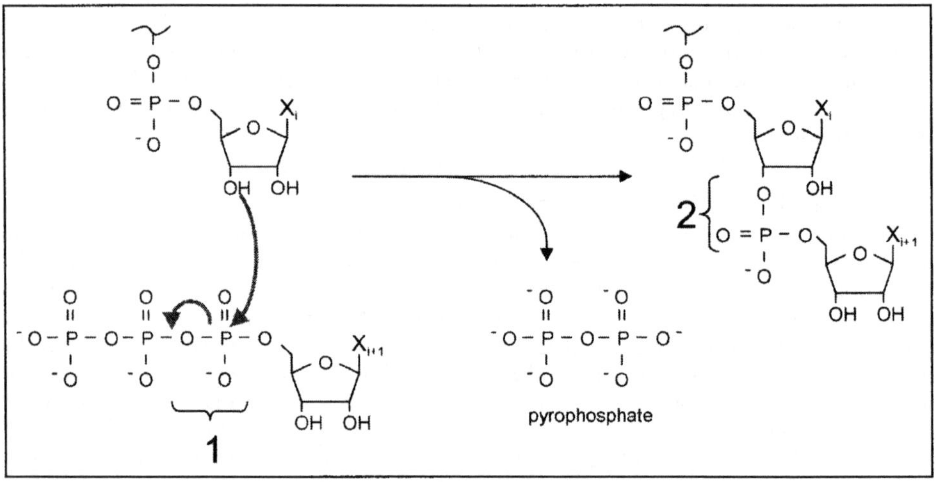

Figure 1. During nucleotide polymerization, a phosphoanhydride bond (bracket 1) is consumed in order to produce a phosphate ester (bracket 2). Electron movement in this and subsequent figures is indicated by gray arrows. Oxyanion transition state intermediates are not shown.

incorporation of metal ions into active sites, and general acid-base catalysis. The next few paragraphs outline these mechanisms and how they shape ribozyme evolution. Later sections address how these and other chemical mechanisms influence the acyltransfer reactions of protein synthesis.

Substrate Binding

The molecular details of binding interactions govern the distances and orientations among substrates and active site residues. These, in turn, dictate much of an enzyme's specificity and its ability to accelerate a chemical reaction. Thus, it will be extremely important to determine how substrate recognition contributes to catalysis in polymerases and other newly isolated ribozymes. A population of substrate-binding RNAs may even be more richly endowed with catalytic species than would be found in a random pool,[12] although such "preadaptation" has not been explored in detail. A generic polymerase must bind both a primer-template junction and activated mononucleotides. The challenges of binding to the primer-template junction represent some of the strongest limitations to existing polymerase ribozymes. Generic features of the helix, such as overall shape and hydrogen bonding to ribose hydroxyls are more important in polymerase ribozymes than are specific structural elements that stabilize large natural RNAs, such as tetraloop-receptor interactions[10] and adenosine platforms.[11] In contrast to the challenges of generic recognition of helices, nucleotide triphosphates offer many potential interaction surfaces through aromatic stacking, hydrogen bonding to sugars and bases, and metal-mediated charge-charge interactions through the phosphates. While nucleic acid aptamers that recognize nucleotides with a wide range of specificities and affinities have been identified and characterized at atomic resolution, there is little known about how polymerase ribozymes bind their respective NTPs.

Coordinated Metal Ions

Specifically bound divalent metal ions play important structural and catalytic roles in several ribozymes through inner- or outer-sphere coordination to water or to phosphate oxygens or to ribose hydroxyls.[13-15] Indeed, specific metal-binding sites have been identified in active sites for several ribozymes through crystallographic analysis, Mn(II) rescue of sulfur-substituted substituents, and other approaches. In addition to general acid/base catalysis, bound metals can accelerate reactions by orienting substrates for attack, stabilizing the developing negative charges on the transition states and leaving groups, and providing strain by distorting a ground state structure more towards that of the transition state.[16-18] Although it was believed for some time that the natural ribozymes all required divalent metal ions for their activity, two groups have recently shown that high concentrations of monovalent ions are sufficient for near maximal activity in the hammerhead, hairpin, and VS ribozymes.[19,20] Thus, many unanswered questions remain as to how the natural ribozymes catalyze reactions at phosphates. Less still is known about metal ion utilization by the many ribozymes recently derived from in vitro selections. Elucidating their mechanisms will keep RNA biochemists busy for years to come.

Acid-Base Catalysis

Reactions at ribose-phosphate bonds require protonation and deprotonation events, and can thus benefit from acid-base catalysis. Ribose hydroxyls are poor nucleophiles unless they are deprotonated to the oxyanion. The relevant leaving groups (generally another ribose hydroxyl) are unstable until they acquire a proton. The cleavage rates for several ribozymes vary in a log-linear fashion with pH, implying a single deprotonation event at the rate-limiting step. For many years, the substituents responsible for proton transfers were thought to be metal hydrates, and the evidence is strong that this is the case for at lease some ribozymes. A metal-bound water is more acidic than free water (e.g., pKa = 11.42 for $Mg(H_2O)_6$ vs. 15.7 for H_2O), allowing proton exchange to occur more readily near neutral pH. For the hammerhead ribozyme, this relationship is retained for a variety of metal ions, with the net reaction rate shifted according to the pKa of the hydrated metal.[21] In recent years nucleotide bases have been recognized to have a role in proton transfer, even though their pKa's in solution are far from neutrality. It has long been recognized from NMR studies that specific structural contexts can markedly shift the pKa's of nucleobases.[22,23] In the hepatitis delta virus (HDV) ribozyme, the protonated N3 of an active site cytosine (C75) is in position to donate a proton as the general acid in that ribozyme's self-cleavage reaction, and a ribozyme-bound hydrated metal hydroxide is in position to abstract a proton in the basic function.[24,25] Specifically, transfer of a proton from C75 to the ribose 5' hydroxyl stabilizes the leaving group. The pKa of the cytosine N3 is normally near 4.2, but the pKa of C75 is perturbed to neutrality within the structural context of the ribozyme, making it "histidine-like".[25] Thus, ribozymes can access at least two catalytic strategies to effect proton transfers.

Evolution of a Ligase Ribozyme into a Polymerase

Polymerase ribozymes have not yet been isolated directly from a pool of random sequences, but remarkable progress has been made through step-wise evolution from ligase ribozymes isolated in the Bartel lab. The original ligases, comprising a diverse collection isolated from random sequence pools, condense a small oligoribonucleotide onto their 5' ends through attack by the 2' or 3' hydroxyl of the oligo on the 5' terminal α-phosphate of the ribozyme (Fig. 2A). Random mutagenesis and further cycles of selection of the "Class I ligase" produced ribozymes with ligation rates around 1 min^{-1}.[26,27] The oligo binds to the ligase ribozyme through base pairing to an internal guide sequence and shows optimized, templated ligation rates as high as 100 sec^{-1}.[27]

Remarkably, the ligase ribozymes also catalyzed limited template-directed primer extension (Fig. 2B). In this reaction, the primer/IGS helix is provided in trans as a primer-template junction, held in place through base-pairing to an adjacent site in the ribozyme.[28] The reaction is exactly analogous to that of the original ligation and almost certainly uses the same active site, with nucleotide triphosphates serving the same role as the 5' terminal triphosphate of the original ribozyme. The rate of polymerization is not significantly diminished compared to ligation by the non-optimal, parental ribozyme (k_{cat} = 0.3 min^{-1} for addition of GTP), although the affinity for NTP substrates is low (K_m^{GTP} = 5 mM). The ribozyme shows 85-95% fidelity when presented with competing nucleotides, depending on NTP concentration.[28] This is comparable to the fidelity of Pol η, an error-prone DNA polymerase associated with repair of

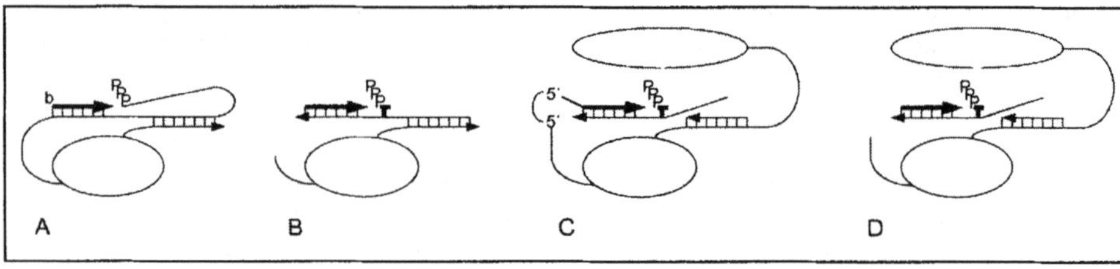

Figure 2. Strategy used in selecting a polymerase ribozyme. In each figure, the primer that is extended by the ribozyme is shown by a thick line, and the 5' triphosphate or NTP that is attacked by the primer is indicated.

DNA damage in humans.[29] The primary limitation of the ligase ribozyme as a polymerase lies in its interaction with the primer-template, which must still base pair with the enzyme, though it must be emphasized that this polymerase is really just a ligase that has been forced into an expanded role.

Johnston et al made two fundamental changes to the ligase that made possible the direct selection of a generic polymerase. First, they appended 76 random nucleotides to the 3' end of the ribozyme and inserted 8 random nucleotides at each of two locations within the ribozyme. Second, they fused the 5' end of the primer oligonucleotide to the 5' end of the ribozyme in a 5'-5' junction, so that any templated primer extension would covalently modify the ribozyme (Fig. 2C). The NTPs used in this step included affinity tags that allowed specific capture of extended species. Furthermore, the sequence of the primer was changed during each cycle of the selection to avoid a requirement for sequence-specific interactions. Finally, the original internal guide sequence was sequestered through addition of an exogenous oligonucleotide, forming a stem that had previously been shown to be required for ribozyme activity. The resulting polymerase ribozymes no longer required base pairing interactions of any kind to hold the primer-template junction onto the ribozyme (Fig. 2D), which extends the primer by more than one full helical turn with approximately 99% fidelity (12 errors observed in 100 isolates, each carrying 11 newly synthesized nucleotides).[30] Observed rates of extension for individual addition steps range from 0.004 hr^{-1} to 6 hr^{-1}, reflecting fast chemistry (on the order of 2 min^{-1} for k_{cat}) and low affinity for both NTPs and primer-template substrates (M Lawrence and DP Bartel, personal communication). Overall processivity is limited because the polymerization rate is on the order of the same as the rate at which the primer/template duplex dissociates from the ribozyme, since processivity is given by the ratio of k_{pol} to k_{off}. Future versions of this ribozymes with improved affinities for NTPs (lower K_m^{NTP}) and for the primer-template duplex (lower K_m^{duplex}) may therefore show greater processivity.

Significance of Polymerase Ribozymes with These Kinetic and Fidelity Parameters

Considerable effort, intelligence, and design have led to a halting polymerase ribozyme. Sequence requirements were carefully evaluated such that extraneous nucleotides could be pruned and new random tracts could be introduced at sites where they were most likely to be beneficial. It is a staggering accomplishment, yet it does not seem reasonable to view these ribozymes as evidence that self-replicating species could have resulted from random, non-enzymatic polymerization of prebiotic nucleotides. There is a considerable functional gap between the current polymerase ribozymes and a useful replicase.[4,31] It currently requires about 24 hours to synthesize one turn of a helix. Three weeks would be required for the nearly 200-nucleotide ribozyme to synthesize its own complement, and another three weeks to copy the complement back into the active plus strand. Still more time would be required for production of ribozymes for other functions and RNAs that serve as structural components of the cell, notwithstanding the time wasted synthesizing inactive RNAs carrying fatal errors (approximately one misincorporation per hundred nucleotides). Hydrolysis of the RNA backbone could easily destroy the ribozyme or its incipient products faster than new ribozymes can accumulate, removing the opportunity for exponential growth and Darwinian selection. Low affinity for the primer-template duplex and for NTPs currently limits reaction conditions to unrealistically high concentrations of each. A useful replicase will require greater speed, processivity, and fidelity. Enhanced substrate affinity could allow reactivity at reasonable substrate concentrations. An active proofreading mechanism could enable the net fidelity to exceed the energetic limits afforded by Watson-Crick base pairing (approximately 99% accuracy). A useful replicase would also require a means of separating the product strands to serve other active roles within the ribocyte or to participate in further replication cycles. Engineering these or other polymerase ribozymes into credible replicases would go a long way towards solidifying theoretical models of RNA-based organisms, irrespective of the assumed providence of such cells. Efforts are no doubt underway to arrive at such improved ribozymes.

Bountiful Ligase Ribozymes As Evolutionary Fodder

Polymerization in some form might be able to evolve from other ligases along evolutionary paths similar to those outlined above. If so, then there could be many different evolutionary starting points from which a replicase could arise. The naturally occurring Group I and Group II self-splicing ribozymes use phosphoester exchange to ligate RNA strands, and were among to first to show limited polymerization activity (see discussion in ref. 31). The Group I intron from the *sunY* gene of bacteriophage T4 has been engineered to carry out template-directed polymerization of sorts. The "monomers" added during each step were trinucleotide fragments, and the leaving group for the reaction was a 5' guanosine residue, for a net reaction of $G\text{-}X_n + G\text{-}Y_n \rightarrow G\text{-}X_nY_n + G\text{-}OH$, with n=10 [32] or n=3.[33]

Even the small, classic, endonucleolytic ribozymes such as the hammerhead, hairpin, VS and hepatitis delta ribozymes, catalyze both the forward cleavage reaction and the reverse ligation reaction, using a 2',3' cyclic phosphate to activate the reaction. Freezing out large-scale RNA motions —either through compact tertiary structure or through the formation of engineered crosslinks— is thought to determine where the ligation/cleavage equilibrium lies. The ligation reaction for the hairpin ribozyme is favored 6-30 fold over the cleavage reaction,[34-36] and ligation by the HDV and VS ribozymes is also notable.[37-40] Introduction of a disulfide crosslink into the hammerhead ribozyme was recently shown to accelerate the rate of ligation without altering the cleavage rate, thereby shifting the equilibrium to favor ligation over cleavage [41]. It might be possible to re-engineer the abundant new small RNA-cleaving ribozymes into ligases by similarly freezing out their large-scale motions.

In vitro selections have yielded an abundance of bona fide ligase ribozymes. Some of these generate 2'-5' linkages (rather than 3'-5'), and nearly all bind the primer strand through an internal guide sequence rather than binding exogenous primer-template duplexes. Nevertheless, the diversity of sequences that catalyze RNA strand ligations suggest bountiful starting points for their evolutionary conversion into polymerases. Some in vitro selected ligase ribozymes are described below, and their ligation rates are summarized in Table 1.

Derivatives of the Canonical Class I Ligase

The Class I ligase that gave rise to the polymerase described above has also spawned other evolutionary derivatives. Wright and Joyce used it to develop a continuous in vitro evolution system, in which a "culture" of replicating ribozymes (aided by a few protein enzymes) can be propagated indefinitely while introducing new mutations.[42] Rogers and Joyce derived a Class I ligase variant that is devoid of any cytosine residues, forming all of its structure using A, G, and U.[43] The C-less ribozyme is 2500-fold slower than the parent from which it was derived, with k_{cat} dropping from 20 min^{-1} in the parental version to 0.008 min^{-1} in the C-less version, but these results showed that complex, functional species can arise even with a reduced alphabet of nucleotides. The activity of the original Class I ligase is sharply reduced at low pH. Miyamoto et al. optimized the low-pH activity through in vitro evolution. Their Class I variant is 250-fold

Table 1. Ligation rates of in vitro-selected ligase ribozymes

Ribozyme	k_{cat}, min^{-1}	Rate Acceleration
original Class I ligase	1	10^7
optimized Class I	100	10^9
C-less Class I	0.008	8×10^4
Class I at pH 4.0	0.0000005	[a]
optimized for reaction at pH 4.0	0.0001	[a]
Class I cogeners	1	10^7
"Spartan Spandrel"	0.0006	6×10^3
TyrS-activated ligase		
- TyrS	0.0000003	3
+ TyrS	0.035	3.5×10^5
Lysozyme-activated ligase		
- Lysozyme	0.000003	30
+ Lysozyme	0.01	10^5
Group I intron core	0.00003	3×10^3
Optimized with 85 nt insertion	0.26	2.6×10^6
Ligase with "lysine-A" analogs	0.000025	2.5×10^2
Ligation from 5'-linked AMP- or 5'-phosphorimidazolide-activated oligonucleotides	0.007	[b] 7×10^4
Uncatalyzed templated ligation	$\sim 10^{-7}$	1

[a] Background reaction at pH 4.0 not determined. [b] Background reaction assumed to be same as with triphosphate activating group.

more active at pH 4.0 than the parental sequence at that pH, but the derivative did not actually prefer acidic buffers, retaining a positive correlation between activity and pH.[44]

Congeners of the Class I Ligase

The selection that yielded the Class I ligase also yielded several other structural classes.[26] Some of these have been shown to catalyze ligations at respectable rates (approx 1 min^{-1}),[26,27] though most have not yet been subject to optimization of their ligase function.

Dual-Function Ligase/Nuclease Ribozymes: The "Spartan Spandrel"

Landweber and Pokrovskaya found a ligase ribozyme that, in the presence of Mn(II), also catalyzes self-cleavage at a site that is distinct from the ligation site. This unintended dual function is referred to by those authors as a "spandrel" to emphasize how evolution for one function can simultaneously preadapt a species for a second function. While the ligation rate is slow (0.0006 min^{-1}), its structure is among the most Spartan of the selected ribozymes, requiring only a few specific nucleotides between paired helices at the ligation junction.[45] The simplicity of the "Spartan Spandrel" suggests that sequence space may be riddled with low-activity ligase ribozymes.

Allosteric Ligases

Robertson and Ellington identified a ligase ribozyme, dubbed "L1," that is activated 1,000 to 10,000-fold by addition of an exogenous oligonucleotide (separate from the ligation substrate).[46] By mutagenizing the catalytic core and applying further cycles of selection, they identified variants of the L1 ligase that were activated nearly 10^5-fold by a tyrosyl transfer RNA (tRNA) synthetase (TyrS) encoded by the Cyt18 gene.[46] Allosteric ligases are being developed in several labs for potential biotechnology applications (e.g., sensors).

Modular Assembly of Ligase Ribozymes

Another set of structurally complex ligase ribozymes was identified by inserting 85 random nucleotides into a group I self-splicing intron core of 225 nucleotides and selecting variants that condensed an oligonucleotide onto its end at the expense of the 5' triphosphate.[48] The intron core alone catalyzed the desired ligation with a rate of 3×10^{-5} min^{-1} (300-fold above uncatalyzed rate), while after selection this rate was improved to 0.26 min^{-1}.

Ribozymes Containing Lysine Analogs

Nucleotide analogs that carry added chemical functionality offer a possible route by which to increase the catalytic activity of ribozymes in general. With this is mind, nucleotides carrying the side chains of histidine (imidazole), lysine (n-alkylamine), and other substituents (pyridine) have been used in selections for various ribozymes. In the ligase arena, Teramoto et al. identified ribozymes containing alkylamino side chains at the N6 position of adenosines to mimic the positively charged side chain of lysine.[47] However, the modification did not appear to confer any advantage to the population, as only a 250-fold rate enhancement was observed over the uncatalyzed reaction rate of approximately 10^{-7} min^{-1}.

Ligation Using Other Chemistries

Polyphosphorylated nucleotides are hypothesized to have been used by ribocytes prior to protein-catalyzed chemistry.[49] However, because polyphosphorylation of the 5' hydroxyl is challenging for prebiotic chemistries, alternative leaving groups have been explored, such as cyclic phosphates, imidazolides, and adenylates[3,50,51] (Fig. 3). Ligase ribozymes using each of these reagents have been isolated. The Szostak group described a ligase in which the 3' OH of the RNA condenses with a donor oligonucleotide activated with 5'-linked AMP[52] (the same adenylate intermediate generated by phage T4 DNA ligase). The same group later found a ribozyme that formed 5'-5' tetraphosphate, triphosphate and pyrophosphate linkages from 5'-phosphorimidazolide-activated oligonucleotides.[53] Neither reaction is rapid —maximal observed ligation rate = 0.4 hr^{-1}, or about 0.007 min^{-1}— but the two activities suggest that one ribozyme could synthesize the substrate needed by another ribozyme, constituting a coupled, ribozyme-catalyzed, two-step reaction mechanism. Along similar lines, the Breaker group has described deoxyribozymes that cap DNA with 5' adenylates, as well as other deoxyribozymes that catalyze templated ligation.[54,55] DNAzymes could therefore catalyze similar multistep reactions.

RNA-Catalyzed Genetics II: Protein Synthesis

The joining of two amino acids by a ribozyme launched the Protein Revolution and began the end of the RNA World. Protein synthesis is a progression of aminoacyl transfer reactions, wherein an amino acid (the acyl group) is passed from one acyl donor to the next (Fig. 4). In each step a nucleophile attacks a phosphate or carbonyl center with displacement of progressively less reactive leaving groups. In the first step of the cascade, the carboxylate of the amino acid joins to the α-phosphate of ATP, displacing pyrophosphate and forming a high-energy, mixed phosphoanhydride between the amino acid and AMP. In the second step, the amino acid is transferred to the 2' or 3' ribose hydroxyl of tRNA to make the less reactive ester, and displacing AMP. Both of these reactions are catalyzed by aminoacyl tRNA synthetase proteins (ARS), which are discussed in detail in other chapters of this volume. On the ribosome, the acyl group (initiator amino acid or growing peptide chain) is transferred to the alpha amino of another amino acid, forming a new peptide bond at the expense of an ester bond with concomitant displacement of tRNA. Ribozymes that catalyze each of the individual reactions of this cascade have been isolated. This section first discusses the underlying chemistry of acyltransfers, then describes the acyltransfer ribozymes that have been isolated in vitro for the three individual steps.

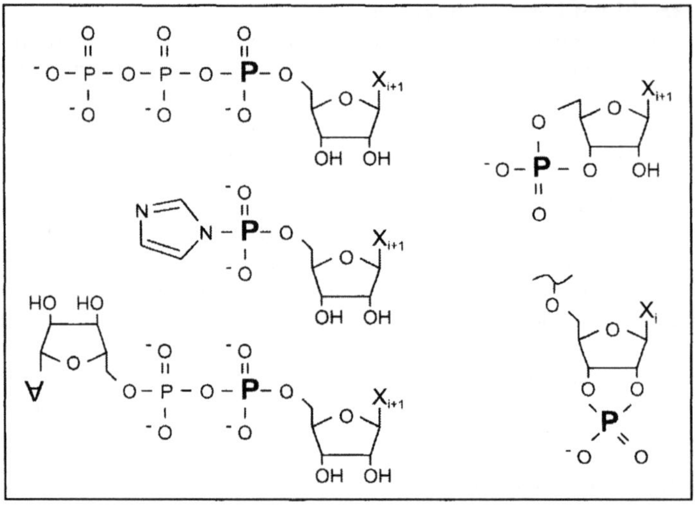

Figure 3. Alternative activated nucleotide monomers that can be used in polymerization and ligation reactions. From top to bottom, left side: triphosphate, phosphorimidazolide, adenylate; right side, 3',5'-cyclic phosphate, 2',3'-cyclic phosphate. The 2',3'-cyclic phosphate is produced upon self-cleavage by several small, natural ribozymes. Reversal of the cleavage reaction re-ligates the two termini.

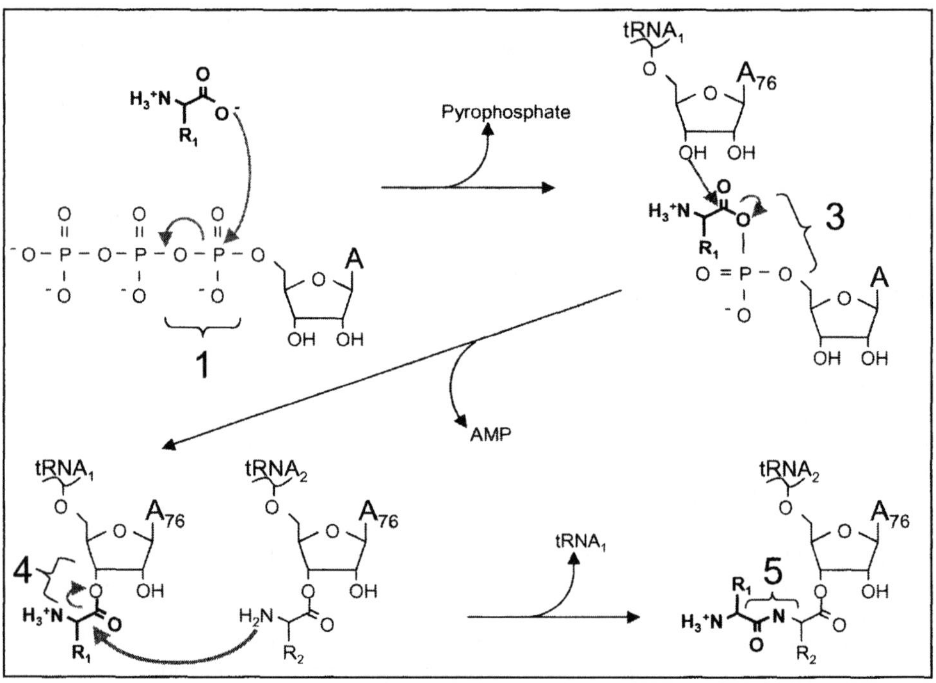

Figure 4. Stepwise acyl transfer in peptide synthesis. The initial activation step converts a phosphoanhydride (indicated by bracket 1) into a mixed carboxy-phosphoanhydride (bracket 3). Reaction with tRNA produces a ribose ester (bracket 4), then condensation with another amino acid produces the peptide product (bracket 5). A_{76} indicates the 3' terminal nucleotide of aminoacyl-tRNA.

Table 2. pKa's and relative activation of several leaving groups relevant to the reactions of transmitting genetic information

Conjugate Acid of Leaving Group	pKa	Activation
RNH_2	≈35	no
ROH	15-20	weak
H-OH	15.7	weak
ribose hydroxyls	12.35	moderate
$HP_2O_7^{-2}$	12.32	moderate
$HOCH_2CN$	11.0	good
RSH	10.6	good
RNH_3^+	≈10	good
$H_2P_2O_7^{-1}$	7.09	very good
H-AMP	6.2	very good
$H_3P_2O_7$	2.15	very good

The Chemistry and Enzymology of the Reactions of Protein Synthesis

The underlying chemistry of natural protein synthesis is dominated by leaving group reactivity, nucleophilicity, electrophilicity, acid-base chemistry, proximity effects, and charge stabilization. Although the mechanistic enzymology of aminoacyltransfer ribozymes is in its infancy, the known mechanisms of protein acyltransferase and those of phosphoesters-manipulating ribozymes let us postulate several catalytic strategies that ribozymes could use to accelerate acyltransfers.

Leaving Group Reactivity

To say that an acyl donor is highly reactive usually means that it carries a good leaving group. Leaving groups whose conjugate acids have low pKa's are generally more reactive than those with high pKa's (Table 2). This trend is part of what makes amides and peptides so stable against hydrolysis. In the absence of any additional reaction, hydrolysis of an amide or peptide bond would displace RNH^- as a leaving group. The RNH^- species has a conjugate acid, RNH_2, whose pKa is very high (e.g., ≈36 for isopropylamine), making the RNH^- a lousy leaving group. Proteases often protonate the nitrogen to facilitate peptide bond hydrolysis (Fig. 5). The resulting RNH_2 is a good leaving group because its conjugate acid, RNH_3^+, has a relatively low pKa (around 10). The significance of using one activating group vs. another takes at least three forms: the need to protect against water hydrolysis (an acyl group that is more reactive with the desired substrate is also more sensitive to spontaneous hydrolysis), the existence of any requirement to protonate the leaving group (which would increase the functional demands on the active site residues), and the geometry of the active site (vs. steric constraints and charge distribution of leaving group; e.g., AMP vs. inorganic phosphate leaving groups).

Four classes of leaving groups are especially important for discussions of RNA-catalyzed protein synthesis. ***Phosphoanhydrides***, such as adenylates, are the most reactive species used during normal protein synthesis. Spontaneous hydrolysis is especially pernicious for the aminoacyl adenylate ($t_{1/2}$ ≈ 10 min at 0°C, pH 7.0; ≈1-3 min at 37°C[56,57]). To prevent their premature hydrolysis or reaction with unintended nucleophiles, ARS enzymes keep the adenylate hidden in the active site until they react with the 2' or 3' hydroxyl of the cognate tRNA. ***Thioesters***, such as acyl coenzyme A (acyl CoA), are also highly activated, as the pKa values of sulfhydryls are near 9-10. The polarizability of the RS^- thiide ion further stabilizes it as a leaving group. Thioesters of coenzymeA and the related pantotheine are used in a plethora of acyltransfers, including non-ribosomal peptide synthesis. ***Sugar esters***, such as aminoacylated

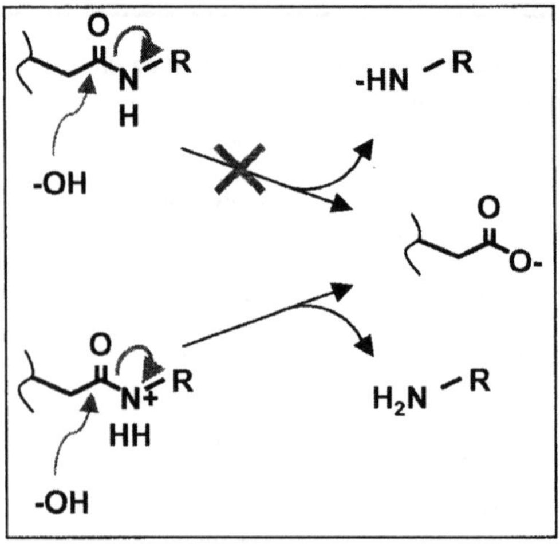

Figure 5. Effect of protonation on leaving group pKa and acyl group reactivity. Direct amide hydrolysis (top) is disfavored because the RNH^- leaving group is unstable; hydrolysis of a protonated amide (bottom) occurs much more readily with the RNH_2 leaving group.

tRNA, are considerably less reactive than phosphoanhydrides or thioesters, as the pKa's of ribose hydroxyls are 12.35.[58] Even so, the amino acid-tRNA ester bond is labile, hydrolyzing rapidly at 37°C when not shielded from solvent by elongation factor EF-Tu ($t_{1/2}$ for hydrolysis ≈20 min at 37°C, pH 7.5[59]). ***Cyanomethyl esters*** show intermediate reactivity (pKa of the conjugate acid, $HOCH_2CN$ = 11). They are more reactive than sugar esters, but less susceptible to hydrolysis than adenylates. While not employed in biological reactions, cyanomethyl esters of amino acids have figured prominently in some ribozyme-catalyzed aminoacylation studies.

Nucleophilicity of the attacking nucleophile also governs reaction kinetics. As above, the pKa of the conjugate acid of a given species is an important determinant of reactivity, since electron-rich atoms are generally better nucleophiles that electron-poor atoms, but the correlation with pKa is not precise. More accurately, good nucleophiles are good Lewis bases (electron donors). Thus nucleophilicity increases with anionic charge (RO^- > ROH) and according to position in the periodic table (RS^- > RO^-; RNH_2 > ROH). The steric environment of the attacking nucleophile is another important determinant of reactivity, as a crowded environment limits the productive angles of approach during reactive center collisions.

General acid-base catalysis serves to activate the nucleophile (deprotonation) and stabilize the leaving group (protonation). As noted above, hydrated metals and nucleobases with shifted pKa's fulfill this role in some ribozymes. For recently selected acyltransferase ribozymes, there are few mechanistic data available that would suggest specific moieties as general acids or bases.

Electrophilicity of the Carbonyl Carbon

The electronegative oxygen withdraws electrons from the carbonyl carbon atom through the double bond, creating a partial positive charge on the carbon. Hydrogen bonding to the oxygen or association with a positive charge further polarizes the C=O double bond, increasing the partial charge on the carbon and making it more reactive with electron-rich nucleophiles. RNA molecules bristle with hydrogen bond donors and acceptors, and they are adept at positioning metal ions and cofactors in specific spatial arrangements. Acyltransferase ribozymes are thus expected to incorporate electrophillic enhancement into their catalytic strategies.

Charge Stabilization

Negative charge accumulates on the carbonyl oxygen in the acyltransfer tetrahedral transition state and on the leaving group as the transition state resolves into products. Depending on the reaction and on the timing of protonation and deprotonation steps, negative charges may also occur on either the attacking or leaving groups at different times in the course of the reaction. Neutralizing these charges could prevent unfavorable burying of charged species while helping to position them within the active site. Oxyanions are integral to the mechanisms of the naturally occurring ribozymes that act on phosphate centers, which, as noted above, are generally believed to be stabilized by specifically bound divalent cations. However, as previously mentioned, several of the natural ribozymes are now known to operate in the absence of divalent metals if provided instead with high concentrations of monovalents,[19,20] suggesting that monovalent ions may offer a functionally equivalent route to oxyanion stabilization. Monovalent-stimulated self-cleavage by the hammerhead ribozyme is log-linear with pH —a property also seen with the divalent-dependent reaction— and with the ionic radius of the cation.[20]

Proximity Effects

If both the nucleophile and acyl donor are sufficiently reactive, an increase in their respective local concentrations can yield significant rate accelerations. For an evolving population of ribozymes, juxtaposition of substrates could be attained, in principle, through recombination among substrate binding pockets.[12] This is particularly true for the highly reactive adenylates (see below). However, a higher local concentration is not always enough for acyltransfers. For example, in both chloramphenicol acetyltransferase (CAT)[60-62] and dihydrolipoyl acetyltransferase (E2p),[63] the attacking hydroxyl nucleophile is normally deprotonated by a conserved His residue. The oxyanion of the transition state in CAT is then stabilized by a hydrogen bond from a conserved Ser while the protonated His is stabilized by a conserved Asp. Mutations in the His-Asp-Ser catalytic triad reduce or eliminate acetyltransfer activity[62,63] even though Km for the substrates is unaffected. Thus, binding and juxtaposition alone may not be enough for reactivity in these enzymes. Establishing the relationships between substrate binding by RNA and the catalytic potential of ribozymes promises to stimulate mechanistic and evolutionary studies of macromolecular catalysis.

Applying These Principles to the Ribosomal Peptide Bond Formation

Proximity effects and acid-base catalysis have both been proposed to operate in ribosome-catalyzed peptide bond formation. In the 2.4 Å structure of the *Haloarcula marismortui* large ribosomal subunit, the peptidyl transferase active site is composed entirely of RNA. The nearest protein components are too far away (≈16 Å) to contribute meaningfully to catalysis.[64] The N3 of a universally conserved adenosine residue in the active site (A2451) is within hydrogen bonding distance of the alpha amino group on the A-site-bound tRNA. From chemical modification by dimethyl sulfate (DMS), the pKa of this adenosine in the E. coli ribosome appears to be shifted from the normal 3.8 (< 1 for the N3 position!) to around 7.6, possibly due to a charge relay to neutralize a buried phosphate.[65] These observations prompted models in which this adenosine acts as a general base to deprotonate the attacking amino group. However, no pKa shift is observed at this position in the ribosomes of three other bacterial species, and mutant ribosomes with base substitutions at the conserved adenosine are still functional.[66,67] These authors of these last two studies postulate that pH-dependent structural rearrangements may have accounted for the previous observations,[66] and that the primary function of the peptidyl transferase active site in ribosomes is simply to juxtapose the reactants.[67] Indeed, poly(U) can direct peptide bond formation when amino acids are supplied as 2'(3') adenosyl esters.[68] Furthermore, there is an A•C base pair immediately behind A2451 (A2450•C2063) in the high-resolution crystal structure of the *Haloarcula* large ribosomal subunit. A•C pairs are only stable when the N1 of adenosine is protonated, and the pKa of N1 in such pairs is often shifted to near neutrality. Deprotonation of the A2450•C2063 pair above neutral pH could act as a structural switch that increases the reactivity of A2451 to DMS.[69,70] The exact mechanism by which the ribosome accelerates peptide bond formation —and the precise contribution of 23S rRNA— remain controversial.

Ribozyme Catalysis of Step 1, Amino Acid Activation

Kumar and Yarus isolated ribozymes that mimic the first step of aminoacylation by forming a mixed carboxylate-phosphate anhydride at the expense of a nucleotide triphosphate. A surrogate for cysteine, 3-mercaptopropionic acid (3MPA), was used in place of an amino acid. During the selection, RNA transcripts were incubated with 3MPA. Any RNAs that joined the 3MPA to their own 5' end were recovered through their ability to form a disulfide between thiopropyl sepharose and the unique sulfur introduced by the 3MPA. Since these transcripts carry GTP at their 5' ends, formation of the mixed anhydride with the terminal α-phosphate was expected to produce a guanidylate (the formal chemical equivalent of an adenylate), with concomitant displacement of pyrophosphate. The product was confirmed by showing that when radiolabeled transcripts were incubated with various amino acids, digested to mononucleotides and separated by HPLC, the radiolabel comigrated with genuine, chemically synthesized aminoacyl guanidylate. The reaction requires Ca(II), and at pH 4.0 it proceeds at a rate of 1.1 min^{-1}, with Km =48 mM for 3MPA. (This Km value is close to the 50 mM concentration used during the selection.) The alpha amino group of an amino acid makes its aminoacyl anhydride much more labile to hydrolysis than the mixed anhydride formed from simple organic acids (such as 3MPA). As a result, the product decays rapidly ($t_{1/2}$ = minutes for aa-adenylates vs. hours for organic acid adenylates at 0°C, pH 7). Using 3MPA in place of a normal amino acid was thus a necessary precondition for success of this selection, as it greatly increased product lifetime.

This RNA catalyzes the formal equivalent of the first step of protein synthesis by loading amino acids onto a mixed phosphoanhydride. The ultimate goal of such a ribozyme is to initiate a series of reactions that result in aminoacylation of a tRNA-like species. Because of the rapid hydrolysis without stable aminoacylation, this ribozyme, for now, only provides a complex means by which to convert a 5' triphosphate into a 5' monophosphate. It is anticipated, however, that an engineered or evolved descendent of the Kumar and Yarus ribozyme will present the mixed anhydride quickly to another substrate (e.g., its own or another RNA's 3' hydroxyl) to complete the reaction, and that they may show greater substrate affinity and specificity.

The ribozymes of Kumar and Yarus are the latest in a series selected to condense various substrates onto the alpha phosphate of their 5' terminal NTP, with concomitant release of pyrophosphate. The "Iso6" ribozyme first isolated by Huang and Yarus[71] catalyzes attack by a variety of phosphorylated compounds[72] to yield transcripts capped by nucleotide cofactors, by the normal eukaryotic mRNA cap (GpppG), by expanded and contracted caps (GpppppG and GppG), and by other phosphorylated organic compounds.[71-73] Iso6 also possesses decapping and pyrophosphatase activities, albeit at rates that are >1000-times slower than the condensation reaction.[74] Iso6 derivatives were the first ribozymes to demonstrate multiple-turnover kinetics between two exogenous, small-molecule substrates. When one, two, or all three 5' terminal nucleotides were omitted from the transcript and instead supplied as exogenous substrates (pppG, pppGpG, and pppGpGpG), each truncated ribozyme catalyzed the formation of (5'—>5') polyphosphate-linked oligonucleotides in trans.[75] Substrate nucleotides are bound with Km values around 10-30 μM, while the pyrophosphate product is a powerful inhibitor of the reaction with Ki around 0.2 μM. Like the amino acid activating ribozymes above, Iso6 is a Ca(II)-dependent metalloribozyme that prefers acidic pH. While Iso6 and the aminoacylguanylate-forming ribozymes share little if any sequence identity, it will be interesting to determine whether they use similar structural frameworks or catalytic mechanisms.

Ribozyme Catalysis of Step 2, Making an Ester from an Activated Amino Acid

In vitro selections can employ either the natural adenylate or some other activation strategy to aminoacylate ribose. Pre-activated amino acids or organic acids in the form of adenylates,[76,77]

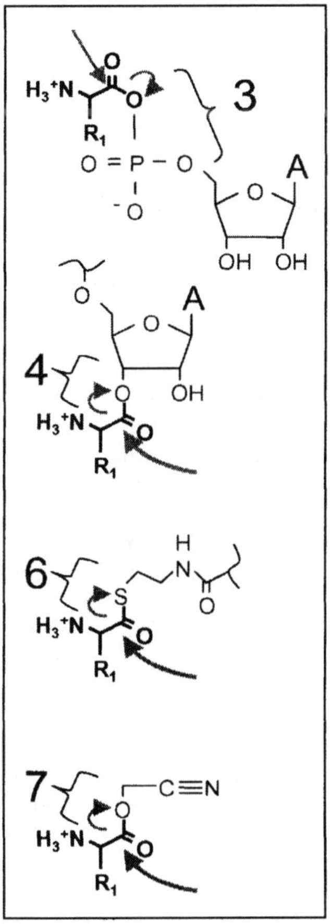

Figure 6. Activated amino acids (in bold) used in biology or in selections in vitro: Adenylate, or phosphoanhydride (bond #3), ribose ester (bond #4), the thioester of CoA (bond #6), cyanomethyl ester (bond #7). Electron movement during nucleophilic attack and displacement of the activating group is indicated by gray arrows.

3'(2') ribose esters,[78,79] and cyanomethyl esters[80] have been presented to RNA pools in the course of selecting acyltransfer ribozymes (Fig. 6).

Activation As Aminoacyl Adenylates

Illangasekare and Yarus incubated RNA transcripts with Phe-AMP, then tagged the Phe-RNA products with the N-hydroxysuccinimido (NHS) ester of naphtoxyacetate.[76] The naphtoxyacetate and the phenyl ring of Phe decreased the overall polarity of the modified RNAs and shifted their HPLC mobilities, allowing recovery of the desired products. Since HPLC is cumbersome for kinetic analysis, these authors later monitored electrophoretic gel shifts on low-pH gels. The original ribozyme, designated R29, was slow (second order rate constant of 70 $M^{-1}min^{-1}$) and non-specific, reacting equivalently with various aminoacyl adenylates.[57,76] Substrate recognition is primarily through the adenosine portion of the aa-AMP, as shown by the fact that AMP competes with the substrate, but free Phe does not.[57] These results provided the first proof that RNA could catalyze this class of condensation reaction, but also showed that naïve ribozymes can be poor reagents for rebuilding biology from scratch. However, further screens and additional engineering produced a rapid, highly specific, aminoacylating ribozyme designated RNA77.[81] This 90 nucleotide species shows second-order rate constants of around 50,000 $M^{-1}min^{-1}$ for Phe-AMP and Tyr-AMP (nearly 10^8-fold above the uncatalyzed rate), but only 1 to 5 $M^{-1}min^{-1}$ for other aa-AMP species.

Activation As Ribose Esters and Cyanomethyl Esters

The activated amino acids used by the Suga lab's ribozymes to acylate the 5' hydroxyl of a substrate RNA are cyanomethyl esters (CME). The CME group is stable enough against hydrolysis to facilitate its use in the selections. It nevertheless efficiently activates the amino acid to react with RNA substrates while offering few binding interactions that might subvert substrate specificity. Energetically neutral exchange reactions then transfer the amino acid from one hydroxyl group to another (Fig. 7). Like the Bartel polymerases, the Suga acyltransferases were built from simple ribozymes to produce successively more elegant elaborations.

The first ribozyme in this series catalyzed only the exchange reaction (the reverse of the second step shown in Fig. 7).[78] RNA pools bearing 5' hydroxyl groups were incubated with short oligonucleotides that carried the aminoacyl ester of N-biotinylated methionine on their terminal 3' oxygen. Those RNAs that catalyzed transfer of the amino acid from the oligonucleotide onto their own 5' hydroxyls were recovered by streptavidin (StrAv) affinity chromatography. These ribozymes stabilize the transition state through outer sphere coordination to a divalent metal ion in the active site.[82] Several follow-up studies defined the nucleotides required in the active structure and in forming the binding site for the metal ion.[83-85] Because the exchange reaction is energetically neutral, these ribozymes are expected to catalyze the

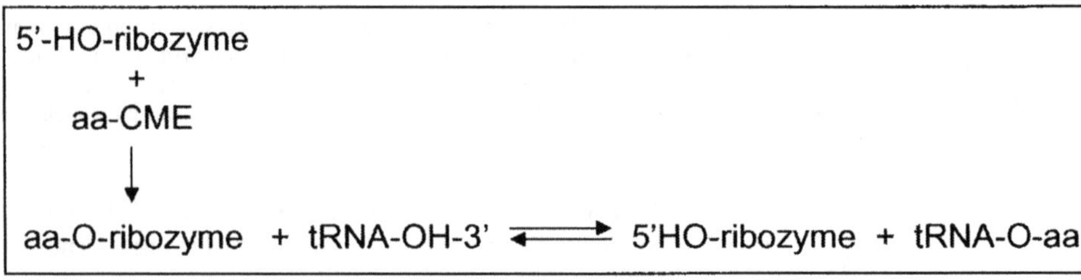

Figure 7. Ribozyme-catalyzed aminoacylation of tRNA starting from CME-activated amino acid and proceeding through an acyl-ribozyme intermediate.

forward and reverse reactions with roughly equivalent proficiency. Furthermore, the 2'(3') esters mimic the acylated 3' end of tRNAs used during peptide bond formation.

The second generation in this series added new substrate recognition activity.[80] Seventy random nucleotides were appended to the 3' end of a proficient, first-generation exchange ribozyme, and the resulting pool was subjected to selection for its ability to transfer biotinylated glutamine onto itself from a CME-Gln-bio substrate (the first step in Fig. 7). The winners from this selection were also required to retain their original exchange activity (second step) by accepting Gln from the 3' end of a small oligonucleotide. The final product RNAs are "ambidextrous" in that they can acquire Gln either from the CME-Gln-bio or from the 3' glutaminylated oligonucleotide. In a leap towards biological relevance, the "AD02" ribozyme showed the ability to perform a sequential 2-step reaction, in which it first acquires an amino acid from an activated source (CME-Gln-bio), then transfers it onto tRNA (both steps in Fig. 7). This second reaction is slow and operates at low yield (k_{cat} = 0.00195 min^{-1}; $Km^{Glu\text{-}CME}$ = 158 µM; 4% conversion of tRNA to aa-tRNA after four cycles of thermal denaturation/renaturation). However, it is highly specific for Gln over the unrelated amino acids Phe and Ala, and the product is a genuine aminoacylated tRNA. This portion of the ribozyme makes inner-sphere contacts with a required metal ion, and a 29 nucleotide stem-loop domain within the ribozyme can be provided in trans to assemble the active species.[86]

The third generation in this series partially removes the requirement for intramolecular tRNA recognition.[80] Lee and Suga appended a 70 nucleotide random segment onto the 5' end of a tRNA to generate a hybrid pool. By incubating the hybrid pool with CME-Phe-bio followed by recovery on immobilized streptavidin, they identified an RNA that directly aminoacylated the tRNA domain. The ribozyme-tRNA hybrid is a substrate for RNaseP, which cleaves off the 5' domain to release mature tRNA aminoacylated on its terminal 3' oxygen.[87] The reaction is not restricted to operating in an intramolecular fashion, as a 57 nucleotide miniribozyme derived from the 5' domain proficiently aminoacylates tRNA in trans.[88] While the ribozyme distinguishes between CME-Phe and unrelated CME-activated amino acids, it is not specific for the activating group, reacting equivalently with Phe-thioesters and Phe-adenylates. Further advances in this area may facilitate expansion of the genetic code and the incorporation of non-natural amino acids into proteins.

One additional ribozyme falls within this class, although it was intended to model a very different reaction. Jenne and Famulok sought peptidyltransferase ribozymes using a 2'(3') AMP ester of biotinylated Phe, which mimics the terminus of aminoacyl tRNA. The RNA pool they used was modified at the 5' end with the amino acid citrulline. Any species capable of condensing the alpha-amino group of citrulline with the activated Phe could have been recovered through the biotin-streptavidin (bio-StrAv) interaction; however, species that adduct the bio-Phe at any other position, such as a terminal or internal hydroxyl, could also be isolated. Indeed, the selected species was shown to self-aminoacylate at an internal 2' OH. As with the original Illangasekare and Yarus self-aminoacylating ribozyme, substrate recognition occurs largely through the AMP moiety.[79]

Ribozyme Catalysis of Step 3, Making an Amide/Peptide from Activated Amino Acids

Mimicking the third reaction of peptide synthesis was an early target of in vitro selections, and four independent sets of ribozymes have been described that form amide or peptide bonds. The first of these was the Lohse and Szostak acyltransferase described above. This ribozyme forms an amide linkage if the 5' OH of the ribozyme is replaced with a 5' NH_2, although the rate of bond formation is decreased about 3-fold by making the substitution (1.8 vs. 0.58 min^{-1} for ribozyme, 0.00083 vs. 0.00029 min^{-1} for the uncatalyzed reaction).[78] This ribozyme is the only one of the four that indisputably uses a ribose ester as the substrate (bond #4 in Figs. 4 and 6). The other three amide- or peptide-bond forming ribozymes all use (or at least appear to use) a mixed phosphoanhydride with AMP as the activated amino acid donor (bond #3 in Figs. 4 and 6).

The second species in this series is that of Wiegand, Janssen and Eaton, which forms an amide bond with biotin if the biotin is provided in a mixed anhydride as the adenylate (bio-AMP).[77] During selection, the attacking amine was on the end of an aliphatic chain covalently attached to the ribozyme. This strategy is similar to the use of the tethered citrulline noted above for the Jenne and Famulok ribozyme. The resulting catalysts are unusual among ribozymes in their requirement for Cu^{2+} ions and for their dependence on 5-imidazole-substituted uridines in place of normal U in the RNA chain. The catalyzed reaction proceeds at a rate of 0.04 min^{-1}, which is 10^4-fold above the uncatalyzed rate. Like the self-aminoacylating ribozymes of Illangasekare and Yarus, the Eaton ribozyme mimics natural tRNA aminoacylation by using AMP as a leaving group. The catalytic mechanisms employed by the Eaton and Yarus ribozymes are not yet known, but their substrates are so reactive that proximity effects may be sufficient to drive their respective reactions.

Third, a 29 nucleotide derivative of RNA29 from the Yarus lab noted above not only self-aminoacylates, it also forms di- and tripeptides from Phe-AMP as side reactions.[89] These RNAs first append an amino acid onto their 3' ends (k_{cat}/Km = 154 $M^{-1}min^{-1}$) and then add one or two additional amino acids, with k_{cat}/Km = 10-30 $M^{-1}min^{-1}$ for first addition. Amino acid adenylates condense into peptides in solution at a rate of approximately 0.3 $M^{-1}min^{-1}$, so RNA29 accelerates the condensation reaction by 30- to 100-fold[90] over the uncatalyzed rate.

Fourth, a ribozyme intended to mimic natural peptide bond formation was isolated by Zhang and Cech.[91] During the selection, the attacking amino acid was a phenylalanine attached to the 5' end of the RNA chain through a flexible linker that carried a disulfide bond. The analog of peptidyl tRNA was N-biotinylated Met, ostensibly supplied as a 3'(2') ester of AMP (but see below!). Condensation of the two amino acids allowed partition of reactive from unreactive molecules on StrAv affinity beads. Cleavage of the disulfide with dithiothreitol released the RNA into solution, providing an extra measure of specificity in recovering active species.[91] It now appears that these ribozymes use the aa-adenylate (*phosphoanhydride*) form of the substrate rather than the 3'(2') aminoacyl *esterified* adenosine. In tandem HPCL/activity assays, the adenylate and the ester are well separated. The adenylate reacted with k_{cat} close to those originally published (Km actually improved slightly), while the 3'(2') aminoacyl esterified adenosine failed to react. Some isolates from the original selection also show moderate stereopreference for the L-Met-containing phosphoanhydride substrate over the D-isomer. (RL Gottlieb, Z Cui, L Sun et al, in preparation). A family of ribozymes related to isolate 27/71 can join other pairs of amino acids into dipeptides, forming at least 30 different dipeptides from the appropriate adenylates. The majority of the catalytic rates are within 5-fold of the rate observed for the original Met-Phe combination.[92] Species RBZ180 derived from the 27/71 family is especially rapid, with k_{cat} = 4.05 min^{-1} and $Km^{aa\text{-}AMP}$ = 210 μM, for an overall acceleration of more than 10^5 over the uncatalyzed rate.

Significance of Ribozymes with These Kinetic and Fidelity Parameters

Acyltransfer ribozymes will need to become faster and more specific if they are to take over the synthesis of cellular aminoacyl tRNAs or peptides. They remain addicted to substrates that are pre-activated as esters, thioesters, or mixed phosphoanhydrides (e.g., adenylates). They are not yet able to generate and utilize the activated species from readily available, kinetically stable, thermodynamically activated reagents, such as ATP. The Kumar and Yarus ribozyme begins to break this chemical dependence by converting amino acids into the highly activated adenylates, but it does so at the expense of its own 5' triphosphate, making this RNA a single-use reagent. An alternative strategy that has yet to be realized is to utilize an exogenous NTP to generate activated amino acids in multiple turnover reactions. An obvious benefit from operating as a multiple turnover enzyme is the increase in overall efficiency for how the ribocyte's resources are used. A multiple-turnover enzyme would also be in a better position to acquire an editing function; a mis-acylated species could be hydrolyzed without requiring resynthesis of the transcript, and of course the editing function would permit increased fidelity. As they stand, ribozymes for peptide synthesis reactions show varying degrees of substrate affinity and specificity. The second-generation Bartel and Suga ribozymes demonstrate that substrate-recognition domains can be grafted onto existing acyltransferases. Similar efforts could yield ribozymes that synthesize di- and tri-peptides of specific sequence, perhaps through direct RNA-templated polymerization of activated amino acids.[93]

One additional prerequisite for inventing translation in an RNA-based metabolism is the necessity for a triplet code by which to link mRNA sequence with amino acid identity. Several mechanistic models have been suggested as to how this may have come about based on biophysical interactions between amino acids and specific binding sites in RNA.[93,94] Many aptamers that recognize amino acids contain the corresponding codons, although the significance of the correlation is controversial.[95] Various models for the origin of the Code are considered elsewhere in this volume.

These challenges aside, once activated amino acids became available to RNA World ribozymes, it would have been a small step to begin using them to make specific oligopeptides, thereby planting the seeds of the Protein Revolution. That ribozymes for uncoded peptide synthesis are relatively frequent in random populations is suggested by the successive aminoacyl- and peptidyl-RNA synthesis by Yarus's RNA77, by this ribozyme's small size, and by the diversity of dipeptides assembled by the Zhang ribozyme. Small size (low information content) implies that relatively few nucleotides need to be specified, thereby increasing the probability that such sequences are encountered in random searches through sequence space and that they could have arisen from within an RNA-catalyzed metabolism.

Towards an RNA-Catalyzed Metabolism: What's Missing?

Replication and translation are necessary components of any RNA World that might have preceded our own evolution —one to propagate the genome and one to launch the Protein Revolution— but it is difficult to envision a viable organism with only these activities. What else is required to build a cell from scratch? Biosynthesis may have been required early, before supplies of abiotically synthesized organic compounds were exhausted completely. Where biosynthesis runs counter to free energy gradients, energy extraction through catabolic reactions could have helped to power those reactions. A sufficient set of RNA-catalyzed activities, then, may need to include a minimal, recognizably modern metabolism. Benner et al used chemical intuition and the phylogenetic distribution of extant pathways to infer that the last ribocytes "had a complex metabolism that included dehydrogenations, transmethylations, C-C bond formation, and an energy metabolism based on phosphoesters," and that it synthesized porphyrins and terpenes.[49,96] Ribozymes for a constrained sort of nucleoside biosynthesis have been described,[97] and several labs are in the early stages of exploring RNA-catalyzed polyphosphorylation and the synthesis of lipids and amino acids (Table 3).

Table 3. Some targets for new ribozyme reactions and chemical activities

Biosynthetic Targets for Ribozymes:
cofactors
nucleotides
amino acids
lipids
Target Reactions for New Ribozymes:
cofactor-dependent electron transfer
Claisen condensation
hydration/dehydration
reactions with cationic intermediates
hydrogenation/dehydrogenation
radical chemistry
mutases

How many of these reactions can RNA actually catalyze? While all of the reactions in the transmission of genetic information are substitutions at esters or at phosphate esters, catalysis of additional reaction classes is required for an expanded metabolism. Confirmed ribozyme-catalyzed reactions encompass many additional reaction types (reviewed in refs. 1,6,7). Some of these include S_N1 displacement at the C1 of ribose to form a glycosidic bond,[97] S_N2 displacement of halogenated carbons to form N-C and S-C bonds,[98-100] C-C bond formation through Diels-Alder chemistry[101-103] and Michael addition,[104] porphyrin metallation,[105] and even very weak redox activity.[106,107] Many of these ribozymes may use catalytic strategies similar to those described above for reactions at carbon and phosphate esters (proximity effects, charge stabilization, acid/base catalysis, etc.). The next several years are likely to see advances in ribozyme catalysis of new classes of reactions, incorporation of nucleotide cofactors, amino acids and other small molecule prosthetic groups into ribozymes, and integration of multiple ribozymes into cells and multistep pathways.

New Classes of Reactions

The opportunities for expanding ribozyme activities are enormous (Table 3). Lipid synthesis makes use of redox reactions, dehydrations, and Claisen condensations—each of which includes an anionic intermediate, like the ester and phosphate reactions above. Some of the mechanistic strategies used in acyltransfer reactions may also be important for ribozyme-catalyzed electron transfers. Redox reactions using flavin or nicotinamide cofactors are often initiated by (de)protonation events and involve anionic intermediates. Reactions that require stabilization of cationic intermediates should be especially facile for RNA, with its polyanionic backbone, yet none of this ilk has yet been described. Ribozymes with new activities will continue to appear in the literature. Some of these activities may never have existed in an RNA World, or may not have a counterpart in extant cells, yet they could still find a niche in the metabolism of artificial cells, in the metabolic engineering of normal or diseased cells, or in practical synthetic applications.

Direct selection for catalysis requires a partition method that distinguishes active RNAs from the rest. A popular tool in recent years has been to couple a substrate covalently to an oligonucleotide, ligate this construct to the nucleic acid pool, allow the reaction to proceed, and purify active species on the basis of the chemical properties of the covalently tethered product (Fig. 8). This strategy has been fruitful for condensation with biotinylated,

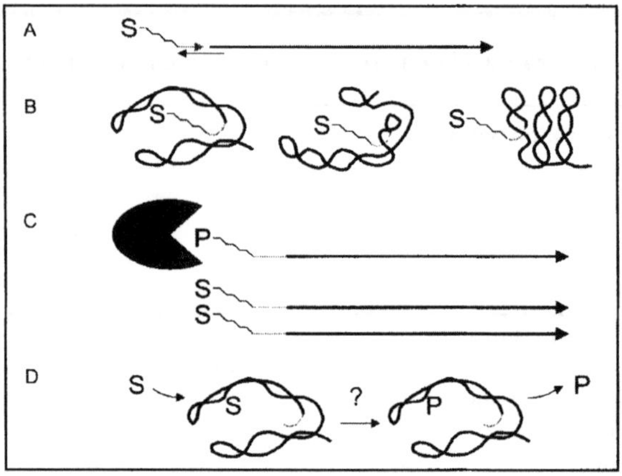

Figure 8. Ribozyme selection in vitro for reactions involving small molecule substrates. A) Substrate (S) is covalently tethered to a DNA or RNA oligonucleotide (thin gray line) through a flexible linker, such as polyethylene glycol (jagged line). The modified oligo is then ligated to each member of the RNA pool (thick line) with the help of a bridging DNA oligonucleotide. Strand polarity 5' to 3' is indicate by arrows. B) Upon folding, the sequence diversity library becomes a shape diversity library, wherein the flexible linker allows the substrate to access most of the folded RNA surface. C) Active species are recovered through the unique properties of the tethered product (P) such as through interactions with streptavidin or mercury if the product contains biotin or sulfer, respectively. D) Reactivity is often maintained when the tethered substrate is instead provided in trans, a critical feature for biological relevance, though some bind the substrate at low affinity and require high concentration.

sulphur-containing, or otherwise tagged substrates, but it does not readily extend to reduction of a C=C double bond, rearrangements (mutases), radical-mediated reactions, (de)hydrations, (de)hydrogenations, or methyltransfers. Ribozymes for these reactions are likely to exist in accessible sequence space —particularly if the RNA can exploit reactive cofactors— though their identification through in vitro selection may require highly creative new strategies.

It is our goal and that of several other labs to move ribozyme biochemistry toward biological relevance. In most cases, new ribozymes catalyze model reactions without producing biologically useful products, or they exploit biologically unreasonable reactants (such as cyanomethyl esters or tethered substrates). Some of the challenges for constructing a ribozyme-based metabolism include improving substrate recognition, exploiting reactive nucleotide cofactors and peptides, and integrating several activities into coordinated metabolic pathways.

Improved Substrate Recognition

Most new ribozymes are identified through their ability to promote a reaction between one substrate that is free in solution and another that is covalently tethered. In most cases, specific binding interactions are observed for the free substrate. In a few cases, reactivity is still notable when the tethered substrate is provided in trans at high concentrations,[75,108] but severing the attachment destroys reactivity in others.[97] Apparently the affinity for free substrate in those latter cases is insufficient to form a productive E•S (Michaelis-Menten) complex. Target recognition per se does not seem to be an intrinsic limitation, as RNA aptamers with high affinity have been isolated that recognize a wide variety of molecular targets. The Bartel polymerase described above is intermediate in this sense. While the tethered primer template duplex used during its selection can be provided exogenously, the limitations to its processivity are derived as much from rapid product dissociation as from slow polymerization chemistry. Many ribozymes bind one or more substrates non-covalently, although little is known about the details of these interactions. Small molecule binding by aptamers, on the other hand, is much better understood, and numerous aptamer modules have been described that bind nucleotide cofactors.

While ribozymes could in principle be developed from such modules, there are not yet any examples of this engineering/evolutionary path. The interrelation between small molecule binding and catalytic activities is fertile ground for future work.

A Role for Nucleotide Cofactors

The chemical versatility of unadorned RNA is generally regarded as being less than that of unadorned proteins, but this comparison is inappropriate. Protein enzymes make extensive use of prosthetic groups such as metal ions and nucleotide cofactors that ribozymes could similarly exploit. The seemingly superfluous inclusion of nucleotide components in cofactors such as CoA, FAD, NAD^+, and SAM has been interpreted as molecular fossils of an RNA world.[49,109,110] (The tRNA portion of glutaminyl-tRNA used in the C-5 pathway for porphyrin synthesis is also cited in this category as evidence that ribocytes made porphyrins,[49] although in this case it is equally probable that this reaction is not primitive; rather, a primitive cell that had already invented translation simply usurped a convenient, readily available source of activated Glu en route to inventing porphyrin synthesis.) In general, if a given cofactor can be made abiotically, or if its synthesis by ribozymes does not violate established RNA-catalyzed chemistry, it is reasonable that it should be available to ribozymes in RNA World models. Adenine forms readily upon condensation of ammonia and hydrogen cyanide.[111] The nicotinamide of NMN^+ and NAD^+ appears in some abiotic chemical reactions,[112] as does the pantotheine portion of CoA.[113-115] In vitro evolution studies of cofactor-mediated catalysis by ribozymes is therefore justified from an RNA world perspective. As of this writing, phosphoryl transfer from ATP remains the only significant nucleotide cofactor-assisted catalysis by nucleic acids,[54,116,117] though this may soon change.

Cofactor-dependent ribozymes could bind their cofactors non-covalently as in aptamer complexes, covalently through a self-capping reaction or other adduction, or by incorporating them into the RNA chain during transcription. The world of protein enzymes is replete with precedent for covalently attached cofactors. Various flavoproteins are covalently tethered to their flavins. In the fatty acid synthesis pathway, acetyl CoA carboxylase carries an attached biotin, and acyl carrier protein carries a covalently attached pantotheine. The self-capping ribozyme of Huang et al appends the phosphorylated cofactor precursors FMN, NMN^+, and phosphopantotheine onto its 5' end, releasing pyrophosphate in each case.[73] Jadhav and Yarus used this ribozyme —augmented with additional random sequence— to identify RNAs that catalyze synthesis of biotinyl CoA and acetyl CoA from the acyl adenylates and ribozyme-tethered CoA.[118] The ability to synthesize acetyl CoA would have been a seminal event in the evolution of RNA World metabolism. While covalent attachment prevents the cofactor from diffusing away, it does not obviate the need for non-covalent binding interactions, both with the substrates that react with the activated form of the cofactor (e.g., acylation substrate) and with the substrates that re-activate the cofactor (e.g., acetyl-AMP).

A Role for Amino Acids and Peptides

Amino acids are abundant in prebiotic chemical reactions. If there is an RNA World in our history, it surely arose in a chemical environment that included some amino acids and small peptides. These reagents are thus legitimate tools for augmenting ribozyme activities, either through direct chemical functionality or by allosteric activation. Indeed, a commonly cited model for takeover of catalytic functions by proteins is that the protein portions of ribonucleoprotein enzymes gradually took on greater importance while the RNA portions diminished, leaving behind only the nucleotide cofactors in the active site (Fig. 9).

Roth and Breaker isolated a deoxyribozyme that requires L-histidine or a closely related analog to catalyze RNA phosphoester cleavage.[119] The pH dependence of the reaction suggests that the rate-limiting step includes protonation of the histidine imidazole. These data are interpreted as indicating that the histidine serves as a general base catalyst similar to the first step of the reaction catalyzed by RNAseA, although it is also possible that the protonated histidine allosterically stabilizes the active structure of the ribozyme. Small peptides could also serve

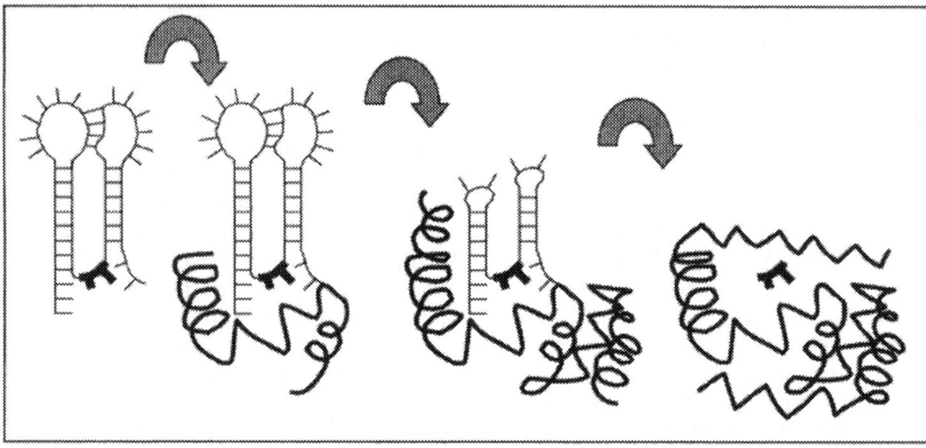

Figure 9. Protein takeover of catalytic function from a ribozyme (left side), through ribonucleoprotein enzymes (middle two), to a modern protein (right side). The dinucleotide cofactor in the active site is the only portion of the original RNA left behind in this model, making it a "molecular fossil." Redrawn based on White, 1982.

chemical roles for ribozymes. Like the nucleotide cofactors, such peptides could be bound to ribozymes through covalent or non-covalent interactions. Baskerville and Bartel found ribozymes that form a stable phosphoamide bond between their 5' termini and a specific polypeptide at the expense of pyrophosphate.[120] An optimized version of the ribozyme recognizes and reacts with the substrate even when the peptide is embedded within a fusion protein. Peptides and proteins could also augment ribozyme activity by inducing allosteric regulation; the nucleoprotein ligase enzymes of Robertson and Ellington are activated as much as 10,000-fold upon binding either the TyrS protein encoded by Cyt18 or the lysozyme of bacteriophage T4.[121] Ribonucleoprotein complexes similar to those shown in Figure 9 are therefore legitimate targets for RNA World research, and they are beginning to show up in the experimental literature. Furthermore, to the extent that small peptides were utilized by ancient ribozyme to augment their activities, any requirement for specific sequences would have provided selective pressure towards increased translational fidelity.

A Role for Modified Nucleotides

More than one hundred different modified nucleotides are found in modern rRNA and tRNA. Some of these have been proposed to have been present during the RNA World to augment stability against hydrolysis or catalytic reactivity, while others have been exploited in ribozyme selections in vitro. Ribozymes have been isolated that carry 5-pyridyl-U or 5-imidazole-U in place of normal uracils,[77,103] or that carry alkylamino side chains at the N6 position of adenosines.[47] There are at least three difficulties in applying these results to RNA World physiology. First, the current set of modified ribozymes catalyze reactions for which unmodified ribozymes have also been identified; thus, it is not inevitable that there are advantages to including such modifications. The second difficulty lies in choosing the appropriate set of modifications from among the hundreds of possibilities. As the proposed ancestral set of nucleotides becomes increasingly dissimilar from the present set of A, C, G, and U, fewer clues to the nature of the RNA World are available from modern biology. For addressing basic chemical questions, modifications can be chosen based solely on how useful they might be for selected catalysts, but some other criterion must be applied for the modification to be relevant to RNA world. Once such criterion could be the demonstration of an abiotic or RNA-catalyzed pathway to its synthesis. Messenger RNA is largely devoid of significant modifications (A→ I and C→U deaminations in double-stranded mRNA do not introduce new chemical functional-

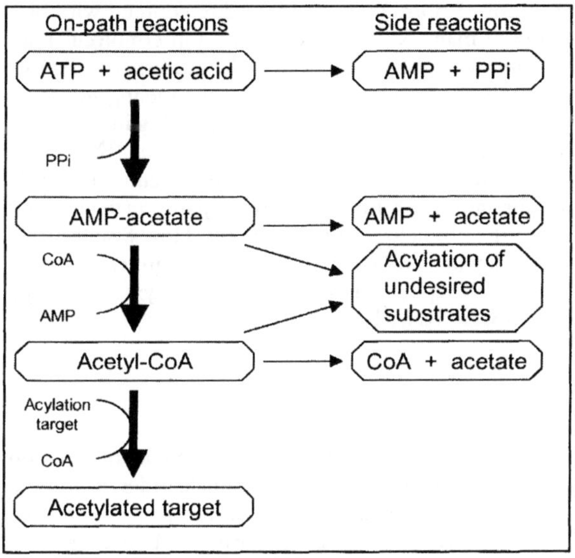

Figure 10. Kinetic control of reaction outcome, illustrated by a simple scheme for acetylating a desired substrate. A useful catalyst must accelerate the on-path reactions (thick arrows) well above the combined rates of competing reactions, such as hydrolysis and reaction with alternative substrates (thin arrows).

ity). The third difficulty lies in removing the modifications from the cellular lexicon once mRNA-encoded protein synthesis developed. It is conceivable that early ribocytes depended upon modified nucleotides, and that losing the ability to synthesize them during the Protein Revolution made the transition from RNA World to Protein World irreversible. Future developments in the theory and application of ribozymes with modified nucleotides may dispel these concerns, though for now it is difficult to know how to incorporate modified nucleotides appropriately into RNA World biochemistry.

Integration of the Individual Activities into a Working Metabolism

Achieving this objective requires both increased catalytic vigor of existing ribozymes and kinetic control of reaction outcomes. The significance of the latter is that biological catalysts are required not merely to accelerate an inevitable reaction, but also to channel reactive intermediates into particular products before they have a chance to participate in side reactions (Fig. 10). It is by guiding this kinetic competition among possible chemical outcomes that enzymes earn their keep within a cell. In principle, a set of ribozymes closely related to those currently available could attach CoA onto the 5' end of a transcript, activate an organic acid such as acetate in the form of an adenylate, transfer that acid onto the CoA to generate a thioester, and from there proceed to substrate acylation. All of these individual activities are available in the laboratory, but at present they do not adequately compete with the alternative reaction pathways, such as hydrolysis.

It would also be useful to have quantitative estimates of the affinities and kinetic rate constants needed to achieve kinetic control for each reaction, and computational models of cells may help in this regard (for review of theoretical cell models, see ref. 122). The rate enhancements of most ribozymes are many orders of magnitude below those observed for protein enzymes. The catalytic prowess of natural enzymes is a tribute to billions of years of optimization, but a more appropriate comparison might be made with catalytic antibodies, which often display rate enhancements similar to those of in vitro selected nucleic acid catalysts.[106] In this light it is not the polymer type (protein vs. nucleic acid), but the evolutionary naiveté of the catalyst, that limits the catalytic activity observed in ribozymes. Additional cycles of engineering and evolution, such as those that produced the polymerase ribozyme, may yet increase the catalytic vigor of these diverse ribozymes to levels sufficient for integration and for practical and biological application.

A cytoplasm carrying high concentrations of ribozymes would need to contend with biophysical and spatio-temporal coordination of many RNA species to avoid intermolecular aggregation. At the same time, productive intermolecular interactions as the quaternary assembly of multi-subunit enzymes could benefit metabolic integration. Substrate channeling and coordinated levels of enzyme activity, for example, could exploit RNA-RNA docking interactions, while minimizing such interactions could prevent such aggregation and keep each RNA maximally accessible to the cell's polymerases for replication. Two experimental avenues currently being explored along these lines include fabricating an artificial, multistep pathway in vitro and integrating in vitro-derived ribozymes into the physiology of modern cells. How would such a system respond and adapt to the complex demands of an intracellular environment? Success along either front, or fabrication of artificial, self-replicating cells based on ribozymes would go a long way in moving RNA biochemistry toward RNA-based life.

Concluding Remarks

Begging for Phosphates

The one essential reaction not yet demonstrated for RNA catalysis of genetic information flow is nucleotide polyphosphorylation. Given suitably activated mononucleotides, ribozymes can catalyze all of the subsequent reactions required for the transmission of genetic information, as well as a scattered sampling of other reaction types. They can bind an exogenous primer-template substrate for polymerization and extend the substrate more than one complete helical turn. They can use the energy of an NTP to activate an organic acid or amino acid, then use the activated substrate in acyl transfers to generate thioesters, ribose esters, amides and peptide bonds.

Proof of Principle?

The plausibility of RNA World theories accrues incrementally as one objection after another falls to experimental observation. Some authors now consider its plausibility to be fully established, but this is an overstatement of the case. At present none of these reactions proceeds with the vigor needed to form a working metabolism. It is tempting to wave the magic wand of Evolution and proclaim that if ribozymes can limp through a given reaction now, they could leap through it once a few favorable mutations are introduced. Experimental demonstration of this postulate will be more satisfying than a quasi-religious extrapolation. The RNA-mediated activities observed to date are to real RNA-based life what Goddard's rockets were to manned space flight: a very long way from the ultimate goal, but extraordinarily important milestones along the way. RNA World theories have risen in stature from enthusiastic conjecture to a hypothetical system that can be approached experimentally. The concept of an RNA World, whether in our own evolutionary past or in the frozen oceans of Jupiter's moon Europa, is now worth taking very seriously.

There are detractors from RNA World theories who seem to want the entire theory proven or discarded at once, with no tolerance for the intervening ambiguity. While it is not yet known whether ribozymes have what it takes to sustain life, it would be simplistic pedantry to dismiss the theory for lack of data before the requisite experiments have been carried out. The experimental goal is not to prove whether our own evolution ever passed through an era in which the world was populated with ribocytes, as this question cannot yet be addressed experimentally. Instead we seek to define both the chemical limits of catalysis by ribozymes and the inherent features of living systems built around RNA catalysis. The 21st Century will see the discussion move towards issues that link the chemical properties of individual macromolecules with the activities required to maintain cellular function. Then sophomoric rhetoric, whether from undisciplined cheerleaders or from jaded detractors, can be replaced by relevant experimental data.

References

1. Joyce GF. The antiquity of RNA-based evolution. Nature 2002; 418:214-221.
2. Joyce GF. RNA Evolution and the origins of life. Nature 1989; 338:217-224.
3. Orgel L. The origin of life—A review of facts and speculations. Trends Biochem Sci 1998; 23:491-495.
4. Gesteland RF, Cech TR, Atkins JF. The RNA World. 2nd ed. Cold Springs Harbor: Cold Springs Harbor Laboratory Press, 1998:49-77.
5. Szostak JW, Bartel DP, Luisi PL. Synthesizing life. Nature 2001; 409 Suppl:387-390.
6. Jaschke A. Artificial ribozymes and deoxyribozymes. Curr Opin Struct Biol 2001; 11:321-326.
7. Wilson D, Szostak J. In vitro selection of functional nucleic acids. Ann Rev Biochem 1999; 68:611-647.
8. Doudna JA, Cech TR. The chemical repertoire of natural ribozymes. Nature 2002; 418:222-228.
9. Bada JL, Lazcano A. Origin of life. Some like it hot, but not the first biomolecules. Science 2002; 296:1982-1983.
10. Pley HW, Flaherty KM, McKay DB. Model for an RNA tertiary interaction from the structure of an intermolecular complex between a GAAA tetraloop and an RNA helix. Nature 1994; 372:111-113.
11. Cate JH, Gooding AR, Podell E et al. RNA tertiary structure mediation by adenosine platforms. Science 1996; 273:1696-1699.
12. Burke DH, Willis J. Recombination, RNA evolution, and bifunctional RNA molecules isolated through Chimeric SELEX. RNA 1998; 4:1165-1175.
13. Pyle AM. Ribozymes: A distinct class of metalloenzymes. Science 1993; 261:709-714.
14. Gesteland RF, Cech TR, Atkins JF. The RNA World. 2nd ed. Cold Springs Harbor: Cold Springs Harbor Laboratory Press, 1998:287-319.
15. Strobel SA, Doudna JA. RNA seeing double: close-packing of helices in RNA tertiary structure. Trends Biochem Sci 1997; 22:262-266.
16. Shan S, Yoshida A, Sun S et al. Three metal ions at the active site of the Tetrahymena group I ribozyme." Proc Natl Acad Sci USA 1999; 96:12299-12304.
17. Yarus M. How many catalytic RNAs?: Ions and the Cheshire Cat conjecture. FASEB Journal 1993; 7:31-9.
18. Steitz T,Steitz J. A general two-metal-ion mechanism for catalytic RNA. Proc Natl Acad Sci USA 1993; 90:6498-6502.
19. Murray J, Seyhan A, Walter N et al. The hammerhead, hairpin and VS ribozymes are catalytically proficient in monovalent cations alone. Chem Biol 1998; 5:587-595.
20. Curtis E, Bartel D. The hammerhead cleavage reaction in monovalent cations. RNA 2001; 7.
21. Dahm SC, Derrick WB, Uhlenbeck OC. Evidence for the role of solvated metal hydroxide in the hammerhead cleavage mechanism. Biochem 1993; 32:13040-13045.
22. Legault P, Farmer BT, Mueller L et al. Through-bond correlation of adenine protons in a C-13-labeled ribozyme. J Am Chem Soc 1994; 116:2203-2204.
23. Wang C, Gao H, Jones RA. Nitrogen-15-labeled oligodeoxynucleotides 3. Protonation of the adenine N1 in the A•C and A•G mispairs of the duplex $\{d[CG(^{15}N^1)AGAATTCCCG]\}_2$ and $\{d[CGGGAATTC(^{15}N^1)ACG]\}_2$. J Am Chem Soc 1991; 113:5486-5488.
24. Shih I, Been M. Involvement of a cytosine side chain in proton transfer in the rate-determining step of ribozyme self-cleavage. Proc Natl Acad Sci 2001; 98:1489-1494.
25. Nakano S, Chadalavada D, Bevilacqua P. General acid-base catalysis in the mechanism of a hepatitis delta virus ribozyme. Science 2000; 287:1493-1497.
26. Bartel DP, Szostak JW. Isolation of new ribozymes from a large pool of random sequences. Science 1993; 261:1411-1418.
27. Ekland EH, Szostak JW, Bartel DP. Structurally complex and highly active RNA ligases derived from random sequences. Science 1995; 269:364-370.
28. Ekland EH, Bartel DP. RNA-catalysed RNA polymerization using nucleoside triphosphates. Nature 1996; 382:373-376.
29. Washington MT, Johnson RE, Prakash S et al. Fidelity and processivity of Saccharomyces cerevisiae DNA polymerase eta. J Biol Chem 1999; 274:36935-36938.
30. Johnston W, Unrau P, Lawrence M et al. RNA-catalyzed RNA polymerization: Accurate and general RNA-templated primer extension. Science 2001; 292:1319-1325.
31. Gesteland RF, Cech TR, Atkins JF. The RNA World. 2nd ed. Cold Springs Harbor: Cold Springs Harbor Laboratory Press, 1998:143-162.
32. Doudna JA, Cech TR. Self-assembly of a group I intron active site from its component tertiary structural domains. RNA 1995; 1:36-45.

33. Green R, Szostak JW. Selection of a ribozyme that functions as a superior template in a self-copying reaction. Science 1992; 258:1910-1915.
34. Fedor MJ. Tertiary structure stabilization promotes hairpin ribozyme ligation. Biochemistry 1999; 38:11040-11050.
35. Esteban JA, Banerjee AR, Burke JM. Kinetic mechanism of the hairpin ribozyme. Identification and characterization of two nonexchangeable conformations. J Biol Chem 1997; 272:13629-13639.
36. Rupert PB, Ferre-D'Amare AR. Crystal structure of a hairpin ribozyme-inhibitor complex with implications for catalysis. Nature 2001; 410:780-786.
37. Hegg LA, Fedor MJ. Kinetics and thermodynamics of intermolecular catalysis by hairpin ribozymes. Biochemistry 1995; 34:15813-15828.
38. Berzal-Herranz A, Joseph S, Burke JM. In vitro selection of active hairpin ribozymes by sequential RNA-catalyzed cleavage and ligation reactions." Genes Dev 1992; 6:129-134.
39. Saville BJ, Collins RA. RNA-mediated ligation of self-cleavage products of a Neurospora mitochondrial plasmid transcript. Proc Natl Acad Sci USA 1995; 88:8826-8830.
40. Beattie TL, Collins RA. Identification of functional domains in the self-cleaving Neurospora VS ribozyme using damage selection." J Mol Biol 1997; 267:830-840.
41. Stage-Zimmermann T, Uhlenbeck O. A covalent crosslink converts the hammerhead ribozyme from a ribonuclease to an RNA ligase." Nat Struct Biol 2001; 8:863-867.
42. Wright MC, Joyce GF. Continuous in vitro evolution of catalytic function. Science 1997; 276:614-617.
43. Rogers J, Joyce G. A ribozyme that lacks cytidine. Nature 1999; 402:323-325.
44. Miyamoto Y, Teramoto N, Imanishi Y et al. In vitro adaptation of a ligase ribozyme for activity under a low-pH condition." Biotechnol Bioeng 2001; 75:590-596.
45. Landweber LF, Pokrovskaya ID. Emergence of a dual-catalytic RNA with metal-specific cleavage and ligase activities: The spandrels of RNA evolution. Proc Natl Acad Sci USA 1999; 96:173-178.
46. Robertson MP, Miller SL. Prebiotic synthesis of 5-substituted uracils: A bridge between the RNA world and the DNA-protein world. Science 1995; 268:702-705.
47. Teramoto N, Imanishi Y, Ito Y. In vitro selection of a ligase ribozyme carrying alkylamino groups in the side chains. Bioconjug Chem 2000; 11:744-748.
48. Jaeger L, Wright M, Joyce G. A complex ligase ribozyme evolved in vitro from a group I ribozyme domain." Proc Natl Acad Sci USA 1999; 96:14712-14717.
49. Benner SA, Ellington AD, Tauer A. Modern metabolism as a palimpsest of the RNA world. Proc Natl Acad Sci 1989; 86:7054-7058.
50. Joyce GF, Orgel LE. Non-enzymatic template-directed synthesis on RNA random copolymers. J Mol Biol 1988; 202:677-681.
51. Ertem G, Ferris JP. Template-directed synthesis using the heterogeneous templates produced by montmorillonite catalysis. A possible bridge between the prebiotic and RNA worlds. J Am Chem Soc 1997; 119:7197-7201.
52. Hager A, Szostak J. Isolation of novel ribozymes that ligate AMP-activated RNA substrates. Chem Biol 1997; 4:607-617.
53. Chapman K, Szostak J. Isolation of a ribozyme with 5'-5' ligase activity. Chem Biol 1995; 2:325-333.
54. Li Y, Breaker RR. Phosphorylating DNA with DNA." Proc Natl Acad Sci USA 1999; 96:2746-2751.
55. Li Y, Breaker RR. In vitro selection of kinase and ligase deoxyribozymes. Methods 2001; 23:179-190.
56. Lacey JC, Senaratne N, Mullins, Jr. DW. Hydrolytic properties of phenylalanyl- and N-acetylphenylalanyl adenylate anhydrides. Origins of Life 1984; 15:45-54.
57. Illangasekare M, Yarus M. Small-molecule-substrate interactions with a self-aminoacylating ribozyme. J Mol Biol 1997; 268:631-639.
58. Izatt RM, Hansen LD, Rytting JH et al. Proton ionization from adenosine. J Am Chem Soc 1965; 87:2760-2761.
59. Rudenger-Thirion J, Giegé R, Felden B. Aminoacylated tmRNA from Escherichia coli interacts with prokaryotic elongation factor Tu. RNA 1999; 5:889-992.
60. Barsukov IL, Lian LY, Ellis J et al. The conformation of coenzyme A bound to chloramphenicol acetyltransferase determined by transferred NOE experiments. J Mol Biol 1996; 262:543-538.
61. Leslie AGW. Refined crystal structure of Type III chloramphenicol acetyltransferase at 1.75 Å resolution. J Mol Biol 1990; 213:167-186.
62. Lewendon A, Murray IA, Shaw WV et al. Replacement of catalytic histidine-195 of chloramphenicol acetyltransferase: Evidence for a general base role for glutamate. Biochem 1994; 33:1944-1950.
63. Hendle J, Mattevi A, Westphal AH et al. Crystallographic and enzymatic investigations on the role of Ser558, His610, and Asn614 in the catalytic mechanism of Azotobacter vinelandii dihydrolipoamide acetyltransferase (E2p). Biochem 1995; 34:4287-98.

64. Nissen P, Hansen J, Ban N et al. The structural basis of ribosome activity in peptide bond synthesis. Science 2000; 289:920-930.
65. Muth G, Ortoleva-Donnelly L, Strobel S. A single adenosine with a neutral pKa in the ribosomal peptidyl transferase center. Science 2000; 289:947-950.
66. Xiong L, Polacek N, Sander P et al. pKa of adenine 2451 in the ribosomal peptidyl transferase center remains elusive. RNA 2001; 7:1365-1369.
67. Polacek N, Gaynor M, Yassin A et al. Ribosomal peptidyl transferase can withstand mutations at the putative catalytic nucleotide. Nature 2001; 411:498-501.
68. Weber A, Orgel LE. Poly(U)-directed peptide bond formation from the 2'(3')-glycyl esters of adenosine derivatives. J Mol Evol 1980; 16: 1-10.
69. Katunin VI, Muth GW, Strobel SA et al. Important contribution to catalysis of peptide bond formation by a single ionizing group within the ribosome. Mol Cell 2002; 10:339-346.
70. Green R, Lorsch J. The path to perdition is paved with protons. Cell 2002; 110:665-668.
71. Huang FQ, Yarus M. 5'-RNA self-capping from guanosine diphosphate. Biochemistry 2997; 36:6557-6563.
72. Huang F, Yarus M. Versatile 5' phosphoryl coupling of small and large molecules to an RNA. Proc Natl Acad Sci 1997; 94:8965-8969.
73. Huang F, Bugg C, Yarus M. RNA-Catalyzed CoA, NAD, and FAD Synthesis from Phosphopantetheine, NMN, and FMN. Biochemistry 2000; 39:15548-15555.
74. Huang F, Yarus M. A calcium-metalloribozyme with autodecapping and pyrophosphatase activities. Biochemistry 1997; 36:14107-14119.
75. Huang F, Yang Z, Yarus M. RNA enzymes with two small-molecule substrates. Chem Biol 1998; 5:669-678.
76. Illangasekare M, Sanchez G, Nickels T et al. Aminoacyl-RNA synthesis catalyzed by an RNA. Science 1995; 267:643-647.
77. Wiegand TW, Janssen RC, Eaton BE. Selection of RNA amide synthases. Chem Biol 1997; 4:675-683.
78. Lohse PA, Szostak JW. Ribozyme-catalyzed amino-acid transfer reactions. Nature 1996; 381:442-444.
79. Jenne A, Famulok M. A novel ribozyme with ester transferase activity. Chem Biol 1998; 5:23-34.
80. Lee N, Bessho Y, Wei K et al. Ribozyme-catalyzed tRNA aminoacylation. Nat Struct Biol 2000; 7:28-33.
81. Illangasekare M, Yarus M. Specific, rapid synthesis of Phe-RNA by RNA. Proc Natl Acad Sci USA 1999; 96:5470-5475.
82. Suga H, Cowan JA, Szostak JW. Unusual metal ion catalysis in an acyl-transferase ribozyme. Biochemistry 1998; 37:10118-10125.
83. Suga H, Lohse PA, Szostak JW. Structural and kinetic characterization of an acyl transferase ribozyme. J Am Chem Soc 1998; 120:1151-1156.
84. Vaidya A, Suga H. Diverse roles of metal ions in acyl-transferase ribozymes. Biochemistry 2001; 40:7200-7210.
85. Flynn-Charlebois A, Lee N, Suga H. A single metal ion plays structural and chemical roles in an aminoacyl-transferase ribozyme. Biochemistry 2001; 40:13623-13632.
86. Lee N, Suga H. A minihelix-loop RNA acts as a trans-aminoacylation catalyst. RNA 2001; 7:1043-1051.
87. Saito H, Suga H. A ribozyme exclusively aminoacylates the 3'-hydroxyl group of the tRNA terminal adenosine. J Am Chem Soc 2001; 123:7178-7179.
88. Saito H, Watanabe K, Suga H. Concurrent molecular recognition of the amino acid and tRNA by a ribozyme. RNA 2001; 7:1867-1878.
89. Illangasekare M, Yarus M. A tiny RNA that catalyzes both aminoacyl-RNA and peptidyl-RNA synthesis. RNA 1999; 5:1482-1489.
90. lewinsohn R, Paecht-Horowitz M, Katchalsky A. Polycondensation of amino acid phosphoanhydrides. III. Polycondensation of alanyl adenylate. Biochim Biophys Acta 1967; 140:24-36.
91. Zhang B, Cech TR. Peptidyl bond formation by in vitro selected ribozymes. Nature 1998; 390:96-100.
92. Sun L, Cui Z, Gottlieb RL et al. A selected ribozyme catalyzing diverse dipeptide synthesis. Chem Biol 2002; 9:619-628.
93. Yarus M. Amino acids as RNA ligands: A direct-RNA-template theory for the Code's origin. J Mol Evol 1998; 47:109-117.
94. Knight R, Landweber L. Guilt by association: The arginine case revisited. RNA 2000; 6:499-510.
95. Ellington A, Khrapov M, Shaw C. The scene of a frozen accident. RNA 2000; 6:485-498.
96. Gesteland RF, Atkins JF. The RNA World. Plainview: CSHL Press, 1993:27-70.
97. Unrau PJ, Bartel DP. RNA-catalysed nucleotide synthesis. Nature 1998; 395:260-263.

98. Wilson C, Sostak JW. In vitro evolution of a self-alkylating ribozyme. Nature 1995; 374:777-782.
99. Wecker M, Smith D. Selection for RNA: Peptide recognition through sulfur alkylation chemistry. Methods Enzymol 2000; 318:229-237.
100. Wecker M. Molecular, Cellular, and Developmental Biology. Boulder: University of Colorado, 1995.
101. Seelig B, Keiper S, Stuhlmann F et al. Enantioselective ribozyme catalysis of a bimolecular cycloaddition reaction. Angew Chem Int Ed Engl 2000; 15:4576-4579.
102. Seelig B, Jäschke A. A small catalytic RNA motif with Diels-Alderase activity. Chem Biol 1999; 6:167-176.
103. Tarasow TM, Tarasow SL, Eaton BE. RNA-catalyzed carbon-carbon bond formation. Nature 1997; 389:54-57.
104. Sengle G, Eisenfuhr A, Arora P et al. Novel RNA catalysts for the Michael reaction. Chem Biol 2001; 8:459-473.
105. Li YS. A catalytic DNA for porphyrin metallation. Nat Str Biol 1996; 3:143-147.
106. Travascio P, Bennet AJ, Wang DY et al. A ribozyme and a catalytic DNA with peroxidase activity: Active sites versus cofactor-binding sites. Chem Biol 1999; 6:779-787.
107. Wilson C, Szostak JW. Isolation of a fluorphere-specific DNA Aptamer with weak redox activity. Chem Biol 1998; 609-617.
108. Stuhlmann F, Jaschke A. Characterization of an RNA active site: Interactions between a Diels-Alderase ribozyme and its substrates and products. J Am Chem Soc 2002; 124:3238-3244.
109. White HB, III. Coenzymes as fossils of an earlier metabolic state. J Mol Evol 1976; 7:101-104.
110. Everse J, Anderson B, Yu K-S, eds. The Pyridine Nucleotide Cofactors. New York: Academic, 1982:1-17.
111. Oró J. Mechanism of synthesis of adenine from hydrogen cyanide under plausible primitive earth conditions. Nature 1961; 191:1193-1194.
112. Dowler MJ, Fuller WD, Orgel LE. Prebiotic synthesis of propiolaldehyde and nicotinamide. Science 1970; 169:1320-1321.
113. Miller SL, Schlesinger G. Prebiotic syntheses of vitamin coenzymes: II. Pantoic acide, pantothenic acid, and the composition of coenzyme A. J Mol Evol 1993; 36:308-314.
114. Miller SL, Schlesinger G. Prebiotic syntheses of vitamin coenzymes: I. Cysteamine and 2-mercaptoethanesulfonic acid (coenzyme M). J Mol Evol 1993; 36:302-307.
115. Keefe AD, Newton GL, Miller SL. A possible prebiotic synthesis of pantetheine, a precursor to coenzyme A. Nature 1995; 373:683-685.
116. Lorsch JR, Szostak JW. In vitro evolution of new ribozymes with polynucleotide kinase activity. Nature 1994; 371-31-36.
117. Lorsch JR, Szostak JW. Kinetic and thermodynamic characterization of the reaction catalyzed by a polynucleotide kinase ribozyme. Biochem 1995; 34:15315-15327.
118. Jadhav VR, Yarus M. Acyl-CoAs from coenzyme ribozymes. Biochem 2002; 41:723-729.
119. Roth A, Breaker RR. An amino acid as a cofactor for a catalytic polynucleotide. Proc Natl Acad Sci USA 1998; 95:6027-6031.
120. Baskerville S, Bartel DP. A ribozyme that ligates RNA to protein. Proc Natl Acad Sci USA 2002; 99:9154-9159.
121. Robertson MP, Ellington AD. In vitro selection of nucleoprotein enzymes. Nat Biothechnol 2001; 19:650-655.
122. Mendes P, Kell D. Non-linear optimization of biochemical pathways: Applications to metabolic engineering and parameter estimation. Bioinformatics 1998; 14:869-883.

Chapter 5

The Scope of Selection

Michael Yarus and Rob D. Knight

Abstract

We have estimated the maximal size for an RNA motif recoverable from selection-amplification for new RNA activities, under conditions that span those in present laboratory use. The number of sequence pieces from which an active site is folded (the modularity) is a crucial variable. Routine laboratory experiments might isolate RNAs of modularity 4 containing ≤ 33 specified nucleotides. The probability of recovering shorter motifs increases rapidly, but the likely maximal motif size declines 1.66 nucleotides per 10-fold decrease in experimental scale. In such experiments, randomized tracts of 80-120 nucleotides extract most of the benefit of longer initially randomized pools. The same methods also permit extrapolation to conditions more plausible during the initiation of an RNA world. Under these conditions, active RNAs were likely highly modular, even more so than in modern experiments. Strikingly, several lines of evidence converge on the conclusion that 15 to 35-mer active sites would be the working material for an early RNA world. If initiation of an RNA world is synonymous with emergence of active structures from randomized sequences (the Axiom of Origin), populations containing only zeptomoles of RNA (hundreds to hundreds of thousands of molecules) might yield RNAs at the lower end of this size range. This makes the RNA world much more accessible than previously suspected.

Introduction

It is rare that a new technique makes possible a type of experiment not feasible before, but this is true of selection-amplification or SELEX.[4,20,22] This procedure consists of cycles of alternating selection (biochemical fractionation) and amplification (replication), applied to RNA or DNA containing randomized tracts of nucleotides. Because nucleic acids are uniquely able to replicate, any usable fractionation can be applied to a starting population, then repeatedly re-applied to the replicated output from the fractionation. The population of molecules increases in purity. When repetitive selection for an initially rare molecule yields sufficient purity, the population is cloned (and active molecules thereby purified to homogeneity).

On one hand, such cyclic fractionation-replication is well suited for specific questions like "what is the sequence of the nucleic acid bound by protein A"? Protein A is repetitively used to sequester molecules for which it has affinity. After multiple selection-amplification cycles, a substantial plurality of molecules have a protein A binding site, revealed as conserved sequences among independently-derived clones. As for other specific types of questions, selection-amplification supplies an alternative to (for example) cloning and sequencing natural protein A binding sites, and clarifies such an experiment by reducing bias due to effects other than binding.

However, this chapter is primarily about the nonspecific, open-ended use of selection-amplification. In such an experiment, one asks "is an RNA (DNA) with property X possible?" As long as a selection (fractionation) exists that specifically concentrates molecules

The Genetic Code and the Origin of Life, edited by Lluís Ribas de Pouplana.

with property X, one can determine their existence and study them in pure form. No prior information is required about the potentially interesting molecules, and no prior example ever need have been observed. Such questions are pressing because of the prediction that many practical RNA activities are presently extinct. This unused RNA potential, imputed relic of an RNA world,[26] can usually be demonstrated in no other way than by open-ended selection-amplification.

There are notable findings using this approach. The four chemical sub-reactions required for translation have been shown to be within RNA capabilities.[29] Three of these; amino acid activation,[19] aminoacyl-RNA synthesis,[10] and direct coding interactions[27] are not extant RNA capabilities. RNA should have once replicated itself, and a pure RNA RNA replicase that uses free primed RNAs as template has been selected.[15] Finally, the existence of an RNA-mediated metabolism has been supported. RNAs can synthesize[9] and utilize the enzymatic cofactor CoA,[13] long thought, because of its structure, to be a molecular remnant of an RNA world.[26]

However, selection-amplification is not infinitely capable. It isolates only molecules that meet its constraints, and addresses the existence of a new RNA activity only within limits. We understand selection's outcome only if we know the scope of selection.

In order to define what sort of molecules are within reach, below we estimate how large a nucleotide motif can be derived from a randomized RNA sequence of specified size, under conditions that span those in current use. This discussion concerns only what might be present in the initial randomized pool, but models of the subsequent selective processes are also available.[23]

Calculations

What follows is only counting, though obscured by notation (details have been placed in an Appendix). The need for counting is fundamentally simple, and can be appreciated from a rough example. Suppose we pick single nucleotides blindly from a hat containing a large number of the standard four. In order to reduce the probability of missing one of A, G, C or U to some small value, we must pick several times 4 of them. In order to be similarly certain we don't miss a dinucleotide, we must pick about 4 times more, since there are 16 kinds rather than 4 kinds. Note the "about": actually, the statistics for these small numbers are a bit different from picking huge numbers of nucleotide sequences (see the Appendix), but close enough for now. We usually wish to reason conversely; that is, for any particular number of sequences from my hat (or in my experiment), motifs of a particular size are present with high probability. What is that size?

This crucial question can be refit for finding motifs of l nucleotides within a pool of RNAs randomized at n positions. How many RNA folds involving l nucleotides (nt) divided into m indivisible sequence modules (the modularity) are present in a randomized sequence n nucleotides long ($l \leq n$)? Each fold is a sample of l-mer sequences. If we multiply folds by the number of randomized molecules used, we know the effective number of sequences we have tested for some new function in a selection-amplification. As in the paragraph just above, we can then estimate the size of the l-mer we might recover. In the discussion below, l, the likely motif size, is our index for the capability of selected RNAs. The implicit assumption is—the larger the motif potentially selected, the more capable the selected molecule is likely to be.

Results

The Importance of Being Modular

Figure 1 shows the total number of motifs containing 20 fixed nucleotides (l = 20) in random regions of different lengths. The results vary widely. One can construct only a single occurrence (O = 1) for a motif that just fits: 1 module of 20 nucleotides in a randomized region 20 nucleotides long (n = 20). But in a second, equally realistic situation, about 10^{10} folds exist for motifs broken into 4 pieces (m = 4) and allowed to find as many positions as they can in random regions 150 long (n = 150).

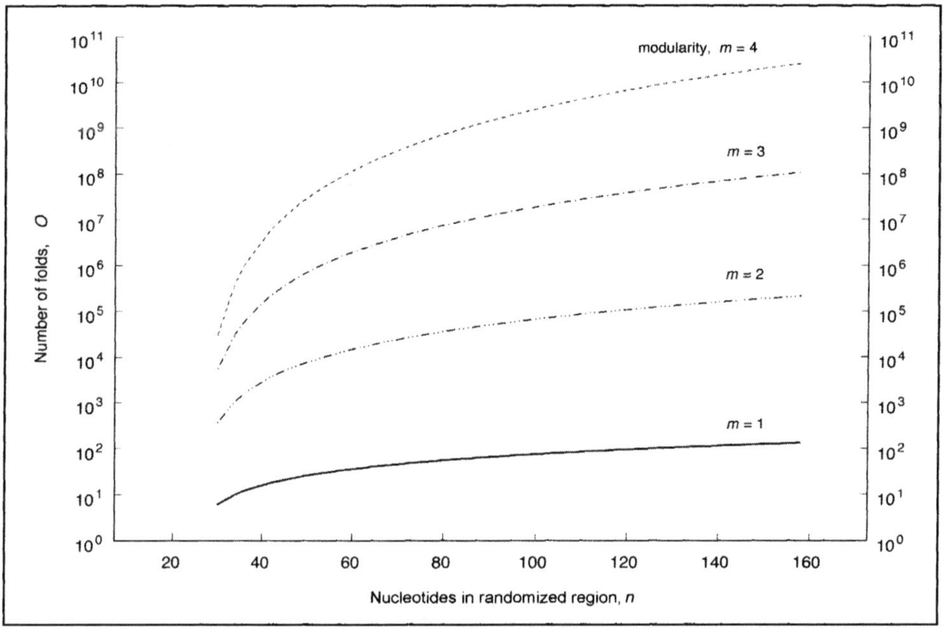

Figure 1. The number of folds (O, eqn. 4) for a 20-mer (l) in randomized regions of varied size (n). The effect of modularity (m), or the number of separable sequences comprising the motif, in increasing the variety of folds is shown. No sampling corrections were used here.

Figure 1 therefore strikingly emphasizes the importance of the modularity, m. The number of folds available for selection increases two or three orders of magnitude with each increase in m, the number of sequence pieces allowed. Therefore, in the absence of special considerations, selection-amplifications isolate molecules whose active sites are folded from as many separated sequence tracts as possible. This has a profound effect on the nature and analysis of selected molecules, to which we return below.

The Importance of Random Region Size, n

What is the most effective size for a randomized region? Figure 2 is directed at this fundamental experimentalist's question, encountered by everyone who has performed a selection-amplification.

In the following, the "presence" of a motif (the probability of occurrence of an l-mer) is discussed. It seems at the outset that one might equally reasonably count a motif present if the probability of its occurrence is 0.5 or with similar justification, 0.99. Throughout the text below, motifs are said to be present when the probability of an l-mer is 0.5, that is, when 50% of all l-mers are present. This is not arbitrary, but chosen to maximize the accuracy of the calculations (see Appendix for details).

The scope of a practical selection is limited by the mass of RNA present. Some number of RNA molecules is always convenient; more is hard. The maximum number of molecules used may be limited in various ways; by economics, by the capacity of PCR machines, by the solubility of macromolecular RNAs in the presence of divalent ions, or by some combination of considerations. In Figure 2 the effect of total initial pool absorbance is shown using different strategies for the size of the randomized region, n, and seeking structures with a fixed, realistic modularity, $m = 4$.

It clearly pays to increase n, making longer random regions on fewer initial molecules. However, this strategy becomes less effective as randomized regions get longer. One adds about

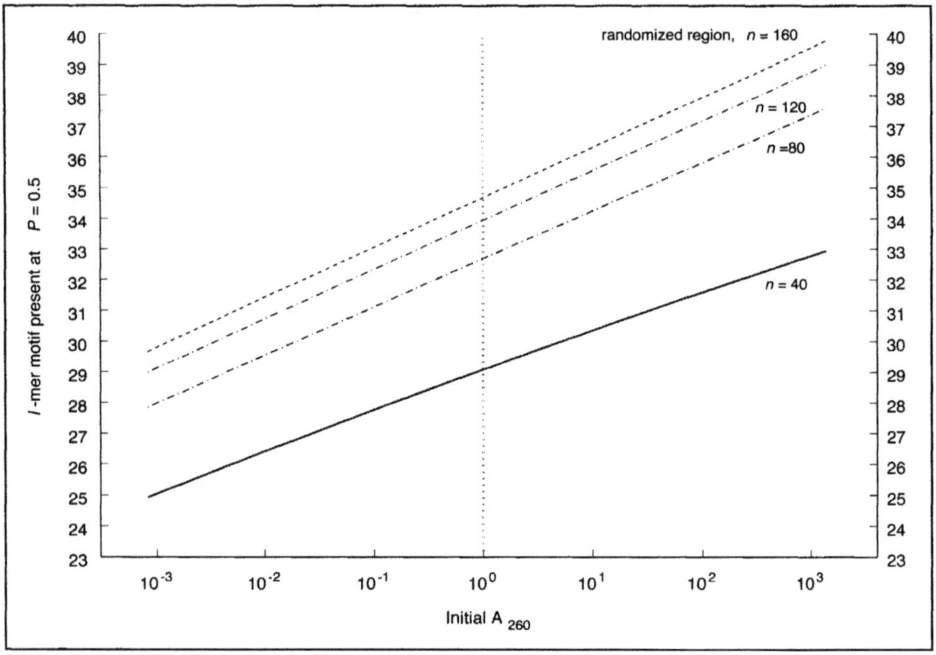

Figure 2. The size of a motif present (with probability = 0.5) for random regions of different lengths, starting with varied A_{260}. Modularity is fixed at $m = 4$.

3.6 fixed nucleotides to the likely motif by increasing the randomized region from 40 to 80, but going from 80 to 120 adds only 1.2 more. Adding another 40 randomized positions (to $n = 160$) potentially enlarges the accessible motif by only 0.7 nucleotides. There are three reasons for this behavior. One has fewer molecules as the mass of each one increases. In addition, Figure 1 shows that the rapidity of the increase in the number of folds coming from each molecule decreases as the randomized tract is lengthened. In addition (Appendix) sampling of sequences becomes less effective (more nonideal) as a randomized region is lengthened, further decreasing effectiveness.

Therefore, if an experiment requires an RNA of unconditionally maximized capability, you should select using randomized regions of maximum attainable size. On the other hand, if ease of analysis, or unbiased replication, or higher stability of the molecules used is of value, most of the benefit of increased n is accessible by using molecules with 80-120 randomized positions. This is particularly true since multiple molecules can sometimes collaborate to form a site,[25] potentially conferring some of the benefits of longer randomized regions via shorter molecules. A numerical point is that experiments of moderate size, conducted with moderate random regions, can contain substantial RNAs. From a reference experiment in Figure 2, 1 mL of $A_{260} = 1$, we may find an active region containing 34 specified (or 68 half-specified) nucleotides within a randomized region 120 long ($m = 4$).

The Importance of Experimental Scale

As we have argued above, the size of experiments is limited. Therefore it is useful to think about the yield from experiments conducted on differing scales.

Figures 3A and 3B depict a set of calculated outcomes. Figure 3A shows the size of the explicit motif present in a population of 1 mL of $n = 80$-mer randomized RNA at starting A_{260}, varied over six orders of magnitude. The vicinity of "typical" experiments (1 mL at an absorbance of 1) is marked with a vertical dotted reference line. Figure 3B rephrases this same relationship in terms of the initial number of RNA molecules, each of unique sequence. The vertical reference line marks another attainable vicinity for experiments (1 nmol of randomized molecules, close to 1 A_{260}).

An experiment at laboratory scale might be conducted in 1000 μL. What is the reward for starting on a million-fold expanded scale, where the initial RNA solution will have to be mustered in many bathtubs, and heated and cooled with some exotic technology? As the Figures show, the answer is largely independent of modularity (at $n = 80$) and uniformly equal to addition of 10 nt, or 1.66 specified nucleotides per ten-fold increment in the experiment. Thus a "typical" experiment, which contained motifs of 33 nucleotides ($m = 4$, $n = 80$) in 1 mL of $A_{260} = 1$ would potentially yield sites of up to 43 specific nucleotides, if the logistics of experimentation in 1000 L could be surmounted. At a more practical level where real experimental decisions are usually confined, there would usually seem to be limited rationale for a ten-fold increase in scale. This analysis is similar from side to side in Figure 3, spanning six orders of magnitude in starting material.

This calculated effect of scale may seem anti-intuitively small, but it is the unavoidable consequence of a simple notion in the first paragraph of CALCULATIONS above. That is to say, we need about 4-fold as much material to recover sequences one nucleotide longer. For 10-fold increases, $4^x = 10$ and $x = 1.66$ nucleotides per order of magnitude, the factor that recurs throughout the calculations in Figure 3. One might say that the complexities of the present calculation (see Appendix) are mainly to show that consideration of folding or sampling nonidealities, for example, do not significantly alter this outcome. And if a 1.66 nucleotide return for ten-fold in magnitude still seems small, then consider the implications for selection-amplification conducted on peptides.[18] The effects of scale arise again below.

The Importance of Motif Size, l

For some purposes we need to know the content of the population as a function of motif size, l. There are $C*4^l$ l-mer motifs in all (C is the number of ways to divide l nucleotides into m modules; see Appendix). Figure 4, showing the number of an average l-mer present versus l, is directed at this question. The horizontal reference line in the plot marks our standard for calculations. When l-mer motifs are present with probability = 0.5, there are 0.693 of the average l-mer amongst the total group of RNAs. For modularity $m = 4$, note that the reference line intersects the plot just below $l = 33$ specified nt, as also shown in Figure 3A.

However, the point of Figure 4 is in the unbroken slope above our previous reference. Shorter motifs are exponentially more present in the population. As we might hope intuitively, the slope of the lines for various modularities is similar, again approximately 1.6 nt/order of magnitude. Therefore a fixed RNA population of randomized sequences selection contains about an order of magnitude more l-mer for each 1.6 nt decrease in motif size, l.

Summing Up

As the first order of business in summing up, we reflect on our approximations. In particular, most errors tend to increase the apparent size of the accessible motif, l. A reckoning has been used (see the Appendix) in which RNA folds are treated as linear abstractions, rather than as real structures in which only certain interactions and certain covalent continuities will be allowed. The number of real structures (and the real l) will be smaller. In addition, in real experiments there is cryptic damage to synthetic DNA that prevents transcription;[4] thus we may often overestimate the number of unique sequences in a selection. Furthermore, the addition of randomized nucleotides to active structures inactivates some or most of them.[21] Thus some motifs counted as being present will be difficult to recover. This is particularly true of less stable (and therefore usually smaller) ones (O. Kovalchuke and M. Yarus, unpublished), which are more easily poisoned by alternative foldings with added sequences. In addition, motifs that exist close to the lines in Figures 2 and 3 exist as one or a few copies in a large population (Fig. 4). Since no real biochemical procedure can be carried out with 100% recovery, these motifs can be lost in stochastic accidents. However, as one backs down from the calculated lines, the

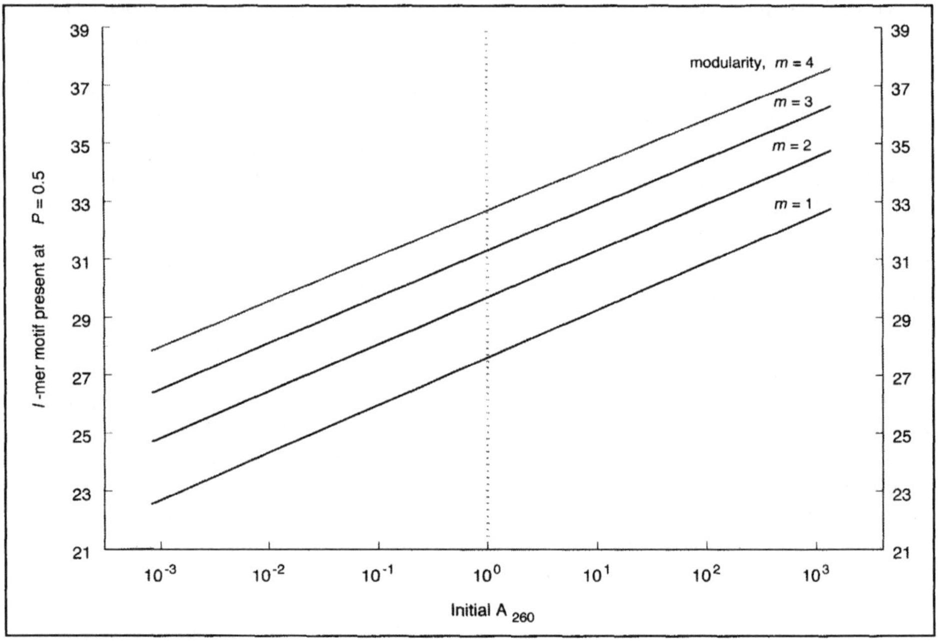

Figure 3A. The size of the motif present with probability = 0.5 in experiments conducted with randomized tracts of 80 nucleotides, and 1 mL of RNA at A_{260}. Modularity of the motif varies from m = 1 to 4.

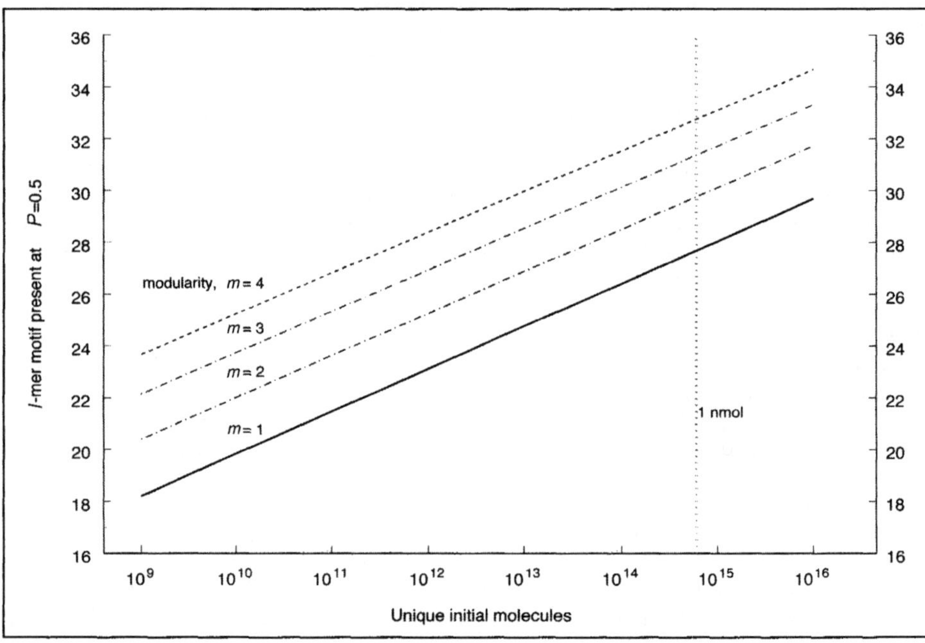

Figure 3B. The size of the motif present with probability = 0.5 in experiments conducted with randomized tracts of 80 nucleotides, and the indicated number of unique RNAs in the initial pool. Modularity of the motif varies from m = 1 to 4.

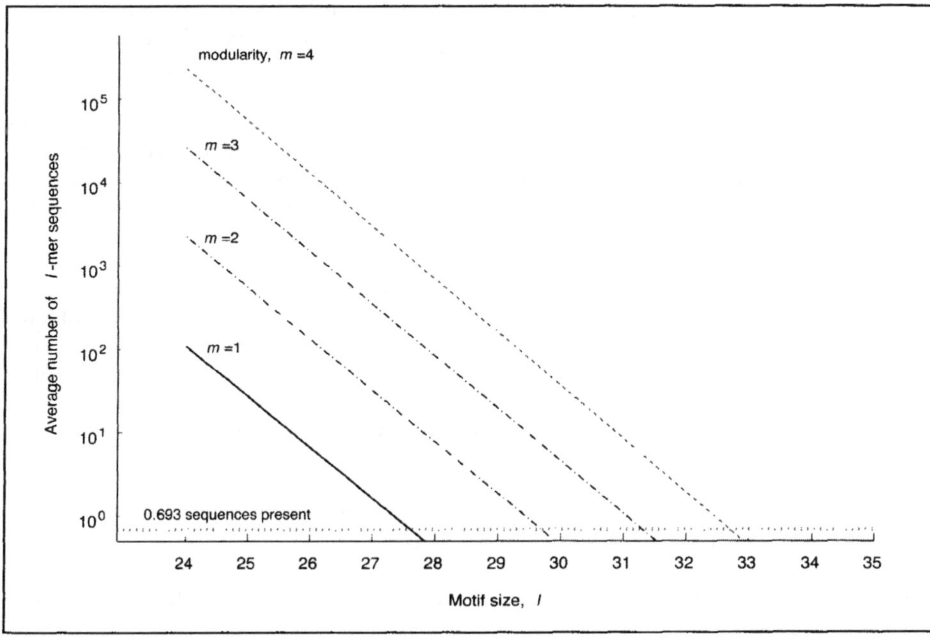

Figure 4. The mean number of specific *l*-mers present in a 1 A_{260} population of RNAs with 80-mer randomized tracts. The modularity of the motif varies from $m = 1$ to 4.

number of predicted copies of a motif increase rapidly, about 4-fold per omitted nucleotide (Fig. 4). Even if the calculated motif is not present, one slightly smaller is likely to be found. If the motif is made slightly smaller yet, its presence is virtually assured. To say the same thing in another way, in eqn 1 when we talk of the size of motifs, we implicitly take the logarithm of the numbers that have these errors. Thus *l*, the size of the motif, is more resistant to error than (for example) the number of folds. Combining all these considerations, calculated lines should be taken as the upper boundary of a region of feasibility. Within this region of feasibility the likelihood of finding a motif increases rapidly as one heads downward in the Figures, and the number of fixed nucleotides in the motif, *l*, decreases.

The present results will now be applied to modern selections, and in addition, to the nature of the first RNAs in an RNA world. The latter extension requires a virtually ubiquitous assumption, therefore worth explicit statement. We assume that the probability of an RNA world is synonymous with the probability of the emergence of active RNA structures from mostly nonfunctional RNA populations having highly varied sequences, and call this the Axiom of Origin.

Modularity

Modularity has large effects. When the fold is composed of a greater number of pieces, this greatly increases the number of folds of a given size, *l* (Fig. 1). Thus the predominant RNAs meeting a selection will quite likely form an active site by folding together separated pieces of its primary sequence. This means that it is likely that there will be spacers with no specific function in a newly selected molecule. It will be difficult to detect and eliminate these, because they will frequently be internal, between active sequences. No good molecular biological method exists for making a selected RNA smaller by random deletion, though deletion during chemical DNA synthesis should be possible.[7] In fact, the facile creation of randomized deletions would usher in selections for the smallest functional units. This would add a useful new dimension to selection-amplification for activities of all kinds (see below).

One might wonder if intercalated sequences between the *m* essential modules would be recognizable in some way. For example, might spacers be less structured? However, even random sequences make about 40-60% base pairs.[6] In addition, inter-module sequences in a selected molecule are not truly random, but have been instead been repetitively chosen to allow the active part of the molecule to fold into its functional configuration. That is, even sequences whose only role is as spacers between functional modules will likely fold to give a purposeful, specific structure, difficult to distinguish from an active site by inspection. In fact, sometimes this tendency can be detected. An initially selected self-aminoacylating 95-mer with a modularity of 4 was reduced to a 29-mer with undiminished activity[10] primarily by internal deletion of what seemed a uniformly structured parent.

Biological selection should resemble selection-amplification in these respects. That is, RNA structures will usually be created with intercalated dispensable regions. These virtually ubiquitous sequences are available raw material for further evolution; for example, for development of other intrinsic functions and interaction with other RNAs. Furthermore, conditions that alter the stability of ribonucleotide folds and therefore the practical modularity are probably crucial to early molecular evolution, though not usually discussed in this context.

Size

Overall confidence in these calculations is somewhat increased by the observation that most selected RNAs are in fact composed of a number of fixed nucleotides smaller than suggested by Figures 2 and 3. However, modularity's effects imply that total apparent size will likely be larger than the real size, making it difficult to know precisely how closely real selections approach calculated boundaries.

Above we played with the notion of bathtubs filled with concentrated RNA solutions. However, the evolutionary significance of these calculations lies in the other direction. At the left in Figure 3B, we read that even in experiments 6 orders of magnitude smaller than typical laboratory regimes, when we have only $\approx 10^9$ molecules to select among ($n = 80$, $m = 4$), active structures containing up to 23 specified nucleotides would be plausible. Since relatively capable ribozymes, for example, the hammerhead[24] and even a self-aminoacylating ribozyme,[10] are significantly below this range, proficient RNA structures can appear in populations containing only femtomoles of RNA. It seems quite likely that this ability to derive structures of substantial size, even from small molecular populations, was crucial to the initiation of an ancient biology based on oligoribonucleotides. For the beginnings of an RNA world, the Axiom of Origin suggests that we need to estimate the minimal size for a productive RNA population, and we return to this topic below.

The effects of modularity reinforce the above conclusion about population size. Modularity becomes more important as the size of the oligoribonucleotide population shrinks. This is readily visible in Figure 3. Because of the form of these results (1.66 motif nucleotides per size factor of 10), modularity adds a similar absolute increase in motif size in populations of every size. Therefore the proportionate impact of modularity grows as populations get smaller. For our index population from a modern selection-amplification (1 nmol RNA), modularity increase from m=1 to 4 adds about 19% to the size of the accessible motif. For a hypothetical ancient RNA population evolving toward biological function (1 fmol RNA), Figure 3B shows that the same change in modularity likely adds about 31% to the motif, and correspondingly to the capability of RNAs that might appear. If only 1 attomole RNA were available (1 amol = 6×10^5 molecules), this same modularity increase would add 42% to the number of nucleotides specified in the accessible motif.

Thus we urge two conclusions—firstly, the ancient RNAs that initiated an RNA world were probably yet more modular in structure than those selected in modern experiments. Secondly, even with the equivalent of only an attomole (6×10^5) of 80-mer RNAs on hand, substantial modular structures (m=4) containing about 18 completely specified nucleotides are near the apparent upper limit of complexity.

A Zeptomole World

How many ribonucleotides are required to specify useful catalytic sites? A convincing general answer to this question is presently beyond us, so we refer to observed structures. Simple activities can be seen in very small molecules. For example, RNA can abet its own hydrolytic instability by binding divalent metals to 7 specific nucleotides, GAAA/UUU.[17] However, a more relevant reference may be the 29-nucleotide self-aminoacylating RNA.[10] This RNA forms a Michaelis complex with a small-molecule substrate and accelerates a reaction not frequent in the natural ribozyme repertoire, involving carbonyl chemistry. Substitution and truncation experiments[11,28] suggest that, among 29 total nucleotides, activity requires ≥ 11 specific nucleotides (this is the minimal number required to create the active structure) but ≤ 19 nucleotides (this is maximal, conserving every nucleotide required in the active structure). Thus we need to calculate the size of an RNA pool that contains motifs of 11-19 nucleotides.

If Figure 3B is extrapolated at 1.66 nucleotides/order, structures of this complexity could be expected from zeptomoles of randomized 80-mer RNA molecules (1 zeptomole = 10^{-21} mol = 602 molecules, ≈ 33 attograms). Thus, surprisingly, catalytic RNAs might appear in unexpectedly tiny RNA populations, beginning at about one thousand times less than that in a modern bacterium. Accordingly, we conclude that sub-attomole RNA populations; that is, only zeptomoles of RNA, may have been sufficient for emergence of an RNA world. As a result, an RNA world is more probable, perhaps much more probable than usually considered. However, we cannot be certain this ribozyme example can be generalized—as pointed out above, selection experiments will not easily find minimal capable RNAs. Further, we need to consider the robustness of the numerical argument that led to this conclusion. There are arguments both for and against its accuracy.

An Argument For

The zeptomole world at first may seem intimately tied to the method and example used for this analysis (CALCULATION, above, and Appendix), and therefore subject to dramatic later revision. Instead, this argument is more independent of numerical details than first appears. It requires only the notion that the accessible motif diminishes by 1.66 nt/order of magnitude in the population size. As pointed out above, this result is deducible from elementary sampling considerations. The CALCULATION arguably only shows that the imposition of other kinds of size-dependence (in the form of folding and sampling) does not submerge the fundamental relationship.

We therefore take the selection-amplification of larger functional RNAs, like the class 1 ligase, as an alternative anchor for the numerical argument. This ligase ribozyme was isolated from a pool with 220 randomized nucleotides.[1] Consistent with the above discussion, it was later reduced to a 112-mer with a 93 nucleotides catalytic region,[3] and also an active 97-mer.[15] This may be a highly modular structure, and it is not certain how many nucleotides are essential in the present sense. However, this number is surely a substantial fraction of 93.[3] If we correct for the 10^{13}-fold difference in the magnitude of these experiments and the zeptomole range (by subtracting 13 x 1.66 ≈ 22 nt), we reproduce the original conclusion. Pools containing zeptomoles of RNA should yield active molecules with tens of essential nucleotides. Correction for the large randomized region leaves the conclusion intact. Accordingly, one needs only the slope in Figures 3 to make a zeptomole world plausible.

An Argument Against

The satisfaction of the Axiom of Origin by zeptomoles of 80-mers depends on realistic estimation of the number of sub-folds from a randomized region. A skeptical view is that a zeptomole world presses our calculation of the number of folds beyond its limits (see the Appendix). We acknowledge above that the present calculations overestimate real folds, and therefore overestimate the versatility of zeptomoles of RNA. Said another way, taking these calculations as upper boundaries for *l* will at some point cease to be useful as the boundary *l*-mer becomes smaller. Perhaps even the 1.66 nt/order of magnitude rule is inaccurate in the extreme limiting cases we need (thereby invalidating the second argument above), though we have confirmed many of the concepts required by computation using small RNA populations (Appendix). The best remedy for these uncertainties seems to be an experimental measurement of the effective number of *l*-mer folds arising from *n* ribonucleotides, and this work is underway.

Comparison with other Size Data

Nevertheless, RNAs in the size range calculated here are likely to be the principal agents of very early evolution. There is a remarkable convergence of independent quantitative evolutionary arguments on molecules of similar size. These present calculations show that structures spanning a few tens of specific nucleotides can arise from small amounts of RNA. This statement acceptably summarizes the results even if the initial number of molecules is stretched to the upper limits of conceivability. In addition, base pairing is error-prone, and this is necessarily reflected in recently selected RNA catalyzed, RNA-templated replication,[15] which has a substantial error rate. Ancient replicators would presumably have begun as unsophisticated nonproofreading catalysts, inescapably limited in this way. Likely error rates limit the plausible size of the RNA accurately reproduced by such replicators to a few tens of nucleotides.[16] Finally, if unordered polymerization of pre-activated nucleotides is carried out on mineral surfaces, RNAs of a few times 10 nucleotides in size can appear.[5] On these several grounds, a primordial RNA cell (a ribocyte) would almost surely begin with RNAs of this size, though it might evolve the capabilities required to maintain larger molecules. Therefore, questions about the origin and initial stages of an RNA world can be focused—is life, even in an initial crude form, possible for an assemblage of 15 to 35-mer active sites? If the answer is "yes", then the Axiom of Origin and the above calculations suggest that the RNA world did not necessarily require protracted gestation, but could have appeared quickly and inevitably.

Appendix

When sampling equally likely sequences for an initial pool for selection-amplification, the Poisson distribution describes the probability that a sequence will be missed. To show this, we first show that the Poisson is appropriate to choice between the relevant numbers of alternative sequences. We then show how we count folds and estimate the largest *l*-mer that is likely to occur. Finally, we correct Poisson statistics for sampling errors in populations of real sequences.

Appropriateness

The calculation of the likelihood of missing a sequence is different for few and many sequences. Thus, we first calculate this probability by a simple magnitude- and distribution-independent method, then show that this is equivalent to the Poisson for relevant cases.

Take k sequences at random from a total of q equally likely choices ($q = 4^l$, where l is sequence length): the probability that one choice yields any given sequence is $1/q$. The probability that one choice misses any sequence is $[1-1/q]$. The probability that we have still missed any sequence even after k tries is $[1-1/q]^k$. Call this quantity Prk:

$$\mathrm{Pr}k(q,k) = [1-1/q]^k$$

For Poisson sampling, for the same process we would predict:

$$\mathrm{Pr}k_{Poisson}(q,k) = e^{-k/q}$$

Prk can be expanded so its limiting behavior is visible:

$$\mathrm{Pr}k(q,k) = 1 + k(-1/n - 1/2n^2 - 1/3n^3..) + (k^2/2)(1/n^2 + 1/n^3 + 11/12n^4..) + (k^3/6)(-1/n^3 - 3/2n^4 - 7/4n^5...) + ..$$

As q, the number of things chosen from) gets larger,

$$Prk(q,k) \rightarrow 1 - k/q + k^2/2q^2 - k^3/6q^3 + ..$$

This is the standard expansion for $e^{-k/q}$, that is, Pr*k* and Prk_{Poisson} will converge as *q* gets larger. The equality is good at even at moderate *q*: Figure 5 shows that, even if we only choose between 64 sequences (trinucleotides; we are usually choosing among $\approx 10^{14}$ sequences), the Poisson is an excellent approximation. In fact, in cases of real interest the two lines in Figure 5 are indistinguishable over tens of orders of magnitude. This result corresponds to expectation for Poisson sampling: that is, it will apply when the events concerned have small probabilities that are constant over the domain of interest. In fact, for real sequences below we will need to correct for departures from equal probabilities.

The Accessible l-mer

For Poisson sampling, we follow Ciesiolka, et al[2] to show that motifs of *l* nucleotides are present within tracts of *n* randomized nucleotides as follows:

$$l - \frac{\ln\left[T(l,m,n)\right] - \ln\left[-\ln(1-P)\right] - \ln\left[C(l,m)\right]}{\ln(4)} = 0 \qquad (1)$$

where *P* is the probability that any motif of *l* nucleotides will be present, *T(l, m, n)* is the total number of folds containing *l* nucleotides in *m* modules within *n*-mer randomized regions of all molecules in the experiment. *C(l, m)* is the number of configurations of the motif (defined below). While *l* appears in eqn. 1 as an implicit transcendental function, such equations are easily and rapidly solved by mathematical software. Mathcad v7 (Mathsoft, Inc) was used for all results in the text above.

We need

$$T = u * O \qquad (2)$$

where *u* is the number of unique molecules and *O* is the number of ways of getting an *l*-mer fold within the randomized *n*-mer in every molecule. Because every molecule in typical selections has a unique sequence, *u* is also the total number of RNA molecules used at the outset. So, in "bench units":

$$T = \frac{A_{260}(6.023x10^{20})v}{(n+50)8500} * O \qquad (3)$$

where *v* ml of a solution of (*n*+50)-mer RNA at absorbance *A260* are used for selection. The RNA (extinction per mol phosphate = 8500) is assumed to have 50 constant nucleotides in addition to its *n* randomized nt. Variations from 50 have little effect on the outcome. The crux of the calculation is the effective number of folds,[21] *O*:

$$O = R * F * C / S \qquad (4)$$

where *R* is the redundancy (the number of variations of an *l*-mer that will satisfy the selection), *F* is the number of folds (ways of disposing *l* nucleotides divided into *m* modules within *n* total nt) and *C* is configurations (possible ways of dividing the *l* motif nucleotides into *m* modules). *S* is a sampling correction that corrects departures from the Poisson expectation to give an 'effective number' of folds (see last section).

R, the redundancy, corrects for a required nucleotide at a given position that may occur as any of the natural 4. A motif containing such a position should have *R* increased 4-fold (satisfactory sequences are 4 times as frequent), compared to a value of 1 for a unique nucleotide at that position. If only a purine will serve at that position, *R* is increased by 2-fold, and so on.

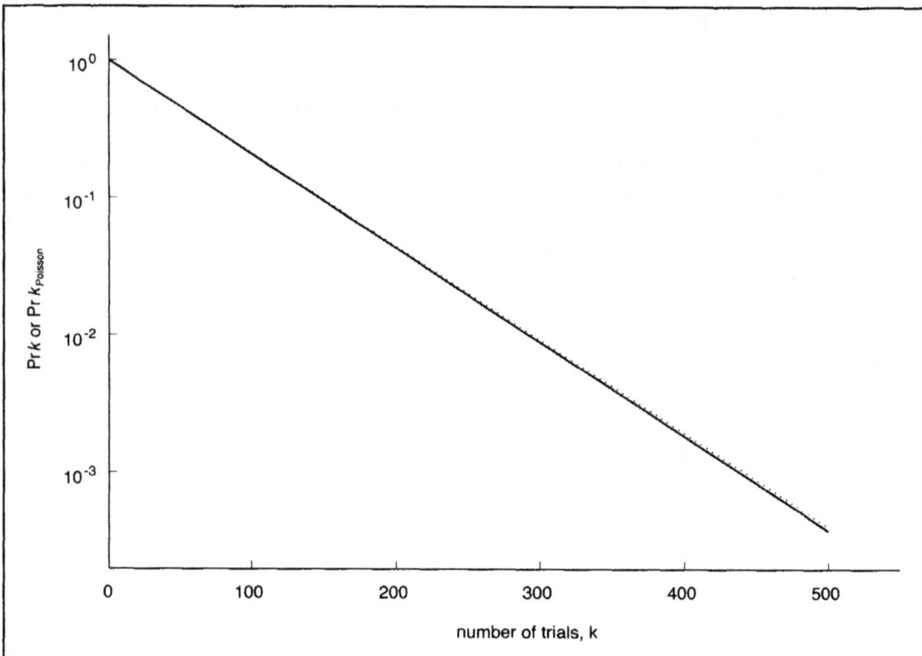

Figure 5. The probability of absence of a sequence after *k* picks among 64 distinct sequences. Pr*k* is the direct calculation (continuous line) and Pr$k_{Poisson}$ is the Poisson approximation (dotted).

With many positions free to vary somewhat, *R* is a multi-term product that can become a large number—Sabeti, et al[21] estimate *R* for the hammerhead ribozyme as 5 x 10^{12}. In addition to nucleotides that are free to vary, there are smaller factors because functional sequences can sometimes be permuted on the linear sequence. Assume that 3 indivisible sequence segments a, b and c come together to make the motif (a structure with modularity $m = 3$). *R* would be multiplied by a permutation factor of 3 if abc, cab and bca are all active molecules.

Having troubled to define *R*, we now set it equal one for all calculations. This convenience frees discussion from details of a motif's tolerance for substitution. We do not know such details in general. Therefore, to make progress we choose to speak in terms of structures composed of nucleotides all of whose identities are unique. As a reminder of this decision, in the text we write of RNAs "composed of fixed nucleotides", or similar language. The smaller factor due to permutation is quite real, because some motifs can be selected in permuted form.[14] However, if a selected RNA makes use of the unique reactivity or unique structure of its 3' or 5' ends or both,[11] permutation becomes implausible. Overall, permutation has been infrequently observed in real selections, and in that respect $R = 1$ with a good conscience.

F is the number of folds. A liberal attitude toward *F* is apt, because the number of folds with a ligand present is the quantity of interest. Therefore, unseen bonds (with a ligand or substrate) connect the *m* sequences in the active site, stabilizing structures not otherwise stable.[8] Consider the molecule as a sequence of nucleotides in the motif and nucleotides in the spacer(s). We multiply the ways of dividing the spacers by the (independent) ways of dividing the modules. First count the number of ways to dispose *m* modules containing a total of *l* nucleotides on a sequence of *n* nucleotides. In order to avoid modules with no nucleotides between them (which would count multiple times those cases where a larger module is the sum of the separate module sizes), we maintain ≥ 1 nucleotide in each spacer. There are three cases.

First: modules may be internal, with spacer sequences at both ends: •—•••———••————•• where the diagram shows 3 modules placed among 8 spacer nucleotides. This yields a molecule with $m+1$ spacer regions; there are l nucleotides in modules and $(n-l)$ nucleotides in spacers. This contributes a term to F:

$$(n - l - 1)!/m!(n - l - \mathrm{m} - 1)! \tag{5}$$

which corresponds to choosing m places for modules from the n-l-1 spaces that will still leave at least one nucleotide in each spacer.

Second: one module may be at an end: — •• —— • ——— ••••• thereby dividing the molecule with m spacers. This term is multiplied by 2 because for every fold, there will be versions with a module at the 5' end: — •• —— • ——— ••••• and the 3' end: ••••• ——— • —— •• —.

$$2(n - l - 1)!/(m - 1)!(n - l - m)! \tag{6}$$

Third: there are folds with modules at both ends: — ••• —— ••••• ——— containing m-1 spacers and being composed by placing m-2 modules in n-l-1 places:

$$(n - l - 1)!/(m - 2)!(n - l - m + 1)! \tag{7}$$

This last term is only relevant when $m \geq 2$. The full equation for F contains the sum of the terms in eqn 5, 6 and 7.

However, this does not yet enumerate all possibilities. The term C, the configurations (eqn 4), accounts for the fact that there are many folds with the same n, l and m, because there are multiple ways to divide l nucleotides into m modules even with fixed spacers. For example, a motif containing $l = 20$ nucleotides in $m = 3$ modules can occur as rather different RNAs constructed by folding together sequence pieces of size 12nt : 3nt : 5nt, or 4nt : 13nt : 3nt, or in yet other ways.

$$C = (l - 1)!/(m - 1)!(l - m)! \tag{8}$$

C counts ways of choosing $(m$-1) places to interrupt the (l-1) linkages between l nt, thereby creating m modules. C must therefore multiply F to account for the different distributions into modules F^*C = (eqn 5 + eqn 6 + eqn 7) (eqn 8). Note that in calculations of the mean number of l-mers or their probability P, C cancels (compare eqn 1) because it is in both O and the total number of l-mers, $C(4^l)$. For actual calculations, the Gamma function (the continuous equivalent of the factorial) was used in order to calculate outcomes with nonintegral n and l.

Relation to Previous Work

Ciesiolka et al[2] did not include modularity greater than 1, and therefore needed neither combinatorics nor sampling correction. Our results are similar where they overlap. Sabeti et al[21] use a simpler form for F and do not include calculation of C or sampling corrections. Their number of folds (that is, folds without configurations) is always larger than the comparable result here. It includes multiply counted adjacent modules with no spacer nucleotides intervening. Therefore our results diverge from theirs most when there are many modules, and when the total nucleotides in modules, l, approaches the size, n, of the randomized region. These conditions maximize module interfaces. Conversely, our results tend to the same limits as does Sabeti, et al[21] at small modularity and large randomized regions, because these conditions minimize folds with module-module interfaces.

Sampling Factors and Folds, the Determination of S

We now consider departures from Poisson sampling in real sequence populations. We have studied these effects using purpose-written software implemented in Perl v5.6, running under Irix on an SGI Octane with a 300 MHz MIPS R12000 processor and 1 GB memory. We explicitly generated random nucleotide sequences using the Math:Random module, made all the combinatorial folds implied by the equations just above, and classified the l-mer motifs that actually occurred. The real probability of motif occurrence was compared to the Poisson expectation. The major finding is that it appears possible to make useful calculations with moderate corrections.

Surprisingly, it is easy to get some motifs, and much more difficult to get others. Not all motifs of size l and modularity m are sampled at the same rate. For example, consider 10-mer $m=2$ folds. Only pentamers must be found to complete a 5:5 nt configuration, while nonamers must be found to complete a 9:1 configuration. This means probability of the absence of decamers as a whole will not decline exponentially with trials (as for the Poisson). Instead it declines as a sum of exponentials, one for each configuration. With greater number of sequences sampled the curve becomes asymptotic to the (possibly much lower) slope for the rarest fold(s). We have used a definition of representation that requires the probability of a fold's absence to be 0.5 ($P = 0.5$). This limits calculations to a range in which we can accurately use an exponential approximation (Fig. 6A).

A second difficulty arises because we repetitively sample existing nucleotide sequences. Combinatorial counting of folds, as above, reuses the same subsection of the sequence many times. Pieces are recombined with other parts of the same sequence to make varied folds, each potentially counted as a separate attempt to find a motif of l nucleotides. Certain nucleotide sequences therefore recur. Accordingly, the probability of obtaining all sequences is not equal. Instead, because particular sequences recur, more than the expected Poisson number of sequences is needed to raise the probability that any l-mer motif can be observed. Said another way, the accessible l-mer is somewhat smaller than for ideal Poisson sampling.

With calculation limited to $P = 0.5$, then over the accessible range of n, l and m, the probability of absence of an l-mer, as the number of sequences sampled increases, can be described with an exponential (Fig. 6A). Therefore, we can define a factor S (the ratio of the Poisson slope to that actually observed) by which the number of sequences must be increased in order to reach the Poisson probability that an l-mer will be absent. The correction S, over the range of our present calculations, implies 1 to 6.4-fold the sequences for Poisson sampling.

The modularity m and the size of the randomized region n are the most influential variables. Departure from the Poisson worsens as modularity increases because the combinatorial use of high modularities re-samples sequences more intensively. Longer randomized regions allow more folds (Fig. 1) so nonideality also increases with n. However, Figure 6B shows that these effects appear regular, and therefore readily predicted. The required correction factor grows linearly with increasing size for the randomized region. The data of Figure 6B were extrapolated, e.g., to $n = 80$, $m = 4$, when needed for calculations.

The departure from Poisson sampling is also worst when l is small, because this also tends to maximize reuse of sequences (Fig. 6C). However, as the Figure shows, when l is an appreciable fraction of n this correction is both small and slowly varying. Therefore we have used the same correction for all l (that implicit in the calculation for n and m, as above). The resulting variation in l over the range of larger motifs produces a variation in the size of the motif estimated as < 3 %, insignificant by comparison with other approximations.

Explicit sampling corrections for larger randomized regions and modularities were beyond the limits of available computer storage and computational speed (compare Fig. 1). As an example related to storage, there are about 10^{10} 20-mer folds of modularity 4 in one randomized 150-mer. To explicitly store the motifs from one randomized molecule and their addresses would therefore require about 340 gigabytes. For these reasons, our corrections were calculated by extrapolation from computable cases, as above. We hope the regularities observed in computation (Fig. 6) will encourage an analytical solution to this interesting sampling problem.

For calculations, O in eqn. 4 is divided by sampling correction factors, S, to give an effective number of folds. With the addition of sampling corrections, all terms in eqn 1 above have an explicit form. Initial RNA populations can now be varied in size and design, and the effects on the presence of given motifs in a starting RNA pool can be estimated (main text above).

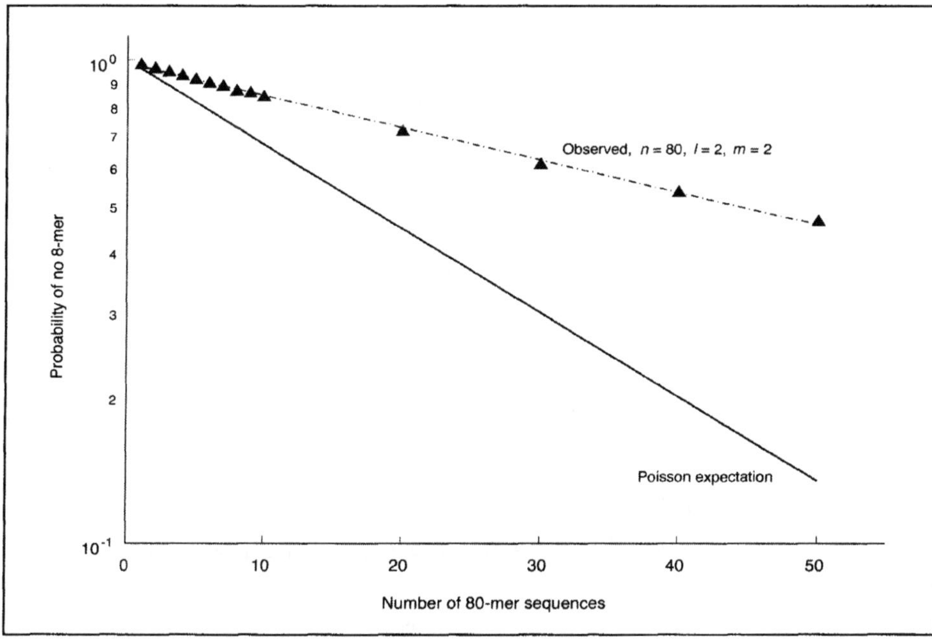

Figure 6A. Approximately exponential decline of the probability of sequence absence for Poisson expectation and for actual sequence sampling in the case $n = 80$, $l = 8$, $m = 2$.

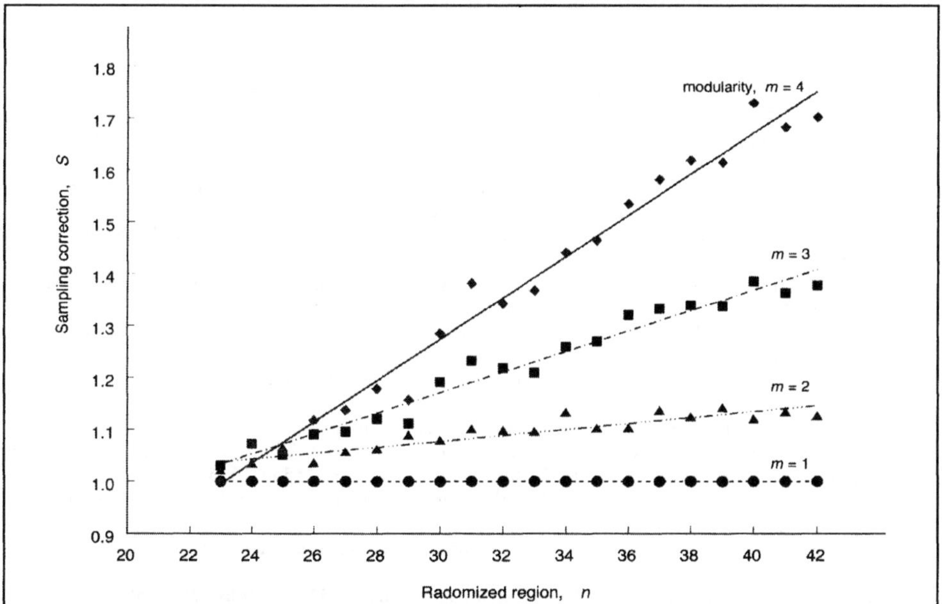

Figure 6B. The correction for departure from Poisson sampling, S, versus the size of the random region for four modularities. The motif length is constant; $l = 20$. For m = 1, there is no recombining sequence parts to get motifs, and no correction is needed. For modularity 2, least squares fitting yields $S = 0.9056 + 0.0057\,n$. For modularity 3, $S = 0.5789 + 0.0198\,n$. For modularity 4, $S = 0.0806 + 0.0398\,n$.

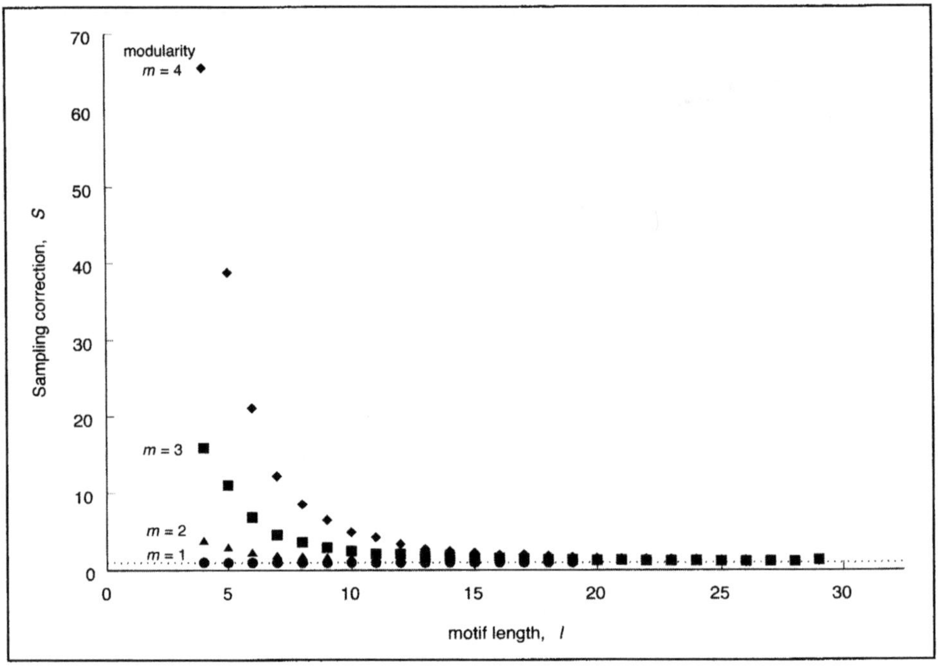

Figure 6C. The correction for departure from Poisson sampling, S, versus the size of the motif *l*. Data are shown for randomized regions $n = 40$, and for modularities 1 to 4. The horizontal dotted line across the bottom of the panel is set at 1.00, the Poisson expectation.

References

1. Bartel DP, Szostak JW. Isolation of new ribozymes from a large pool of random sequences. Science 1993; 261:1411-1418.
2. Ciesiolka J, Illangasekare M, Majerfeld I et al. Affinity selection-amplification from randomized ribooligonucleotide pools. Methods Enzymol 1996; 267:315-335.
3. Ekland EH, Szostak JW, Bartel DP. Structurally complex and highly active RNA ligase derivatives from randomized RNA sequences. Science 1995; 269:374-370.
4. Ellington A, Szostak JW. In vitro selection of RNAs that bind specific ligands. Nature 1990; 346:818-822.
5. Ferris JP, Hill AR, Liu R et al. Synthesis of long prebiotic oligomers on mineral surfaces. Nature 1996; 381:59-61.
6. Fresco JR, Alberts BM, Doty P. Some molecular details of the secondary structure of ribonucleic acids. Nature 1960; 188:98-101.
7. Hecker KH, Rill RL. Error analysis of chemically synthesized polynucleotides. Biotechniques 1998; 24:256-260.
8. Hermann T, Patel DJ. Adaptive recognition by nucleic acid aptamers. Science 2000; 287:820-825.
9. Huang F, Bugg CW, Yarus, M. RNA-catalyzed CoA, NAD and FAD synthesis from phosphopantetheine, NMN and FMN. Biochemistry 2000; 39:15548-15555.
10. Illangasekare M, Yarus M. A tiny RNA that catalyzes both aminoacyl-RNA and peptidyl-RNA synthesis. RNA 1999; 5:1482-1489.
11. Illangasekare M, Kovalchuke O, Yarus M. Essential structures of a self-aminoacylating RNA. J Mol Biol 1997; 274:519-529.
12. Illangasekare M, Sanchez G, Nickles T et al. Aminoacyl-RNA synthesis catalyzed by an RNA. Science 1995; 267:643-647.
13. Jadhav VR, Yarus, M. Acyl-CoAs from coenzyme-ribozymes. Biochemistry 2002; 41:723-729.
14. Jenison RD, Gill SC, Pardi A et al. High-resolution molecular discrimination by RNA. Science 1994; 263:1425-1429.

15. Johnston WK, Unrau PJ, Lawrence MS et al. RNA-catalyzed RNA polymerization: accurate and general RNA-templated primer extension. Science 2001; 292:1319-1325.
16. Joyce GF, Orgel LE. Prospects for understanding the origin of the RNA world. In: Gesteland RF, Atkins JF, eds. The RNA World. Cold Spring Harbor: Cold Spring Harbor Laboratory Press, 1993:1-25.
17. Kazakov S, Altman S. A trinucleotide can promote metal ion-dependent specific cleavage of RNA. Proc Nat Acad Sci USA 1992; 89:7939-7943.
18. Keefe AD, Szostak JW. Functional proteins from a random-sequence library. Nature 2001; 410:715-718.
19. Kumar RK, Yarus M. RNA-catalyzed amino acid activation. Biochemistry 2001; 40:6998-7004.
20. Roberson D, Joyce J. Selection in vitro of an RNA enzyme that specifically cleaves single-stranded DNA. Nature 1990; 344:467-468.
21. Sabeti PC, Unrau PJ, Bartel DP. Accessing rare activities from random RNA sequences: the importance of the length of molecules in the starting pool. Chem Biol 1997; 4:767-774.
22. Tuerk C, Gold L. Systematic evolution of ligands by exponential enrichment. Science 1990; 249:505-510.
23. Vant-Hull B, Payano-Baez A, Davis, RH et al. The mathematics of SELEX against complex targets. J Mol Biol 1998; 278:579-597.
24. Verma S, Vaish NK, Eckstein F. Structure-function studies of the hammerhead ribozyme. Curr Opin Chem Biol 1997; 1:532-536.
25. Vlassov A, Khvorova A, Yarus M. Binding and disruption of phospholipid bilayers by supramolecular RNA complexes. Proc Nat Acad Sci USA 2001; 98:7706-7711.
26. White III HB. Coenzymes as fossils of an earlier metabolic state. J Mol Evol 1976; 7:101-104.
27. Yarus M. RNA-ligand chemistry: A testable source for the genetic code. RNA 2000; 6:475-484.
28. Yarus M, Illangasekare M. Aminoacyl-tRNA synthetases and self-acylating ribozymes. In: Gesteland RF, Cech TR, Atkins JF, eds. The RNA world. 2nd Edition. Cold Spring Harbor: Cold Spring Harbor Laboratory Press, 1999:183-196.
29. Yarus M. On translation by RNAs alone. CSHSQB V66, The Ribosome, 2001:207-215.

Chapter 6

The Evolutionary History of the Translation Machinery

George E. Fox and Ashwinikumar K. Naik

Translation and the Origin of Life

Current theories on the origin of life envision an RNA World as the culmination of chemical evolution. The extent of this RNA World, and the biochemical complexity of the progenotes[1] that populated it, is subject to much debate. It, nevertheless, is likely a point of agreement among workers in the field that the discovery of machinery for the chiral synthesis of defined sequence peptides would have paved the way for transition to the modern protein world. With the discovery of an RNA replicase, which might initially have been a catalytic RNA or an early peptide product, the stage would be set for the development of populations of progenotes that had both of these features in one enclosure. Such advanced progenotes would be the first entities capable of having the genetic couple between replication, transcription and translation that is the hallmark of life, as we know it. The modern day tmRNA[2] at one stage is recognized as a tRNA by the ribosome while it subsequently serves as a mRNA during translation. This unusual RNA might be representative of the types of entities present in the late RNA World.[3] The addition of DNA as a better storage medium for genetic information would finalize the transition from the progenotic world to the living systems that exist in the modern world.

The origins view described above suggests that the cellular protein synthesis machinery began to congeal to its modern form before the emergence of the last common ancestor of modern life. Is this consistent with what we know? Defining ancestors is in principal straightforward; genes that are shared by essentially all descendants are likely to have been genes of the ancestor, if the possibility of polyphyletic gene losses[4] or gains can be reasonably excluded. Several attempts to outline the properties of the last common ancestor in this way were undertaken[5,6] and the list of shared properties is rather extensive with many relating to the translation machinery. The approach has, however, been challenged by Doolittle[7] because of the possibility that extensive lateral gene transfer could confuse the results but defended by Glansdorff[8] who has argued that the significance of lateral gene transfer has been overemphasized.

A somewhat different approach has been to examine the information processing systems in toto. Thus, when complete genome data from all three Domains first became available the information processes in the three major lineages were compared.[9] It was concluded that the (eu)bacterial machinery of DNA replication is unique to that lineage. Subsequently, this issue was reassessed by examining each essential protein and the conclusion reaffirmed.[10] In the case of transcription, core components were found in all three lineages with the archaeal and eukaryal versions appearing more similar.[9] Despite this affinity, it was observed that transcriptional regulation may be more similar between the Archaea and the Bacteria as both have an

The Genetic Code and the Origin of Life, edited by Lluís Ribas de Pouplana.

operon-based organization. Recently, a high-resolution structure was obtained for an eukaryotic RNA polymerase that transcribes mRNA. A comparison with the bacterial enzyme reveals a shared core structure and therefore a conserved catalytic mechanism with bacterial RNA polymerase.[11] In the case of the ribosomal machinery, the core components are clearly very homologous in all three Domains and hence it was argued that translation was likely to have been highly developed by the time the last common ancestor as defined by 16S rRNA data appeared.[9] In summary, consistent with the early view of origins described in the first paragraph above, current thinking is that the modern machinery of both translation and transcription were largely in place by the time the common ancestor defined by 16S rRNA sequence comparisons appeared. Thus, the origins and subsequent history of the translation and transcription machinery may provide a window into the era in which prebiotic entities first made the transition to true living organisms.

Origins of Translation: What Can We Hope to Learn in the Near Future?

There are two critical mysteries associated with translation; the chemical mechanism for the actual synthesis of the peptide bond and the subsequent movement of the mRNA relative to the ribosome (translocation). One can, and probably should, envision that the initial synthesis of peptides was by one of several prebiotic mechanisms, e.g., perhaps a condensation on mineral surfaces.[12] It is, however, the development of the peptidyl transferase center that must mark the beginning of the story of the translation machinery. With the first high-resolution structure of the ribosomal subunits[13,14] came the specific suggestion of an RNA catalyzed acid-base mechanism involving specific residues in 23S rRNA.[15,16] However, this mechanism has been challenged biochemically[17,18] and mutation analysis of the key residues did not prove consistent either.[19,20,21a] In fact, it may well be that the peptidyl transferase center only provides proper positioning and is not a catalyst at all.[3,19,21] Although we are not there yet, it is likely that this core chemistry of the ribosome will be fully understood in the near future. Further down the road is an understanding of the molecular mechanism for the relative movement between the ribosome and the mRNA. It is clear that this must involve coordinated conformational changes, at least some of which are inherent to the most primitive portions of the machinery.[22] In the absence of explicit knowledge of what these changes are, it is impossible to fully understand the translocation step. Some progress has been made, (see next section for a partial discussion) but a final solution will likely require high resolution structural information from ribosomal particles in the different structural forms associated with the various stages of translation.

It has long been appreciated that the original translation machinery was much simpler than that of today.[22-25] As our understanding of the modern ribosome has improved so have models of its early origins.[3] There has also been an increasing interest in developing experimental models of early stages in the history of translation. For example, in vitro evolution has been used to generate aminoacylated tRNAs[26-28] and ribosome-free peptide bond synthesis has been claimed.[29] An earlier report of ribosome free peptide bond synthesis catalyzed by the simple dipeptide Ala-His[30] has however been disputed.[31] The premise that underlies this chapter is that there is now, or soon will be, sufficient information available to deduce the actual specific history of the ribosome and its many components in some detail. There are three extant sources of information that will provide insight to this history. The first is biochemistry; e.g., conserved reactions, pathways, regulation, etc. The second and most commonly relied on is the sequence information that is now widely available from the many genome projects. The third, and to date most under utilized, is macromolecular structure. When preserved by functional requirements, structural features are especially resilient and have the potential to survive even longer than sequence similarities. We will discuss herein how and to what extent the history of the translation machinery might be unraveled.

Timing Information

One particular example of how atomic resolution structural information can and has given important insight to the sequence of evolutionary events in the history of translation will be reviewed in some detail. The elongation factors, EF-Tu and EF-G have long been known to compete for an "overlapping" site on the ribosome. EF-Tu forms a ternary complex with a charged tRNA and GTP, which enters the ribosome at the "A" site. An EF-Tu-GDP complex is released from the ribosome following GTP cleavage. The charged tRNA stays with the ribosome and is moved to the "P" site following peptide bond formation by a reaction known as translocation. This translocation step occurs, though at a greatly reduced rate, in the absence of EF-G.[32,33] The complex of EF-G and GTP mediates the translocation step making it far more efficient. Early topology data showed that the tRNA occupies different physical sites at different times. The structure of EF-Tu[34] and EF-G[35] were studied separately at atomic resolution by X-ray crystallography.

The key insight came with the determination of the structure of EF-Tu when bound to aminoacylated tRNA[36] in the ternary complex. It turns out that the overall three-dimensional shape of EF-G closely mimics the shape of the EF-Tu ternary complex.[36-38] Although the details have not been worked out, it is now clear that the translocation step involves the entry of EF-G (an EF-Tu/tRNA structural analog) into the A site of the ribosome[39] with the result that the tRNA may be mechanically displaced[40] to the neighboring P site. Alternatively, the bound EF-G may simply be preventing the peptidyl-tRNA from moving back to the A site and thus control the correct directionality of events.[15] Following cleavage of GTP the EF-G/GDP complex is released from the ribosome leaving an empty A site where the next tRNA can enter. Cyroelectron microscopy has revealed that the process of translocation is accompanied by significant rearrangements within and between the individual ribosomal subunits.[41-44] Clearly the mechanics of translocation by modern ribosomes are based on far more than RNA chemistry.

From the perspective of unraveling ribosome history, the key point here is the insight to timing that is implicit in the idea of mimicry. It is logical to infer that a mimic is an imitation of an original. Therefore, one of these complexes must be older than the other. In this case it would appear logical that the EF-Tu/tRNA/GTP complex is older than the EF-G/GTP complex on the grounds that one must have tRNAs before there is a need to develop translocation machinery. In this view, the data strongly suggests that EF-G mediated translocation evolved later in the history of translation than the EF-Tu facilitated tRNA entry. With this perception in mind one might then draw further inferences, for example about the relative age of the partially overlapping regions of the ribosomal RNAs that they interact with.

Molecular mimicry nicely illustrates that information about the relative time for the origins of various ribosomal components and properties may be obtainable. Once one starts to think this way it is clear that there are in fact a surprisingly large number of ways in which one might infer timing. Consider the ribosomal proteins. Some have clear homologs in all three Domains of life while others are found only in the Bacteria and are thefore probably more recently acquired. One can also get further insight to timing from the structures of the ribosomal proteins. For example, ribosomal protein L6 has clearly been duplicated.[45] One of the two domains presumably predates the other. Since they bind to two different regions of the 23S rRNA, if one can decide which protein domain is older one might then be able to infer something about the relative age of the interacting regions of the RNA too.

Another obvious example comes from a comparison of the secondary structure of the large ribosomal RNAs. Although the overall structure is very conserved, the eukaryotic rRNAs have several unique expansion segments that account for the larger size of the RNAs. These expansion segments have probably been added later in history. As an extension of this observation one must consider that not all portions of the rRNAs are necessarily of equal age. This is actually as we expect, as it is unlikely that the first RNAs were of the size of the modern 16S and

23S rRNAs. Presumably they grew larger over time. Thus, the history of various secondary structural features has recently been traced[46] and exploited to develop a novel method of deducing phylogenetic relationships.[47] Perhaps even more to the point, conserved features in the RNA primary and secondary structure are being mapped against the three-dimensional structures thereby revealing spatial relationships that have changed or been conserved over time.[48]

In order to reconstruct a reasonable hypothesis for the history of the ribosome it will be necessary to identify many clues that relate the time of origin of one component or component domain to another in some way. The interpretation of one individual timing event will typically be tentative. Thus, a skeptic might argue that "coevolution" kept the shape of the EF-Tu/tRNA/GTP complex and the EF-G/GTP complex similar as each became gradually more complex. It will necessarily require the examination of the larger picture of what we know in order to evaluate the reasonableness of the individual interpretations. For example, one might infer that a ribosomal protein, which interacts primarily with a unique expansion region in the eukaryotic rRNA would be relatively new. If so, it should also turn out that the protein in question is unique to eukaryotic organisms.

Insights to Ribosomal History from tRNA Structure

Transfer RNA is at the core of the modern translation machinery where it facilitates both the specificity of the process via the codon-anticodon interaction and the actual chemistry of peptide bond formation. The structure of tRNA has been known for some time[49,50] and has been recently redetermined at much higher resolution.[51] As has been explicitly pointed out,[52-54] the tRNA molecule consists in essence of two domains; the charging/synthesis domain comprised of the CCA stem and the TΨC stem/loop and the codon/anticodon recognition domain comprised of the D stem/loop and the anticodon stem/loop. It has been persuasively argued that the two domains historically evolved separately; the charging/synthesis domain likely being the older of the two. In this view, the ability to make peptides presumably preceded the role of the ribosome in making peptide bond synthesis a template directed process.[53,55] How might this occur? It was observed that a direct duplication of an RNA hairpin can result in a cloverleaf like secondary structure.[56] Indeed, such a duplication event can readily result in the formation of the key tertiary interactions that characterize the modern tRNA as well.[57]

It was originally shown for alanine[58,59] and is now well established for many tRNAs that a surprisingly small RNA (a "minihelix") which incorporates the portion of the tRNA comprising the acceptor stem and sometimes the TΨC stem/loop can be specifically charged with the cognate amino acid.[60] The minihelix domain alone is in fact recognized by a number of other proteins and enzymes as well, e.g., the ribozyme- RNAse P, T54 methyltransferase, the tRNA nucleotidyl transferase and EF-Tu suggesting this domain may indeed be quite ancient. In contrast, the anticodon is not even required for recognition by many class II tRNA synthetases. Therefore, it seems plausible that the anticodon domain is less ancient. Thus, the structure of tRNA gives us a working hypothesis for the order of early events in translation. The translation machinery would begin with a primitive "half tRNA" charging/synthesis system, followed by a "whole tRNA" charging/template directed synthesis system and finally a whole ribosome mediated translation system. The fundamental conformational changes that facilitate the movement of mRNA relative to the tRNA would likely be present from the earliest stages.[22]

For their part the synthetases being complex proteins could not have existed until late in the process. It has been widely speculated that the original synthetase was a ribozyme and experimental evidence suggests this is feasible.[61-65] The modern synthetases might have evolved in stages.[66] They may have initially facilitated the actions of a ribozymes, added additional useful capability such as the ability to bind the amino acid or affinity for ATP and ultimately displaced the ribozyme altogether.[67] The ability to recognize and transfer the peptide to the minihelix domain of tRNA would have evolved later and finally the ability to recognize identity of tRNA (especially as defined by the anticodon) last.

Individual Protein History

One obvious way in which one can infer the age of a ribosomal protein component is to infer its distribution. Those proteins, which occur in all three Domains of life were presumably present in the last common ancestor whereas those that became part of the machinery at a later time might be restricted to one Domain of life or even a subset of organisms in one Domain. Thus, there are in fact ribosomal proteins such as L3, L4 and L24 that are universally found whereas others such as L28, L31 and L36 are not. There are some caveats with this conclusion. The first is the possibility of lateral gene transfer, which is common in bacteria, could lead to a protein being widely distributed without it actually being old. The extent of lateral transfer of translational apparatus proteins has been studied.[68] Seventy-six components were examined in detail and 21, mostly tRNA synthetases, showed evidence of lateral transfer. In the case of the ribosomal proteins, there typically is little evidence of lateral transfer. Lateral transfer has been detected for L7[68] and especially for S14.[69] Four other proteins, L28, L31, L33 and L36 often have paralogs and may also be subject to lateral transfer.

A second difficulty is the possibility that the equivalent protein has been overlooked due to lack of sensitivity of sequence comparisons. One illustrative example is L19, which lacks an obvious homolog in the Archaea. In the *Deinococcus* 50S subunit,[70] L19 is seen to combine with L14 to form an extended inter-protein beta sheet that is involved in forming an intersubunit bridge. In the *Halococcus marismortui* ribosome, protein L24e, which lacks sequence homology with L19, serves the same purpose. The issue, currently unresolved, is whether this is an example of evolutionary convergence making this particular intersubunit bridge possibly recent, or simply a failure to detect sequence homology due to limited functional need for extensive sequence conservation.

Given that the ribosome is quite old, one might have expected the early ribosomal proteins to have diverged to spawn later ones and possibly even super families of proteins used elsewhere. There are likely to be historical relationships between some of the proteins that will provide timing insights to the development of the subunits. One such example, e.g., S6 and S10, has been uncovered[71] and verified by structural data.[72] It is clear from the structural data that there certainly is not a single ancestor for all of the ribosomal proteins. There are a large variety of different types of folds seen in the ribosomal proteins including but not limited to alpha/beta types, alpha helical bundles, beta barrels and the beta ribbon.[73,74] In the case of the *Deinococcus radiodurans* 50S subunit,[70] it is pointed out that L29 may contain a leucine zipper and the nonuniversal and hence putatively newer proteins, L32 and L36, contain the Zn-finger motif, which may be a later evolutionary discovery. Even the largest ribosomal proteins typically consist of domains of no more than[70] amino acids, possibly reflecting the early genetic events of protein evolution.[75]

Ribosomal Protein S1

Genomic comparisons of the ribosomal proteins have typically not revealed large numbers of nonribosomal proteins with obvious relationship to the ribosomal proteins. One notable exception is ribosomal protein S1. Ribosomal protein S1 is, in the first place, not typical at all. It is substantially larger than all other ribosomal proteins and not integrally part of the ribosome. It is involved in initiation and has been associated with anti-termination and trans-translation as well. It lacks an Archaeal homolog and is sometimes missing in Bacteria and hence likely a relatively recent addition to the ribosomal machinery. Ribosomal protein S1 contains six copies of an RNA binding domain (OB fold) belonging to the cold shock protein superfamily of oligonucleotide binding proteins. This motif contains approximately[70] residues and is characterized by a number of conserved glycine and valine residues. Polynucleotide phosphorylase, a bacterial exonuclease that degrades mRNA from 3' to 5' direction contains a single S1 motif at its C-terminus.[76] NMR was used to elucidate the structure of this S1 motif in polynucleotide phosphorylase. It was found to have a five stranded anti-parallel beta barrel arrangement. Conserved residues are seen on one face of the barrel and adjacent loops form the putative RNA binding site.[77]

A variety of other proteins contain the S1 motif. The most notable of these is the universal translation initiation factor IF1 and its eukaryotic equivalent eIF1a, both of which also have the characteristic five stranded beta barrel arrangement.[78,79] It has also been reported in the alpha subunit of the eukaryotic initiation factor 2.[80] An S1 motif is also found in yeast PRP22, which is an RNA helicase like protein required for the release of the mRNA from the spliceosome.[81] The S1 motif has also been identified in the N-terminal end of ribonuclease E, which is an essential single strand specific endoribonuclease, involved in both 5S ribosomal RNA processing and the rapid degradation of mRNA in *E. coli*.[82] The S1 motif is presumably involved in the endoribonucleolytic activity of the protein. Cytoplasmic axial filament protein from a variety of prokaryotes has an S1 motif at its N-terminus and the protein shows strong similarity to the N terminal domain of RNAse E. Bycroft and colleagues have also reported finding S1 motifs in NusA and EMB-5 protein in yeast and *C. elegans*.[77] The cyanobacterium, *Synechococcus* PCC6301, has been reported to have a novel nucleic acid binding protein with high similarity to S1.[83] Our own searches (Naik and Fox, unpublished data) have revealed several additional examples.

The proteins containing S1 domains can be broadly grouped into three main functional groups of RNA processing, involvement in transcription or translation and chromatin or septum regulation. The S1 motif is found in all three domains of life but only the IF-1/eIF1A type are universally distributed suggesting this might be the original source of the fold. If this is correct, one would likely conclude that ribosomal protein S1 is a late addition to the ribosome, likely derived from the initiation machinery.

Ribosome Subunit Evolution: Does Assembly Recapitulate History?

An especially interesting feature of the modern ribosome is its ability to self-assemble from components in vivo. This assembly process has been replicated in the laboratory for *E. coli*, first for the 30S subunit84 and later for the 50S subunit.[85-87] There is a defined order in which the various components interact in laboratory studies leading to subunit assembly and it is likely, but not certain, that a similar pathway is used in vivo. This information is typically summarized in a diagram known as a reconstitution map. The map illustrates the known dependencies. Some ribosomal proteins bind directly to the RNA whereas others such as L29 are only incorporated later after other proteins have been added (e.g., L24, L4 and L23 in the case of L29).

Although all the ribosomal proteins are involved in the modern assembly process, not all of them were always present. As noted previously, even though many of the ribosomal proteins are universal and hence likely to be present in the last common ancestor, some proteins are unique to each Domain and therefore likely of more recent origin. Given the assumption that the proteins are of different age, it is of interest to see how the two classes fit in the assembly picture. In Table 1, the Bacterial ribosomal proteins are classified as to whether they have recognizable equivalents in all three Domains, i.e., universal, and whether they are terminal in the assembly process. A protein is considered terminal if no other protein is dependant on its presence for assembly. There is in fact, a clear correlation, the recently added proteins are not central to the ribosomal assembly process. Indeed, they are almost always at the periphery where other proteins are not dependent on their presence for assembly. In short, proteins that are believed to be post-common ancestor additions to the genome according to sequence comrisons also appear to be late additions to the assembly process as well.

Not all the putative later additions are peripheral however. Some of the nonuniversal proteins are part of paths in which all subsequent proteins are also nonuniversal. The two most obvious examples are the paths L20-L21–L30 and L17-L28-L33. This suggests that even though all the proteins on these paths are relatively recent, the last ones might be even more recent and hence perhaps not even universal in the Bacterial Domain. When the nonconserved proteins and the connections associated with them are removed from the assembly map, the remaining proteins are all still interconnected with the single exception of L13. Figure 1 shows these proteins and the core assembly interactions that are associated with them.

Table 1. Bacterial ribosomal proteins classified as to whether they have recognizable equivalents in all domains and whether they are terminal in the assembly process

Protein	Terminal	Universal
L1	No[a]	Yes
L2	No	Yes
L3	No	Yes
L4	No	Yes
L5	No	Yes
L6	No[b]	Yes
L7/12	Yes	No
L9	Yes[a]	No
L10	No	Yes
L11	No	Yes
L13	Yes	Yes
L14	Yes[a]	Yes
L15	No	Yes
L16	Yes	Yes
L17	No	No
L18	No	Yes
L19	Yes	No
L20	No	No
L21	No[c]	No
L22	No	Yes
L23	No	Yes
L24	No	Yes
L25	No[c]	No
L27	Yes	No
L28	No[c]	No
L29	No	Yes
L30	Yes	No
L31	Yes	No
L32	Yes	No
L33	Yes	No
L34	Yes	No

The large subunit proteins are classified according to their occurrence[99] and terminality in the assembly map. Special situations are as follows: [a] Protein binds directly to the RNA but because it does not contribute to binding of other proteins is regarded as terminal; [b] This protein may be involved in the addition of L16 and therefore is not regarded as terminal; and [c] This protein is on a path where all subsequent proteins are not universally distributed.

The interdependent universal proteins that account for the core interactions in assembly can be subdivided into four groups, which are to some approximation built upon one another in a linear way, though the cooperativity that is central to the process should not be overlooked. The first group includes the assembly initiator proteins L3 and L24[88] as well as L4. These proteins interact directly with the RNA and facilitate, at least in part, the subsequent binding of the second group of proteins, L2, L22, L23 and L29. Thus, L2 binds in part directly to the RNA but its incorporation into the subunit is strongly facilitated by prior binding of L4. L4 and L24 are very important for incorporation of L22. Ribosomal protein L4 is also very impor-

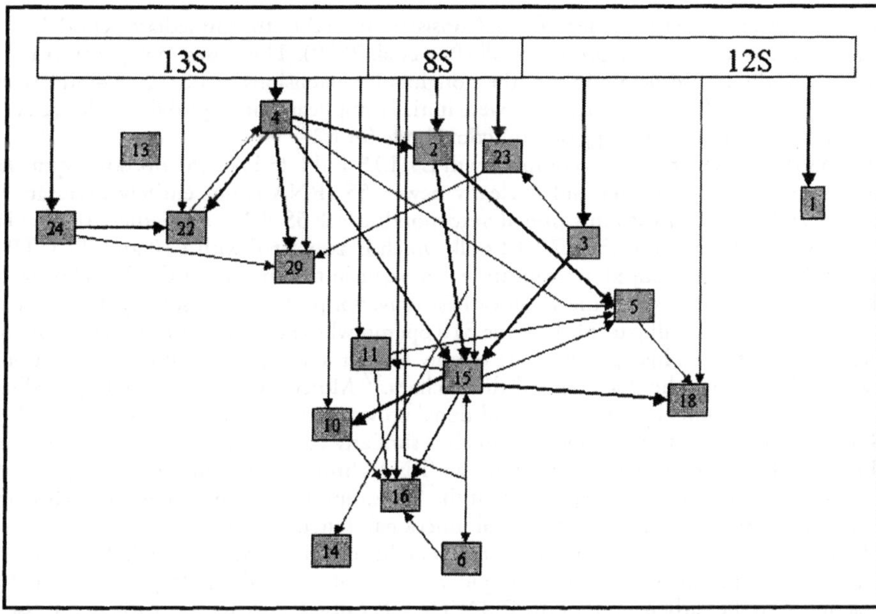

Figure 1. Location of universal proteins in the 50S assembly map. Only the proteins that have equivalents in all three Domains of life are shown in this map as compared to the original assembly map.[86] The 23S rRNA is represented as three fragments of 13S, 8S and 12S. Lines from the RNA indicate strong direct binding of the proteins to the RNA. Dark arrows indicate strong assembly dependence. The map indicates, for example, that protein L29 can be incorporated later, after L24, L4 and L23 have been added.

tant for binding of L29, which is also facilitated by L24 and L23. L23 interacts strongly with the RNA but is also dependent on L3.

Following incorporation of the first two groups of proteins the next major additions appear to be L5 and L15. With L5 and L15 incorporated, the diagrams suggest that the final universal proteins, L6, L10, L11, L16 and L18 appear to be next. Three universal proteins, L1, L13 and L14 are not integral to the assembly process when viewed this way. Although ribosomal protein L15 is significant in the later stages of assembly where it interacts with over ten proteins,[89] it appears that the essence of what it facilitates is innate to the structure of the ribosome that has already been assembled. A mutant strain lacking L15 is viable with a prolonged generation time and active 50S particles can be reconstituted without it.[90] These same authors were also able to create active particles without L16 and L30.

The clear inference from the assembly map is that the earliest proteins in a chain of assembly process are likely earlier and more critical additions to the ribosome machinery than the subsequent proteins. This is fully consistent with the independent conservation data. The proteins that are older based on phylogenetic distribution are more integral to the process than the proteins, which are not universal. The assembly data goes further, however, as it makes specific predictions about relative age of specific proteins. Thus, L21 is likely older than L30 even though both are relatively recent additions because L21 binding facilitates L30 binding. This is in fact consistent with the phylogenetic distribution of these two proteins as L21 is nearly universal and L30 is not. Moreover, L30 is typically not present when L21 is absent. One caveat is that the relative ages of proteins on different chains of assembly can't be readily determined. For example, L2 may predate L24 even though L2 is dependent on prior incorporation of L4.Among the universal proteins, the subgroups discussed above might be interpreted as progressing from oldest to youngest. One would expect that the peptidyl transferase center was the earliest region of the ribosome to form. Consistent with this, the immediate peptidyl transferase center does not contain protein at all (Ban et al, 2000). The closest polypeptides to the active site on the ribosome are in fact the nonglobular extensions of L2, L3, L4 and L10E. Ribosomal proteins L2, L4 and L16[91,92] were initially implicated in peptidyl transferase activity. Later work continued to suggest a functional role for L2.[93-95]

The 5S rRNA, which forms a complex with L5, L18 and L25 joins the ribosome quite late in the assembly process center and is clearly newer. 5S rRNA is not entirely necessary for viability[96] and consistent with this animal mitochondria lack 5S rRNA altogether. Another late addition would appear to be the L7/L12 stalk which is associated with L10 and the GTPase center of the ribosome. Indeed, translocation can occur in the absence of GTPase cleavage.[33]

It is noteworthy that six of the seven earliest universal proteins in assembly, L3, L4, L22, L23, L24, and L29, are all part of the same S10 operon whose gene order is preserved in both Archaea and Bacteria. This operon is one of only four protein operons whose organization is conserved throughout the Bacteria and the Archaea.[97] Moreover, it is regulated by feedback control at the RNA level by L4. Several additional universal ribosomal proteins, L5, L16, L15, L18, L24. are found in the Spc operon, which is similarly conserved in structure. It may very well be that these operons had their origins as "RNA chromosomes" in the late stages of the RNA World.[9,97] In summary, it appears that the 50S subunit assembly process is a rich source of information on the relative age of several functional domains of the ribosome.

If the assembly process does in part represent an historical record then it will be possible to use it to obtain further insight. For example, one can infer that the early proteins would be interacting with the early regions of the RNA. Thus, because L4 occurs very early in the assembly process and is a key element in folding Domain IV of the 23S rRNA, it would be predicted that this region of the RNA is quite old. Consistent with this, most of the contacts between the 50S and 30S subunits are located in this region.[98] With ribosomal structures of increasing resolution from multiple organisms it is already possible to define the regions of interaction between all the proteins and the RNA at the residue level and this is currently being done (Hury, Nagaswamy and Fox, unpublished results). There are of course complexities as well. For example, although some of the proteins are universal they are actually dramatically different in the three Domains, whereas others are highly homologous in all three domains so there is clearly more to the story than what has been presented here. Likewise, at this stage at least the 30S assembly map does not tell an obvious story of that subunit's history.

Order of Events Model for the Development of the Translation Machinery

In order to put the preceding discussion into perspective, we have developed a very general time line encompassing some possible events in the history of the ribosomal machinery (Table 2). It is useful to do this now in order to provoke the type of discussion, data analysis and even experiment that will be needed to produce future proposals that are actually reasonable. At present we are doubly hampered as we don't quite know what the most important events actually were and even if we did, any proposal about their order is necessarily extremely speculative at this early stage. It is also endemic to the essentially informatics approach to ribosome history described here that it is typically more difficult to relate the timing of a sequence of events relative to another sequence of events than it is to determine the relative timing of events within either series. For example, we deduced from the 50S assembly map that L2 likely preceded L15 but we can't use that information to easily determine whether L15 went into use before or after ribosomal protein S9.

Finally, building an order of events model is greatly complicated by the fact that many aspects of translation, e.g., the development of the translocation machinery, likely occurred in stages rather than all at once. Moreover, more than one refinement was likely occurring at any one time. In any event, an outline of the history of the translation machinery, which each reader will undoubtedly find erroneous in different ways is presented below with brief comments. The list is broken up into five time periods within which the various activities should be considered to be possibly overlapping in time.

Table 2. History of events in the evolution of ribosomal machinery

A . Earliest Beginnings: RNA World

1. Initial Formation of Peptidyl Transferase Center in RNA World
   - A. Emergence of RNAs that can aminoacylate RNAs -leads to small charged RNAs (minihelices)
   - B. Emergence of RNAs that can catalyze peptide bond formation between minihelices.
2. Beginnings of coherent 50S subunit
   Portions of Domain V of 23S rRNA likely present. Addition of more RNA or essentially random peptides might have decreased hydrolysis reaction and increase activity by protecting the core reaction.
3. Extension of tRNA to two domains;
   Second tRNA domain allows templating, which significantly increases reaction rate. Characteristic conformational changes associated with translocation are present.
4. Beginnings of coding
   It is now possible to store information. This makes it useful to have a genome (RNA) and hence primitive polymerases might offer a significant selective advantage to progenotes that have them.

B. Beginnings of Transition to Protein World: Late RNA World

5. Ancestors of core proteins such as L3 and L4 are present.
   Emergence of defined sequence peptides means RNA World will soon end.
6. Initial creation of 30S particle
   Further protection of the reaction machinery is possible by stabilizing template. Bridges between subunits were probably initially only RNA. Portions of Domain IV of 23S rRNA that interact with 30S subunit are likely to be present.

C. Early Protein World: Major Refinements Increase Speed and Accuracy

7. EF-Tu ancestor
   Improved control of tRNA access to machinery
8. Addition of 5S rRNA complex
   Further development of 50S subunit underway. Many new proteins present such as L5 and L18 that are associated with 5S rRNA incorporation
9. GTPase Reaction Center Formed
   The emerging protein world allows the development of the translocation machinery. L10 now present. At least portions of 23S rRNA Domain II are present.

D. RNA/Protein World: Complex Proteins Now Possible

10. Protein based tRNA synthetases,
    Possible expansion of genetic code
11. Universal Soluble protein factors; IF1, IF2, EF-G , release factors
    Further development of 30S subunit, Initiation machinery forms
12. Initial regulatory machinery
    Multiple RNA chromosomes probably exist, each coding for a group of ribosomal components that eventually become the major operons that survive in extant organisms.
13. Additional universal proteins added to 50S and 30S subunits.
    Proteins such as L15 and L16 that improve ribosome assembly.

E. Emergence of DNA Genome- Life as We Know It

14. Lineage specific refinements of translation machinery.
    Nonuniversal proteins such as S1 are added in their various lineages.
    Variants of initiation process develop

Implications and Future Work

The evidence suggests that chiral protein synthesis could have been catalyzed by ribosomes lacking such advanced features as the GTPase center, modern synthetases, the 5S rRNA complex, improved rates of assembly etc. Since these features were nevertheless largely in the common ancestor, it appears then that the capability of making defined sequence peptides existed well before the common ancestor. In contrast, the existence of RNA level control for the ribosomal components and the general observation that the DNA machinery appears more recent than that of transcription or translation suggests that DNA was not used as the genetic material until well after efficient ribosomes existed. As a result one must consider two issues that have not been extensively discussed. First, was the nature of evolution different before DNA was implemented as the genetic material? In the RNA/protein world there would be a direct linkage between genome and mRNA with larger error rates probable. In addition, the storage capacity of an RNA genome would likely be limited. The progenotes that populated this world might have relied extensively on lateral transfer. Second, when did DNA genomes actually arise and how rapidly did genes accrete into them? It may well be that the difficulties in defining the earlier branching points on phylogenetic trees relate to the fact that these ancestors had either much smaller genomes or did not yet even have DNA genomes.

With respect to the ribosome itself, the opportunity to learn much more is just around the corner. As we have better models of the mechanisms of peptide bond formation and translocation it will be possible to experimentally test them and ultimately develop laboratory models for early ribosome evolution. Certainly knowing how the ribosome works will be a major step in improving the time line as we will at least know what to put on it. With respect to the bio-informatics approach outlined here, much remains to be done with the data already available. In particular, a detailed analysis of the RNA/RNA, RNA/protein and protein/protein interactions is needed. This will be especially productive when high resolution structural data is available from distinct Domains and species as then the comparative information can be used to determine what is old and what is new in a very detailed way.

Acknowledgements

The work described herein was supported in part by grants from the Exobiology Program of the National Aeronautics and Space Administration (NAG5-12366 and its predecessor NAG5-8140) and the Institute of Space Systems Operations at the University of Houston.

References

1. Woese CR, Fox GE. The concept of cellular evolution. J Mol Evol 1977; 10(1):1-6.
2. Karzai AW, Roche ED, Sauer RT. The SsrA-SmpB system for protein tagging, directed degradation and ribosome rescue. Nature Struct Biol 2000; 7(6):449-455.
3. Brosius J. tRNAs in the spotlight during protein biosynthesis. Trends Biochem Sci 2001; 26(11):653-656.
4. Becerra A, Islas S, Leguina JI et al. Polyphyletic gene losses can bias backtrack characterizations of the cenancestor. J Mol Evol 1997; 45(2):115-118.
5. Benner SA, Cohen MA, Gonnet GH et al. Reading the palimpset: Contemporary biochemical data and the RNA world. In: Gasteland RF, Atkins JF, eds. The RNA World. 1st ed. Cold Spring Harbor: Cold Spring Harbor Laboratory Press, 1993:27-70.
6. Lazcano A. Cellular evolution during the early Archaean: What happened between the progenote and the cenancestor? Microbiologia SEM 1994; 11:13-18.
7. Doolittle WF. The nature of the universal ancestor and the evolution of the proteome. Curr Opin Struct Biol 2000; 10(3):355-358.
8. Glansdorff N. About the last common ancestor, the universal life-tree and lateral gene transfer: A reappraisal. Mol Microbiol 2000; 38(2):177-185.
9. Olsen GJ, Woese CR. Archaeal genomics: An overview. Cell 1997; 89(7):991-994.
10. Leipe DD, Aravind L, Koonin EV. Did DNA replication evolve twice independently? Nucleic Acids Res 1999; 27(17):3389-3401.
11. Cramer P, Bushnell DA, Kornberg RD. Structural basis of transcription: RNA polymerase II at 2.8 angstrom resolution. Science 2001; 292(5523):1863-1876.

12. Ferris JP, Hill Jr AR, Liu R et al. Synthesis of long prebiotic oligomers on mineral surfaces. Nature 1996; 381(6577):59-61.
13. Ban N, Nissen P, Hansen J et al. The complete atomic structure of the large ribosomal subunit at 2.4 A resolution. Science 2000; 289(5481):905-920.
14. Wimberly BT, Brodersen DE, Clemons Jr WM et al. Structure of the 30S ribosomal subunit. Nature 2000; 407(6802):327-339.
15. Nissen P, Hansen J, Ban N et al. The structural basis of ribosome activity in peptide bond synthesis. Science 2000; 289(5481):920-930.
16. Katunin VI, Muth GW, Strobel SA et al. Important contribution to catalysis of peptide bond formation by a single ionizing group within the ribosome. Mol Cell 2002; 10(2):339-346.
17. Barta A, Dorner S, Polacek N. Mechanism of ribosomal peptide bond formation. Science 2001; 291(5502):203.
18. Bayfield MA, Dahlberg AE, Schulmeister U et al. A conformational change in the ribosomal peptidyl transferase center upon active/inactive transition. Proc Natl Acad Sci USA 2001; 98(18):10096-10101.
19. Polacek N, Gaynor M, Yassin A et al. Ribosomal peptidyl transferase can withstand mutations at the putative catalytic nucleotide. Nature 2001; 411(6836):498-501.
20. Thompson J, Kim DF, O'Connor M et al. Analysis of mutations at residues A2451 and G2447 of 23S rRNA in the peptidyltransferase active site of the 50S ribosomal subunit. Proc Natl Acad Sci USA 2001; 98(16):9002-9007.
21. Nissen P, Hansen J, Muth GW et al. Mechanism of ribosomal peptide bond formation. Science 2001; 291:203a.
21a. Beringer M, Adies S, Wintermeyer W et al. The G2447A mutation does not effect ionization of a ribosomal group taking part in peptide bond formation. RNA 2003; 9:919-922.
22. Woese CR. Translation: In retrospect and prospect. RNA 2001; 7:1055-1067.
23. Woese CR. Molecular mechanics of translation: A reciprocating ratchet mechanism. Nature 1970; 226(5248):817-820.
24. Woese CR. Just so stories and Rube Goldberg machines: Speculations on the origin of the protein synthetic machinery. In: Chambliss G, Craven GR, Davies J et al, eds. Ribosomes: Structure, Function and Genetics. Baltimore: University Park Press, 1980:357-373.
25. Maizels N, Weiner AM. Peptide-specific ribosomes, genomic tags, and the origin of the genetic code. Cold Spring Harbor Symp Quant Biol 1987; 52:743-749.
26. Lee N, Bessho Y, Wei K et al. Ribozyme-catalyzed tRNA aminoacylation. Nature Struct Biol 2000; 7(1):28-33.
27. Schimmel P, Kelley SO. Exiting an RNA world. Nature Struct Biol 2000; 7(1):5-7.
28. Saito H, Kourouklis D, Suga H. An in vitro evolved precursor tRNA with aminoacylation activity. EMBO J 2001; 20(7):1797-1806.
29. Tamura K, Schimmel P. Oligonucleotide-directed peptide synthesis in a ribosome- and ribozyme-free system. Proc Natl Acad Sci USA 2001; 98(4):1393-1397.
30. Shimizu M. Detection of the peptidyl transferase activity of a dipeptide, alanylhistidine, in the absence of ribosomes. J Biochem 1996; 119(5):832-834.
31. Larkin DC, Martinis SA, Roberts DJ et al. Do small dipeptides mediate a peptidyl transferase reaction with aminoacylated RNA?" Orig Life Evol Biosph 2001; 31(6):511-526.
32. Gavrilova LP, Kostiashkina OE, Koteliansky VE et al. Factor-free ("nonenzymatic") and factor-dependent systems of translation of polyuridylic acid by Escherichia coli ribosomes. J Mol Biol 1976; 101(4):537-552.
33. Spirin AS. Ribosomal translocation: Facts and models. Prog Nucleic Acid Res Mol Biol 1985; 32:75-114.
34. Kjeldgaard M, Nyborg J. Refined structure of elongation factor EF-Tu from Escherichia coli. J Mol Biol 1992; 223(3):721-742.
35. Czworkowski J, Wang J, Steitz TA et al. The crystal structure of elongation factor G complexed with GDP at 2.7A resolution. EMBO J 1994; 13(16):3661-3668.
36. Nissen P, Kjeldgaard M, Thirup S et al. Crystal structure of the ternary complex of Phe-$tRNA^{Phe}$, EF-Tu and a GTP analog. Science 1995; 270(5241):1464-1472.
37. Moore PB. Molecular mimicry in protein synthesis. Science 1996; 270(5241):1453-1454.
38. Nissen P, Kjeldgaard M, Nyborg J. Macromolecular mimicry. EMBO J 2000; 19(4):489-495.
39. Agrawal RK, Penczek P, Grassucci RA et al. Visualization of elongation factor G on the Escherichia coli 70S ribosome: the mechanism of translocation. Proc Natl Acad Sci USA 1998; 95(11):6134-6138.
40. Abel K, Jurnak F. A complex profile of protein elongation: Translating chemical energy into molecular movement. Structure 1996; 4(3):229-238.
41. Frank J, Agrawal RK. A ratchet-like inter-subunit reorganization of the ribosome during translocation. Nature 2000; 406(6793):318-322.

42. Agrawal RK, Spahn CM, Penczek P et al. Visualization of tRNA movements on the Escherichia coli 70S ribosome during the elongation cycle. J Cell Biol 2000; 150(3):447-460.
43. Wriggers W, Agrawal RK, Drew DL et al. Domain motions of EF-G bound to the 70S ribosome: Insights from a hand-shaking between multi-resolution structures. Biophys J 2000; 79(3):1670-1678.
44. VanLoock MS, Agrawal RK, Gabashvili IS et al. Movement of the decoding region of the 16 S ribosomal RNA accompanies tRNA translocation. J Mol Biol 2000; 304(4):507-515.
45. Golden BL, Ramakrishnan V, White SW. The structure of ribosomal protein L6: structural evidence of gene duplication from a primitive RNA binding protein. EMBO J 1993 12(13):4901-4908.
46. Caetano-Anolles G. Tracing the evolution of RNA structure in ribosomes. Nucleic Acids Res 2002; 30(11):2575-2587.
47. Caetano-Anolles G. Evolved secondary structure and the rooting of the universal tree of life. J Mol Evol 2002; 54(3):333-345.
48. Mears JA, Cannone JJ, Stagg SM et al. Modeling a minimal ribosome based on comparative sequence analysis. J Mol Biol 321(2):215-234.
49. Kim SH, Quigley GJ, Suddath FL et al. Three-dimensional structure of yeast phenylalanine transfer RNA: Folding of the polynucleotide chain. Science 1973; 179(70):285-288.
50. Robertus JD, Ladner JE, Finch JT et al. Structure of yeast phenylalanine tRNA at 3 A resolution. Nature 1974; 250(467):546-551.
51. Shi H, Moore PB. The crystal structure of yeast phenylalanine tRNA at 1.93 A resolution: A classic structure revisited. RNA 2000; 6(8):1091-1105.
52. Schimmel P, Ribas de Pouplana L. Transfer RNA: From minihelix to genetic code. Cell 1995; 81(7):983-986.
53. Schimmel P, Henderson B. Possible role of aminoacyl-RNA complexes in noncoded peptide synthesis and origin of coded synthesis. Proc Natl Acad Sci USA 1994; 91(24):11283-11286.
54. Dick TB, Schamel WA. Molecular evolution of transfer RNA from two precursor hairpins: Implications for the origin of protein synthesis. J Mol Evol 1995; 41(1):1-9.
55. Di Giulio M. On the origin of protein synthesis: A speculative model based on hairpin RNA structures. J Theor Biol 1994; 171(3):303-308.
56. Di Giulio M. On the origin of the transfer RNA molecule. J Theor Biol 1992; 159(2):199-214.
57. Nagaswamy U, Fox GE. RNA ligation and the origin of tRNA. Orig Life Evol Biosph 2002 in press.
58. Francklyn C, Schimmel P. Aminoacylation of RNA minihelices with alanine. Nature 1989; 337(6206):478-481.
59. Shi JP, Schimmel P. Aminoacylation of alanine minihelices. "Discriminator" base modulates transition state of single turnover reaction. J Biol Chem 1991; 266(5):2705-8.
60. Martinis SA, Schimmel P. Small RNA oligonucleotide substrates for specific aminoacylations. In: Söll D, RajBhandary U, eds. tRNA:Structure, Biosynthesis, and Function. Washington, DC: ASM Press, 1995:349-370.
61. Lohse PA, Szostak JW. Ribozyme-catalysed amino-acid transfer reactions. Nature 1996; 381(6581):442-444.
62. Illangasekare M, Sanchez G, Nickles T et al. Aminoacyl-RNA synthesis catalyzed by an RNA. Science 1995; 267(5198):643-647.
63. Illangasekare M, Yarus M. A tiny RNA that catalyzes both aminoacyl-RNA and peptidyl-RNA synthesis. RNA 1999; 5(11):1482-1489.
64. Zhang B, Cech TR. Peptide bond formation by in vitro selected ribozymes. Nature 1997; 390(6655):96-100.
65. Kumar RK, Yarus M. RNA-catalyzed amino acid activation. Biochemistry 2001; 40(24):6998-7004.
66. Francklyn C, Musier-Forsyth K, Martinis SA. Aminoacyl-tRNA synthetases in biology and disease: New evidence for structural and functional diversity in an ancient family of enzymes. RNA 1997; 3(9):954-960.
67. Ribas de Pouplana L, Schimmel P. Aminoacyl-tRNA synthetases: potential markers of genetic code development. Trends Biochem Sci 2001; 26(10):591-596.
68. Brochier C, Bapteste E, Moreira D et al. Eubacterial phylogeny based on translational apparatus proteins. Trends Genet 2002; 18(1):1-5.
69. Brochier C, Philippe H, Moreira D. The evolutionary history of ribosomal protein RpS14: Horizontal gene transfer at the heart of the ribosome. Trends Genet 2000; 16(12):529-533.
70. Harms J, Schluenzen F, Zarivach R et al. High-resolution structure of the large ribosomal subunit from a mesophilic eubacterium. Cell 2001; 107(5):679-688.
71. Jue RA, Woodbury NW, Doolittle RF. Sequence homologies among E. coli ribosomal proteins: evidence for evolutionarily related groupings and internal duplications. J Mol Evol 1980; 15(2):129-148.

72. Brodersen DE, Clemons Jr WM, Carter AP et al. Crystal structure of the 30 S ribosomal subunit from Thermus thermophilus: Structure of the proteins and their interactions with 16 S RNA. J Mol Biol 2002; 316(3):725-768.
73. Ramakrishnan V, White SW. Ribosomal protein structures: Insights into thearchitecture, machinery and evolution of the ribosome. Trends Biochem Sci 1998; 23(6):208-212.
74. Lo Conte L, Brenner SE, Hubbard TJ et al. SCOP database in 2002: Refinements accommodate structural genomics. Nucleic Acids Res 2002; 30(1):264-267.
75. Nakagawa A, Nakashima T, Taniguchi M et al. The three-dimensional structure of the RNA-binding domain of ribosomal protein L2; a protein at the peptidyl transferase center of the ribosome. EMBO J 1999; 18(16):1459-1467.
76. Regnier P, Grunberg-Manago M, Portier C. Nucleotide sequence of the pnp gene of Escherichia coli encoding polynucleotide phosphorylase: Homology of the primary structure of the protein with the RNA-binding domain of ribosomal protein S1. J Biol Chem 1987; 262(1):63-68.
77. Bycroft M, Hubbard TJ, Proctor M et al. The solution structure of the S1 RNA binding domain: A member of an ancient nucleic acid-binding fold. Cell 1997; 88(2):235-242.
78. Sette M, van Tilborg P, Spurio R et al. The structure of the translational initiation factor IF1 from E. coli contains an oligomer-binding motif. EMBO J 1997; 16(6):1436-1443.
79. Battiste JL, Pestova TV, Hellen CU et al. The eIF1A solution structure reveals a large RNA-binding surface important for scanning function. Mol Cell 2000; 5(1):109-119.
80. Gribskov M. Translational initiation factors IF-1 and eIF-2 alpha share an RNA-binding motif with prokaryotic ribosomal protein S1 and polynucleotide phosphorylase. Gene 1992; 119(1):107-111.
81. Company M, Arenas J, Abelson J. Requirement of the RNA helicase-like protein PRP22 for release of messenger RNA from spliceosomes. Nature 1991; 349(6309):487-493.
82. Kaberdin VR, Miczak A, Jakobsen JS et al. The endoribonucleolytic N-terminal half of Escherichia coli RNase E is evolutionarily conserved in Synechocystis sp. and other bacteria but not the C-terminal half, which is sufficient for degradosome assembly. Proc Natl Acad Sci USA 1998; 95(20):11637-11642.
83. Sugita C, Sugiura M, Sugita M. A novel nucleic acid-binding protein in the cyanobacterium Synechococcus sp. PCC6301: A soluble 33-kDa polypeptide with high sequence similarity to ribosomal protein S1. Mol Gen Genet 2000; 263(4):655-663.
84. Mizushima S, Nomura M. Assembly mapping of 30S ribosomal proteins from E. coli. Nature 1970; 226(252):1214.
85. Rohland R, Nierhaus KH. Assembly map of the large subunit (50S) of Escherichia coli ribosomes. Proc Natl Acad Sci USA 1982; 79(3):729-733.
86. Herold M, Nierhaus KH. Incorporation of six additional proteins to complete the assembly map of the 50 S subunit from Escherichia coli ribosomes. J Biol Chem 1987; 262(18):8826-8833.
87. Nierhaus KH. The assembly of prokaryotic ribosomes. Biochimie 1991; 73(6):739-755.
88. Nowotny V, Nierhaus KH. Initiator proteins for the assembly of the 50S subunit from Escherichia coli ribosomes. Proc Natl Acad Sci USA 1982; 79(23):7238-7242.
89. Lieberman KR, Noller HF. Ribosomal protein L15 as a probe of 50 S ribosomal subunit structure. J Mol Biol 1998; 284(5):1367-1378.
90. Franceschi FJ, Nierhaus KH. Ribosomal proteins L15 and L16 are mere late assembly proteins of the large ribosomal subunit: Analysis of an Escherichia coli mutant lacking L15. J Biol Chem 1990; 265(27):16676-16682.
91. Fahnestock SR. Evidence of the involvement of a 50S ribosomal protein in several active sites. Biochemistry 1975; 14(24):5321-5327.
92. Dohme F, Fahnestock SR. Identification of proteins involved in the peptidyl transferase activity of ribosomes by chemical modification. J Mol Biol 1979; 129(1):63-81.
93. Cooperman BS, Wooten T, Romero DP et al. Histidine 229 in protein L2 is apparently essential for 50S peptidyl transferase activity. Biochem Cell Biol 1995; 73(11-12):1087-1094.
94. Wittmann-Liebold B, Uhlein M, Urlaub H et al. Structural and functional implications in the eubacterial ribosome as revealed by protein-rRNA and antibiotic contact sites. Biochem Cell Biol 1995; 73(11-12):1187-1197.
95. Uhlein M, Weglohner W, Urlaub H et al. Functional implications of ribosomal protein L2 in protein biosynthesis as shown by in vivo replacement studies. Biochem J 1998; 331(Pt2):423-430.
96. Dohme F, Nierhaus KH. Role of 5S RNA in assembly and function of the 50S subunit from Escherichia coli. Proc Natl Acad Sci USA 1976; 73(7):2221-2225.
97. Siefert JL, Martin KA, Abdi F et al. Conserved gene clusters in bacterial genomes provide further support for the primacy of RNA. J Mol Evol 1997; 45(5):467-472.
98. Yusupov MM, Yusupova GZ, Baucom A et al. Crystal structure of the ribosome at 5.5 A resolution. Science 2001; 292(5518):883-896.
99. Harris JK, Kelley ST, Spiegelman GB et al. The genetic core of the universal ancestor. Genome Res 2003; 13:407-412.

CHAPTER 7

Functional Evolution of Ribosomes

Carlos Briones and Ricardo Amils

Historical Perspective

Protein synthesis is a complex process that constitutes the last step of gene expression. The first studies that correlated this cellular activity with certain "ribonucleoprotein particles" present in the microsomal fraction are dated from the mid-fifties.[1] Subsequently, extensive research was carried out to determine if those particles were composed of one or two species (with sedimentation coefficients 70S or 30S+50S respectively) of ribonucleoprotein aggregates. One of the first studies was performed on yeast, by measuring the effect of Mg^{2+} cation on the association/dissociation equilibrium of those subunits.[2] In 1958, the term "ribosomes" was proposed for the cellular ribonucleic particles with a sedimentation coefficient ranging from 20S to 100S.[3] That year was also decisive for protein synthesis research, due to the characterization of *Escherichia coli* ribosomes[4] and to the formulation of the "central dogma" of the genetic information flux, where ribosomes played an essential role.[5]

During the following decades, ribosome research made two spectacular advances, the discovery of messenger RNA (mRNA) as an intermediate molecule in the transmission of genetic information[6] and the optimization of an in vitro protein synthesis system.[7] They, in turn, led to the beginning of two fruitful lines of research: the development of systems for in vitro reconstitution of ribosomes,[8,9] which facilitated their structural characterization, and the study of the inhibition of protein synthesis with antibiotics,[10,11] which was very important for the functional characterization of ribosomal particles.

Remarkably, at that time most of the researchers considered the ribosome a multi-enzymatic complex where proteins carried the different catalytic functions, while ribosomal RNA (rRNA) was seen as simply a structural scaffold for the proteins. These ideas, together with greater progress in analytical techniques for proteins than for nucleic acids, focused the research on ribosomal proteins. In this sense, one of the most ambitious projects at the time, developed at the Max-Planck of Berlin, was oriented towards the characterization and sequencing of all *E. coli* ribosomal proteins.

Nevertheless, during the seventies H. Noller's group, following the inspiration of C. Woese's work,[12] finished the sequencing of *E. coli* 16S and 23S rRNAs, and as a result a revolutionary concept started to emerge: the idea that rRNA could play a functional role in ribosomes.[13] In fact, it had been previously proposed that primitive ribosomes could be made only of RNA,[14] as it was highly improbable that, in a prebiotic context, proteins preceded RNA.[15] At the same time, the difficulty of assigning specific functions to any ribosomal protein became evident. The possibility that RNA carried catalytic functions was demonstrated with the discovery of self-splicing introns[16] and the tRNA processing by RNase P.[17] The characterization of different RNA-enzymes (ribozymes) and the development of the "RNA-World" concept influenced ribosome research, with several functions required for protein synthesis being proposed to occur in specific rRNA regions.

The Genetic Code and the Origin of Life, edited by Lluís Ribas de Pouplana.

Since 1990, highly refined structural techniques (among them, X-ray diffraction and cryo-electron microscopy) have allowed the construction of detailed 3-D models of the ribosome and its subunits.[18-20] The combined use of biochemical and genetic techniques are generating interesting results in the characterization of the different functional domains. Nevertheless, we are far from understanding the precise macromolecular interactions among ribosomal proteins, rRNAs and soluble factors that are responsible for the translation of the genetic message.

Ribosomes and Translation

The reactions that take place during the three steps of the translation cycle (initiation, elongation and termination) involve the interaction of more than one hundred macromolecules with elements external to the ribosomal particle, soluble protein factors, mRNAs and aminoacyl-tRNAs playing an important role. From an energetic point of view, protein synthesis is the main biosynthetic activity within the growing cell. It has been determined that in exponential phase *E. coli* cultures, more than 90% of the available energy is consumed in the protein synthesis process.[21] As a consequence, the mechanisms involved in the regulation of translation are fundamental in all organisms.[22] In bacteria there are about 15,000 to 20,000 ribosomes per cell, which corresponds to one quarter of the total cellular mass. rRNA represents 85% of the total mass of the cellular RNA.

Ribosomal RNA

Sequence (primary structure) comparison of rRNAs has shown a high degree of similarity among all the organisms analyzed since 1980. This implies a very slow evolutionary rate, which has been related to restrictions associated to the functional role of rRNA during protein synthesis. Thus, rRNA has become the main molecular clock for studying evolution, and the sequencing of its genes (mainly those of the minor subunit, or SSU rDNAs) constitutes at present the most widely used taxonomic and phylogenetic tool.[23] The proposal of the domain Archaea (formerly Archaebacteria) as the third group of cell systems, showing profound differences with respect to the well-established Bacteria and Eukarya domains was a milestone in molecular evolution.[12,24-26]

Secondary and tertiary rRNA structures give information about its spatial conformation within the ribosomal particle. Comparative sequence analysis, also called "covariation analysis" is very useful for predicting such higher-order structures,[13] as it identifies parts of the sequence which are either conserved in a group of organisms (denominated "signatures", with high taxonomic and ecological value) or are hypervariable. The hypervariable regions of the rRNA sequences are generally associated with promoters of secondary structures (double helices), for which the only requirement is structural stability entailing complementary mutations for the maintenance of base pairing. Most of the predicted structures have been experimentally probed using different methodologies, including: rRNA mutational analysis, chemical or enzymatic rRNA modification, cross-linking studies, 2-D electrophoresis of digested rRNAs, specific oligonucleotide accessability, electron microscopy of rRNA, nuclear magnetic resonance (NMR), and free energy analysis.[27-29] These techniques confirmed that secondary and tertiary rRNA structures are highly conserved in all extant organisms, regardless of their differences in sequence.

The quaternary structure of ribosomes gives information about the interactions among rRNAs and ribosomal proteins, and also rRNA-rRNA and protein-protein interactions. The study of rRNA-protein contacts has been focused on the determination of rRNA "recognition signals" for proteins, the search for common features in rRNA-binding proteins, and the study of conformational changes in rRNA induced by the interaction with proteins.[13] For this purpose, a number of biophysical, biochemical and genetic techniques have been coordinately used: NMR of rRNA-protein complexes, cryo-electron microscopy, footprinting techniques, rRNA-protein cross-linking, reconstitution studies and sequence comparison of rRNA-binding proteins.[20,28,30]

In Vitro Reconstitution of Ribosomes

The synthesis of ribosomes in vivo is a very complex process requiring the precise coordination in the expression of more than fifty genes, and the ordered assembly of its products to allow the formation of functionally active particles. The pioneering work of M. Nomura's group in the sixties showed that when *E. coli* ribosomes were exposed to high ionic strength, some of the proteins were removed, resulting in ribosomal "cores" deprived of different functional properties.[8] It was also shown that the process could be reversed, which constituted the first step towards in vitro reconstitution of ribosomes. Partial in vitro reconstitution involves a previous production of those "cores" lacking a number of ribosomal proteins. The controlled addition of the lost proteins produces particles with their original sedimentation coefficient and fully active in protein synthesis. First partial reconstitution experiments where performed on *E. coli* 30S and 50S subunits.[9]

The advance in rRNA and ribosomal protein purification techniques allowed the development of total in vitro reconstitution systems, that made it possible to dissect the assembly process and uncover the main rules governing the structure/function relationships among the different components. The first optimized systems allowed the total reconstitution of *E. coli* 30S subunit[31] and *Bacillus stearothermophilus* 50S subunit.[32] It was also possible to develop total reconstitution systems for *E. coli* 50S subunit,[33-34] the thermophilic archaeon *Sulfolobus solfataricus* 50S subunit,[35] *E. coli* 70S ribosomes,[36-37] and the extreme halophilic archaeon *Haloferax mediterranei* 30S and 50S subunits and 70S ribosomes.[38-40] Noticeably, reconstitution systems have not yet been developed for eukaryotic ribosomes.

The combination of reconstitution systems with other biochemical techniques allowed the production of ribosomes with modified components, suitable for further structural and functional analysis.[41] From a structural point of view, there are three main variables involved in the in vitro reconstitution process: ionic strength (mainly responsible for the selection of specific interactions), concentration of divalent cations (involved in rRNA conformation), and temperature (used to drive conformational changes and to disassemble nonfunctional abortive particles produced by wrong macromolecular interactions). Therefore, in suitable physicochemical conditions, ribosomal components have sufficient information to promote the formation of fully active ribosomal particles.

In this context, it is important to emphasize that while universal conservation of rRNA sequences has been associated with structures committed to function,[13] several point directed mutagenesis experiments show controversial results.[42] While the introduction of specific mutations into universally conserved rRNA sequences are in general lethal, complementary in vitro reconstitution experiments with mutated rRNA produce functional particles, suggesting that universal rRNA sequences are not necessarily related to function, but to the correct assembly of the particle, a critical step for the generation of functional ribosomes.

Functional Ribosomal Neighborhoods

The ribosome is a complex macromolecular machine that is actively involved in the various steps of protein synthesis. The main functional regions of the ribosome are peptidyltransferase and GTPase (located in the major subunit), and mRNA decoding site (located in the minor subunit). During elongation of the peptide chain and translocation of the ribosome along the mRNA, there is a concerted movement of the mRNA and bound tRNAs relative to the ribosome. Also, the nascent peptide is forced to advance in an unfolded form along a channel until it reaches the surface of the ribosome. These movements must involve switching of alternative ribosomal conformations, in which the components of both ribosomal subunits act in cooperation.[43-45]

The use of some of the structural techniques mentioned above has shown that highly conserved rRNA loops are involved in different ribosomal functional neighborhoods.[46] This fact suggested that those conserved loops represent the most primitive regions of the rRNA.[47] Nevertheless, it has been proved that most of the conserved loops are not functionally active by themselves, requiring the cooperation of other rRNA regions and ribosomal proteins. Therefore, after fifty years of research, it is currently accepted that functional neighborhoods (functional space) of the ribosome are mixed rRNA-protein domains, where components of different macromolecules form the interaction sites for the ligands involved in the protein synthesis process.[20,45]

Protein Synthesis Inhibitors as Functional Markers

The first approaches to the study of the inhibition of protein synthesis in bacteria involved the use of chloramphenicol and chlortetracycline at their minimum inhibitory doses.[48] The development of in vitro translation systems (that avoided experimental problems related to transport and inactivation of antibiotics) allowed the systematic study of the specific inhibition of every protein synthesis step (for a review see refs. 10 and 11). Currently, a large number of protein synthesis inhibitors are known and their structures and mode of action have been determined. Simultaneously, such inhibitors and some of their molecular analogs have been used as probes for the characterization of the functional regions of the ribosome.[49] Therefore, ribosomal research and that of the inhibitors of protein synthesis have been intimately coupled for the last decades.

The taxonomic specificity of protein synthesis inhibitors,[11] which is the base of their medical use, allowed their classification into three groups: group I, specific inhibitors of bacterial ribosomes (some of them are clinically relevant and have been widely used for the last fifty years); group II, specific inhibitors of eukaryotic ribosomes (most of them are very toxic and useful for cancer therapy); and group III, universal inhibitors without specificity, inhibiting both bacterial and eukaryotic ribosomes. This classification does not include archaeal ribosomes, since it is prior to the proposal of Archaea as the third domain of life.[24] However, no protein synthesis inhibitor specific for archaea has been characterized uptown.

Taking into account this taxonomic specificity of protein synthesis inhibitors, a systematic study of ribosomal sensitivity was undertaken by different groups after the proposal by Woese and collaborators that Archaea (formerly archaebacteria) were phylogenetically different to bacteria (formerly eubacteria) to gain insight in this rather controversial issue.[50]

Evolutionary Clocks and Molecular Phylogeny

Before the discovery of the molecular basis of inheritance, morphological, physiological and behavioral diversity provided the only analyzable characters for systematics.[51] However, during the last fifty years, the development of molecular biology has produced a radical change in evolutionary studies. Since a large amount of phylogenetic information is stored in the genomes of organisms, increasingly powerful genome analysis techniques have developed.[52] On the basis of the pioneering work of Zuckerkandl and Pauling,[53] who provided the first indications of a molecular clock, the concept of semantophoretic molecules (nucleic acids or proteins) to be used as molecular chronometers is now widely accepted. Those molecules measure not only evolutionary relationships but also the approximate time of divergence.[12]

As mentioned above, the advances in nucleic acid sequencing techniques, massively applied to ribosomal RNA, have converted the comparison of homologous genes into one of the most powerful molecular approaches for inferring phylogenetic history.[23] Nevertheless, controversies have arisen among evolutionary biologists with respect to several problems related to sequence analysis and its phylogenetic value. First, it is commonly accepted that there must be a close correlation between sequence divergence and time, but it is clear that functional constrains do not allow for evolutionary rates to be constant among all molecules, or even among the domains of a given molecule.[54] Some of the controversies have technical bases: the influence of the alignment procedure on the topology of the tree,[55] or the dependence on the compositional difference among sequences, mainly their G+C content.[56,57] Others are related to the incompatibility of geological data and those obtained using different molecular clocks.[58,59]

But many of the discussions are related to the discovery that not all genes lead to the same three-domain scheme,[60] an observation which is explained on the basis of lateral gene transfer,[61,62] which is currently recognized as an important driving force of prokaryotic evolution.[63] Extensive debate also followed the use of paralogous genes as a way to root the universal tree of life.[64,65] Several studies have revealed contradictions between protein trees and rRNA trees, or even among protein trees themselves. Thus, different families of paralogous genes may produce alternative roots for the universal tree and every possible grouping among the major lineages may appear.[60,66] Presently, it is clear that not all phylogenetic questions can be answered with a method as simple as sequence analysis.

Moreover, the release of complete genome sequences from over 140 organisms, which has already proven to be an extraordinary resource in the understanding of biological evolution, has fueled this debate by showing that phylogenies constructed with many universally distributed genes exhibit major differences with the rRNA-based universal tree. Combined with the need to manage the huge amount of information now available in the public databases, such differences have prompted the need to develop more integrative approaches such as genome trees based on gene content.[67-70] The resulting trees, which are not phylogenies in the classical sense since they are the outcome of a simple hierarchical classification, are nevertheless remarkably similar to the overall topology of SSU rRNA trees.

Such whole-genome assessments should be seen as important tools in the understanding of evolutionary patterns, and their use should not be despised because they are not cladograms but phenograms, which nevertheless allow us to make important inferences about our ancestral states. This approach can be extended to other phenotypic properties corresponding to universal functional structures whose genetic bases are clear and independent of environmental or growth conditions and could be good candidates for use in evolutionary studies.

Almost everyone would agree with the concept that function is the main target of selection, and that functional analysis is not a trivial task, especially given the lack of information on the tertiary and quaternary structures of complex semantophoretic macromolecules. Obviously the growing database of structural biology and the use of powerful computing analytical tools should pinpoint the structural bases of function, to identify them in the gene sequence and eventually give us insight into the rules of the evolution of function. But this type of analysis is growing slowly due to the actual crisis in bench work vocation in biological sciences (isolation and characterization of new organisms, purification of semantophoretic macromolecules, performance of structural studies). Everybody dreams of getting this type of information from the sequence, because companies using robots can produce genomic information at considerable speed (complete prokaryotic genomes in less than a week), but we are facing a severe problem in the quality of annotation, because our data banks are full of poorly edited sequences. Strict rules must be instituted regarding the nomenclature of gene products and their interactions in publications and the edition of sequences in data banks, similar to those required for the clarification of microbial taxonomy and the adequate performance of microbial culture collections. Obviously the support of microbial ecology (search for new organisms) and conventional biochemistry (functional studies) are needed to fill the gaps and improve the quality of our data banks.

Functional Phylogeny of Ribosomes

During the last decade our group has been involved in a systematic study of the phylogenetic relationships among organisms based on the functional analysis of their protein synthesis machinery.[71] We have proposed the term "functiotype" for the singular part of the phenotype that comprises basic universal cellular functions such as translation, replication, transcription or even energy yielding processes.[72,73] The advantages of the translational apparatus over other "functiotypic" systems are diverse. They are mainly based on the large amount of structural, functional and genotypic information available for the ribosomal systems. Indeed, protein synthesis is a basic and universal function performed by all known organisms, and ribosomes are assemblies of a limited number (between 50 and 90) of genetically well-characterized macromolecules. As mentioned, important components of the ribosomal particles, the rRNAs, are the most traditionally accepted molecular clocks. Moreover, the ribosomal function does not seem to be drastically dependent on environmental conditions.

The inhibitory effects produced at different antibiotic concentrations have been studied in optimized cell-free systems (to avoid interference in in vivo experiments by transport or inactivation) using poly(U) as messenger to avoid additional complication in the analysis of the results caused by varying experimental conditions like temperature or ionic strength.[71,74] Our database contains the inhibitory effects of thirty-eight antibiotics on thirty-five ribosomal systems belonging to the three domains of life: Bacteria, Archaea and Eukarya. The inhibitors were selected to represent the most important structural groups, the different functional specificities involved in protein synthesis (interaction of the ternary complex, peptidyltransferase and translocation) and their domain specificity.[11]

Functional Analysis of Archaeal Ribosomes

We first concentrated our functional analysis on archaeal ribosomes in order to clarify their functional relatedness with the bacterial/eukaryotic systems. The conclusion, after the analysis of more than twenty archaeal ribosomal systems belonging to the main groups (halophiles, methanogens and sulfur metabolizing archaea) was that archaeal ribosomes exhibit neither a bacterial nor a eukaryotic sensitivity pattern, but a mosaic of sensitivities probably related to their phylogenetic position. Some archaeal ribosomes, like those from the sulfur-metabolizing archaea, showed extreme insensitivity to all known protein synthesis inhibitors.[75] This insensitivity is not due to their high optimal temperature, since appropriate controls using extreme thermophilic bacteria of the genus *Thermus* showed a pattern of antibiotic sensitivity similar to that of the mesophilic reference bacteria *E. coli*.[74] This control strongly suggested that domain specificity rather than ecological constrains were responsible for the inhibitory patterns observed for sulfur metabolizing archaea.

Methanogens exhibited the most variable range of sensitivities, probably a reflection of their broad phylogenetic diversity.[76] The sensitivities measured for the halophilic archaea exhibited a similar mosaic pattern, although the near-saturation salt concentration in which these systems operate raised doubts about the negative inhibitory values observed for cationic aminoglycoside antibiotics.[77] Elaborate controls performed with the halotolerant bacteria *Vibrio costicola*, whose ribosomes are capable of performing protein synthesis at low and high ionic strength, suggested that competition rather than absence of binding sites is responsible for the lack of sensitivity of halophilic ribosomes to aminoglycosidic antibiotics.[78]

Phylogenetic Value of Ribosomal Functional Analysis

The antibiotic sensitivity data bank has been analyzed using different statistical methods. As a first approximation, a data matrix based on sensitivity or insensitivity (scored as 1 and 0 values when compared with bacterial and eukaryotic reference systems) was constructed.[76,79] Analysis by principal components yielded a two-dimensional distribution that showed good clustering for all the halobacteria and the sulfur-dependent archaeal ribosomes analyzed, while methanogens occupied intermediate sensitivity positions. The cluster analysis of the data revealed three main groups: haloarchaea, methanogens and sulfur-dependent archaea, which were in agreement with the results based on rRNA sequence homology. Also, the relationship between haloarchaea and methanogens was clear in this study.

A careful analysis of the inhibitory curves obtained from different ribosomal systems revealed, in many cases, a gradual change in the antibiotic sensitivity of the ribosomes, suggesting that these differences in sensitivity were the consequence of the evolution of the binding sites for these functional probes. As an example, (Fig. 1) shows the degree of sensitivity of ribosomes from different archaea to thiostrepton, an antibiotic with strong bacterial specificity (group I), and to puromycin, a universal antibiotic able to inhibit all known ribosomes (group III). In both cases it is clear that the affinity of the functional probes for their binding sites changes as a result of the structural modifications of ribosomes. Other protein synthesis inhibitors showed similar effects.[71]

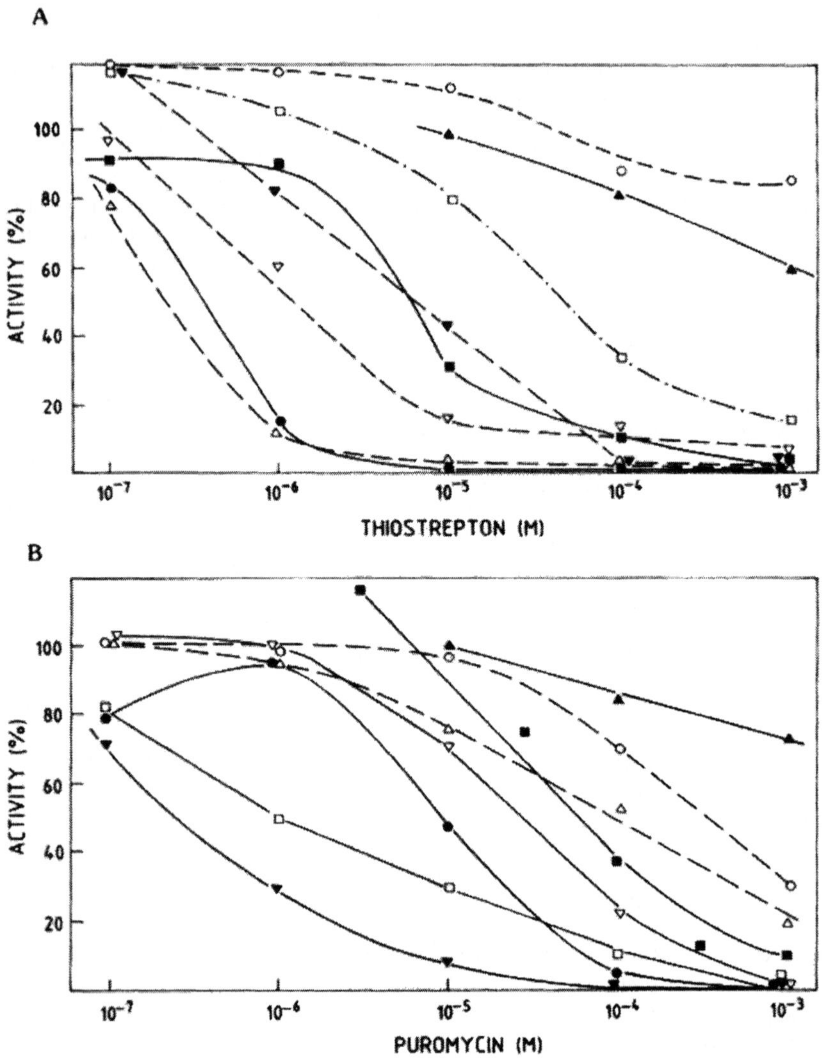

Figure 1. Sensitivity of archaeal ribosomes to the protein synthesis inhibitors thiostrepton (A) and puromycin (B). Adapted from ref. 80.

It was obvious, from the beginning of the work, that the reduction of the wealth of functional information (ribosome affinity, range of inhibition, binding cooperativity, etc.) stored in the inhibition curves to a plus or minus value, was a painful exercise of reductionism, consequence of the lack of appropriate methodologies to perform a comparative analysis of this type of data. To overcome this problem we used a quantification formula that transformed each inhibition curve into a dimensionless value representing the ribosomal sensitivity to a given antibiotic, in which the inhibition efficiencies at low antibiotic concentrations were rewarded. Quantitative antibiotic sensitivity versus ribosome matrix was used to construct a phenetic tree, which basically agreed with the phylogenetic tree obtained with total SSU rRNA sequence comparison.[80]

More recently, a new mathematical method based on fractal analysis, which extracts as much of the functional information contained in the inhibition curves as possible was employed.[73,81] The statistical procedure is based on the principal components analysis and allowed the construction of phenograms which closely resemble those obtained using SSU rRNA sequence comparison. This result underlined the phylogenetic value of functional analysis. Figure 2 shows one of the dendrograms obtained with functional analysis (left) and with the 16S/18S rRNA sequence comparison (right) for a representative number of translational systems. The overall topology of both dendrograms is very similar. The functional phenogram reveals a clear separation among the three cellular lineages and shares the typical major branching order of the clustering according to rRNA.[26,82]

Within Eukarya, although the analysis of more ribosomal systems could further clarify its internal branching order, the appearance of the ciliate *Tetrahymena thermophila* as an outgroup of the other organisms is significant. In Bacteria, a clear separation is observed between Proteobacteria and Cyanobacteria. However, the internal branching order of Proteobacteria depends on the system used. The inclusion of the cloroplast from *Spinacia oleracea* into the cyanobacterial cluster clearly indicates that the ribosomes of the chloroplast maintain the sensitivity pattern of the bacterial group from which they originated. Thus, the endosymbiotic origin of organella[83,84] is also reflected in our functional analysis.

Among Archaea, methanogens, extreme halophiles and sulfur-dependent thermophiles are classified into three separate clusters as occurs in the rRNA dendrogram. The established relation between methanogens and extreme halophiles[23,82] is not observed in the functional phenogram, although in fact our analysis does not clearly define the branching order among the three archaeal groups. Halophiles show a very similar branching topology in both dendrograms. Indeed, this is the group with the closest branching order in every (sequence or fuctional) analysis performed. This illustrates that the high sequence similarity displayed by rRNAs from extreme halophiles[85] is functionally consistent. However, within sulfur-dependent thermophiles a similar pattern does not appear in both dendrograms. This can be related to the functional particularities exhibited by hyperthermophilic archaeal ribosomes, which represent the most refractory group to protein synthesis inhibitors described so far.[75] An important reduction in the number of functional markers (from twenty eight to fifteen) does not change the overall topology of the phenogram, which underlines the robustness and redundancy of the functional data.[81]

When the analysis of the functional data bank is refined and most of the information provided by the inhibition curves is considered, the homogeneous patterns of sensitivities described for the bacterial and eukaryotic ribosomes disappear. The full spectrum of bacterial and eukaryotic diversity has not yet been explored. The possibility of finding bacterial and eukaryotic organisms with heterogeneous sensitivity patterns is illustrated by the abnormal pattern of insensitivity to aminoglycoside antibiotics exhibited by the ribosomes of the extreme thermophilic, anaerobic bacteria, *Thermotoga maritima*,[86] which according to its SSU rRNA sequence homology is close to the root of the bacterial domain.

Phylogenetic Bases of Ribosomal Functiotype

What are the reasons for the strong correlation observed between phylogenetic data based on SSU rRNA sequences (genotype) and the functional data obtained using translational inhibitors? Both systems give similar information, although the data are apparently different. A reasonable assumption is that the most important functional information in the SSU rRNA sequence is stored in specific portions of the linear sequences, namely the highly conserved unpaired regions in the rRNA secondary structure. These sequence signatures correspond, in general, to loops at the end of stretches of helical structures. Even though we recognize their importance, we still do not know much about their position in space and their interaction with other parts of the sequence. Antibiotics are small molecules in comparison to the ribosomal particle, and they act as specific functional effectors for the protein synthesis process. Using these functional-structural probes we are able to detect specific configurations of the ribosomal

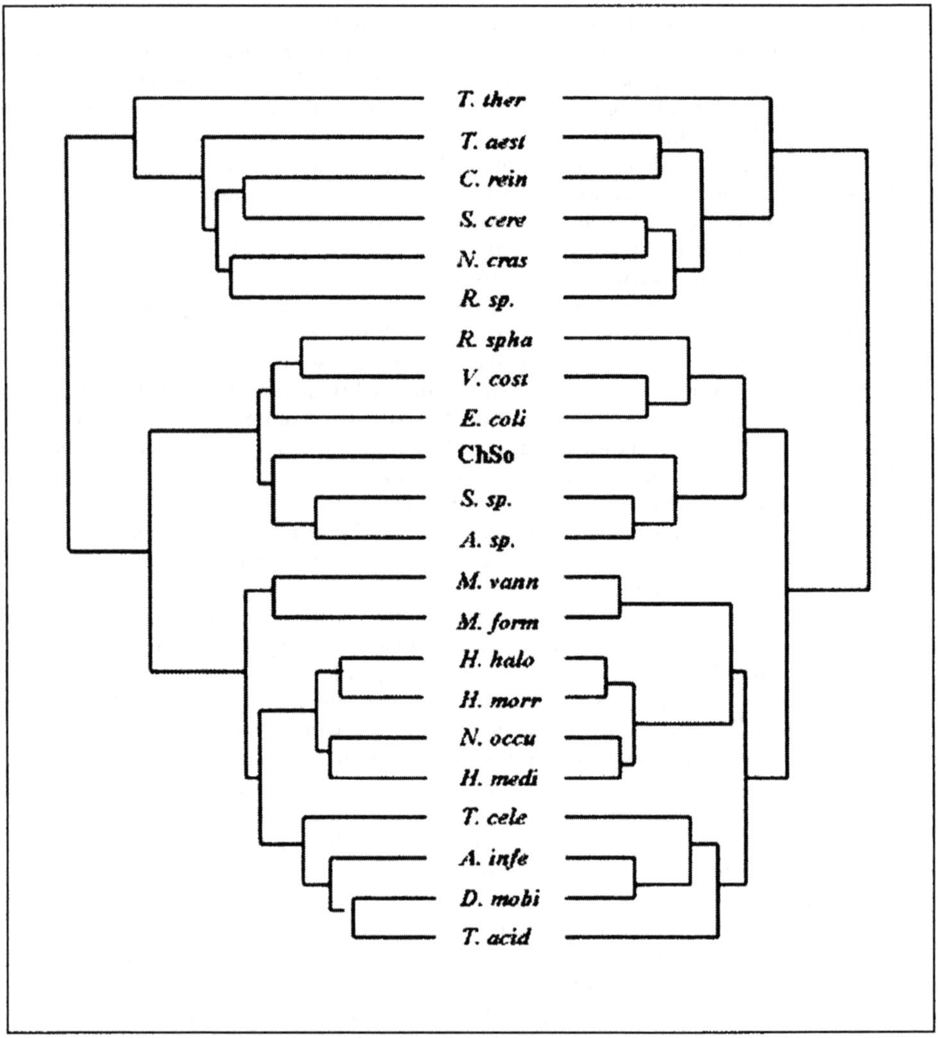

Figure 2. Dendrogram obtained with functional analysis (left) and SSU rRNA sequence comparison (right). Abbreviations of organisms: T.ther (*Tetrahymena thermophila*), T.aest (*Triticum aestivum*), C.rein (*Chlamydomonas reinhardtii*), S.cere (*Saccharomyces cerevisiae*), N.cras (*Neurospora crassa*), R.sp (*Rattus* sp.), R.spha (*Rhodobacter sphaeroides*), V.cost (*Vibrio costicola*), E.coli (*Escherichia coli*), ChSo (Chloroplast from *Spinacia oleracea*), S.sp (*Synechococcus* sp.), A.sp (*Anabaena* sp.), M.vann (*Methanococcus vannielii*), M.form (*Methanobacterium formicicum*), H.halo (*Halobacterium halobium*), H.morr (*Halococcus morrhuae*), N.occu (*Natronococcus occultus*), H.medi (*Haloferax mediterranei*), T.cele (*Thermococcus celer*), A.infe (*Acidianus infernus*), D.mobi (*Desulfurococcus mobilis*), T.acid (*Thermoplasma acidophilum*).

functional space, which in principle must be related to the structures originated from specific positions of bases in the sequence. In this respect, we must recall that in certain cases one base change is sufficient to dramatically alter the affinity of an antibiotic. This phenomena exhibits a domain consistency that reveals the evolutionary importance of this genotypic alteration.[71]

The collection of differential sensitivities exhibited by ribosomes belonging to different domains is the result of the natural selection of different mutations, which drives the evolution of the ribosomal particles. Obviously there are limitations to this type of analysis. The most important is the lack of knowledge regarding the amount of functional space footprinted by the selected antibiotics, although they must be of the same magnitude as those imposed by the comparative sequence analysis of SSU rRNA, which is a disrupted representation of the ribosomal functional space.

The existence of antibiotic-binding sites maintained in all ribosomal systems suggests that the basic components of the translational machinery have been preserved throughout evolution. In general, phylogenetically shared sensitivities should predate the branching of the three lineages. The progressive structural evolution of the ribosome would have promoted the appearance and loss of interaction sites for other effectors in different evolutionary lines. The present-day spectrum of sensitivities reflect the result of a "fine tuning" of the ribosomal function in different organisms, and therefore constitutes a record of their evolutionary history.

We still do not have a clear picture of the ribosomal prototype that gave rise to the existing ribosomal systems, although if we compare the current situation with the ideas discussed few years ago,[76] we see significant progress. Once the phylogenetic consistency of the functional inhibitory data bank is established, it can be used to generate and evaluate functional models. It would be interesting to intersect the ribosomal sequence space (genotype) with the correspondent functional space (functiotype) in order to increase the amount of information about the molecular bases of ribosomal function and its evolution. As a first step, and in collaboration with other groups, we are using comparative rRNA footprinting to generate information on specific nucleotides involved in antibiotic binding with the correspondent specific function inhibited.[87,88] However, several examples of differences in antibiotic sensitivity in ribosomes with no apparent differences in the rRNA sequence suggest that functional sites depend on cooperative interactions among rRNAs and ribosomal proteins.[89] This is an important fact in order to understand the evolution of the translational apparatus at the functional level. Recent data show that X-Ray diffraction patterns of ribosomal-antibiotic complexes can give insight into this area of research.[90]

In conclusion, during the last decade, a number of authors have clearly shown the limitations of molecular phylogeny in understanding the evolution of organisms. As they point out, it is clear that the comparison of sequences may only be used for reconstructing the evolution of genes but not of organisms, and the development of "new evolutionary paradigms where genomes, biochemistry and organisms are all considered in concert" is required.[60] We have discussed the possibilities offered by a comparative functional study of ribosomes using protein synthesis inhibitors (functiotype). The phenograms derived from this study are not only topologically similar to the SSU rRNA phylogenetic trees, but also to the genome-trees based on whole gene content. The structural-functional information provided by this type of analysis constitutes an useful tool for the study of the evolution of organisms.

References

1. Littlefield JW, Keller EB, Gros J et al. Studies on cytoplasmic ribonucleoprotein particles from the liver of the rat. J BiolChem 1955; 217:111-123.
2. Chao FC. Dissociation of macromolecular ribonucleoprotein of yeast. Arch Biochem Biophys 1957; 70:426-431.
3. Roberts RB, ed. Microsomal Particles and Protein Synthesis. New York: Pergamon Press, 1958.
4. Tissières A, Watson JD. Ribonucleoprotein particles from E. coli. Nature 1958; 182:778-780.
5. Crick F. On protein synthesis. Symp Soc Exp Biol 1958; 12:138-163.
6. Brenner S, Jacob F, Meselson F. An unstable intermediate carrying information from genes to ribosomes for protein synthesis. Nature 1961; 190:576-581.
7. Niremberg MW, Matthaei JH. The dependence of cell-free protein synthesis in E. coli upon naturally occurring or synthetic polyribonucleotides. Proc Natl Acad Sci USA 1961; 47:1588-1602.
8. Meselson M, Nomura M, Brenner SM et al. Conservation of ribosomes during bacterial growth. J Mol Biol 1964; 9:696-711.
9. Hosokawa K, Fujimura RK, Nomura M. Reconstitution of functionally active ribosomes from inactive subparticles and proteins. Proc Natl Acad Sci USA 1966; 55:198-204.

10. Gale EF, Cundliffe E, Reynolds PE et al. The Molecular basis of antibiotic action. London: Wiley & Sons, 1972.
11. Vázquez D. In: Kleinzeller A, Springer GF, Wittmann HG, eds. Inhibitors of Protein Biosynthesis. Berlin: Springer-Verlag, 1979.
12. Woese CR, Fox GE. Phylogenetic structure of the prokaryotic domain: The primary kingdoms. Proc Nat Acad Sci USA 1977; 77:5088-5090.
13. Noller HF. Structure of ribosomal RNA. Ann Rev Biochem 1984; 53:119-162.
14. Crick F. The origin of the genetic code. J Mol Biol 1968; 38:367-379.
15. Orgel LE. Evolution of the genetic apparatus. J Mol Biol 1968; 38:381-393.
16. Cech TR, Zaug AJ, Grabowsky PJ. In vitro splicing of the ribosomal RNA precursor of Tetrahymena: Involvement of a guanosine nucleotide in the excision of the intervening sequence. Cell 1981; 27:487-496.
17. Guerrier-Takada C, Gardiner K, Marsh T et al. The RNA moiety of RNAse P is the catalytic subunit of the enzime. Cell 1983; 35:849-857.
18. Wimberly BT, Brodersen DE, Clemons WM et al. Structure of the 30S ribosomal subunit. Nature 2000; 407:327-339.
19. Ban N, Nissen P, Hansen J et al. The complete atomic structure of the large ribosomal subunit at 2.4 Å resolution. Science 2000; 289:905-920.
20. Spahn CMT, Penczek PA, Leith A et al. A method for differentiating proteins from nucleic acids in intermediate-resolution density maps: Cryo-electron microscopy defines the quaternary structure of the Escherichia coli 70S ribosome. Structure 2000; 8:937-948.
21. Ingraham JL, Maaloe O, Neidhardt FC. Growth of the bacterial cell. Sunderland, Mass: Sinauer, Sinauer Associates, Inc., 1983.
22. Smit MH, van Duin J. Control of translation by mRNA secondary structure in Escherichia coli. J Mol Biol 1994; 244:144-150.
23. Woese CR. Prokaryote systematics: The evolution of a science. In: Balows A, Trüper HG, Dworkin M et al, eds. The Procaryotes. 2nd Eds. New York: Springer, 1991:3-18.
24. Woese CR. Archaebacterial and cellular origins: an overview. Zbl Bak Hyg I Abt Orig 1982; C3:1-17.
25. Woese CR. Bacterial evolution. Microbiol Rev 1987; 51:221-271.
26. Woese CR, Kandler O, Wheelis ML. Towards a natural system of organisms: Proposal for the domains Archaea. Bacteria and Eukarya. Proc Natl Acad Sci USA 1990; 87:4576-4579.
27. Turner DH, Sugimoto N. RNA structure prediction. Ann Rev Biophys Biopys Chem 1988; 17:167-192.
28. Christiansen J, Egebjerg J, Larsen N et al. Analysis of rRNA structure: experimental and theoretical considerations. In: Spedding G, ed. Ribosomes and Protein Synthesis: A Practical Approach. Oxford: Oxford Univ Press, 1990:229-252.
29. Chastain M, Tinoco I. Structural elements in RNA. Prog Nucleic Acid Res Mol Biol 1991; 41:131-177.
30. Draper DE. Ribosomal protein - RNA interactions. In: Zimmermann, RA, Dahlberg AE, eds. Ribosomal RNA: Structure, Evolution, Processing and Function in Protein Biosynthesis. Boca Ratón: CRC Press, 1996:171-197.
31. Traub P, Nomura M. Structure and function of E. coli ribosomes. V: Reconstitution of functionally active 30S ribosomal particles from RNA and proteins. Proc Natl Acad Sci USA 1968; 59:777-784.
32. Nomura M, Erdmann VA. Reconstitution of 50S ribosomal subunits from dissociated molecular components. Nature 1970; 228:744-748.
33. Nierhaus KH, Dohme F. Total reconstitution of functionally active 50S ribosomal subunits from E. coli. Proc Natl Acad Sci USA 1974; 71:4713-4717.
34. Amils R, Matthews EA, Cantor CR. An efficient in vitro total reconstitution of the Escherichia coli 50S ribosomal subunit. Nucleic Acids Res 1978; 5:2455-2470.
35. Londei P, Teixidó J, Acca M et al. Total reconstitution of active large ribosomal subunits of the thermoacidophilic archaebacterium Sulfolobus solfataricus. Nucleic Acids Res 1986; 14:2269-2285.
36. Lietzke R, Nierhaus KH. Total reconstitution of 70S ribosomes from Escherichia coli. Methods Enzymol 1988; 164:278-283.
37. Nierhaus KH. In: Spedding G, ed. Reconstitution of ribosomes, en Ribosomes and Protein Synthesis: A Practical Approach. Oxford: Oxford Univ Press, 1990:161-189.
38. Sánchez ME, Ureña D, Amils R et al. In vitro reassembly of active large ribosomal subunits of the halophilic archaebacterium Haloferax mediterranei. Biochemistry 1990; 29:9256-9261.
39. Sánchez ME, Amils R. Absolute requirement of ammonium sulfate for reconstitution of active 70S ribosomes from the extreme halophilic archaeon Haloferax mediterranei. Eur J Biochem 1995; 233:809-814.

40. Sánchez ME, Londei P, Amils R. Total reconstitution of active small ribosomal subunits of the extreme halophilic archaeon Haloferax mediterranei. Biochim Biophys Acta 1996; 1292:140-144.
41. Hill W, Dahlberg A, Garret RA et al, eds. The Ribosome: Structure, Function and Evolution. Washington: ASM.
42. Cunningham PR, Weitzmann CJ, Negre D et al. In vitro analysis of the role of rRNA in protein synthesis: site-specific mutation and methylation. In: Hill W, Dahlberg A, Garret RA et al, eds. The Ribosome: Structure, function and evolution. Washington: ASM, 1990:243-252.
43. Noller HF, Green R, Heilek G et al. Structure and function of ribosomal RNA. In: Matheson AT, Davies JE, Dennis PP et al, eds. Frontiers in Translation: An International Conference on the Structure and Function of the Ribosome. Ottawa: University of Toronto Press, 1995:997-1009.
44. Nyborg J, Kjelgaard M. Elongation in bacterial protein biosynthesis. Curr Opin Biotech 1996; 7:369-375.
45. Nissen P, Hansen J, Ban N. The structural basis of ribosome activity in peptide bond synthesis. Science 2000; 289:920-930.
46. Zimmermann RA, Thomas CL, Wower J. Structure and function of rRNA in the decoding domain and the peptidyltransferase center. In: Hill WE, Dahlberg A, Garrett RA et al, eds. The Ribosome: Structure, Function & Evolution. Washington: AMS, 1990:331-347.
47. Gray MW, Schnare MN. Evolution of rRNA gene organization. In: Zimmermann RA, Dahlberg AE, eds. Ribosomal RNA: Structure, Evolution, Processing and Function in Protein Biosynthesis. Boca Ratón: CRC Press, 1996:49-70.
48. Gale EF, Paine TF. The assimilation of aminoacids by bacteria. The action of inhibitors and antibiotics on the accumulation of free glutamic acid and the formation of combined glutamate in Staphylococcus aureus. Biochem J 1950; 48:298-301.
49. Abad JP, Amils R. The location of the streptomycin binding site explains its pleiotropic effects on protein synthesis. J Mol Biol 1994; 235:1251-1260.
50. Böck A, Kandler O. Antibiotic sensitivity of archaebacteria. In: Woese C, Wolfe RS, eds. Archaebacteria. The Bacteria: A treatise on structure and function, vol VIII. Academic Press, 1985:525-544.
51. Mayr E. The growth of Biological Thought: Diversity, Evolution, Inheritance. Cambridge: Harvard University Press, 1983.
52. Hillis D, Morits C. In: Hillis DM, Moritz C, eds. Molecular Systematics. Sunderland: Sinauer Ass., 1990:502-514.
53. Zuckerkandl E, Pauling L. Molecules as documents of evolutionary history. J Theor Biol 1965; 8:357-366.
54. Britten RJ. Rates of DNA sequence evolution differ between taxonomic groups. Science 1986; 231:1393-1398.
55. Nei M. Relative efficiencies of different tree making methods for molecular data. In: Miyamoto MM, Carcraft J, eds. Phylogenetic analysis of DNA sequences. Oxford University Press, 1991:90-128.
56. Hasegawa M, Hashimoto T. Ribosomal RNA trees misleading? Nature 1993; 361:23.
57. Steel MA, Lockhart PJ, Penny et al. Confidence in evolutionary trees from biological sequence data. Nature 1993; 364:440-442.
58. Doolittle RF, Feng DF, Tsang S et al. Determining divergence times of the major kingdoms of living organisms with a protein clock. Science 1996; 271:470-477.
59. Golding B. Evolution: When was life's first branch point? Curr Biol 1996; 6:679-682.
60. Brown JR, Doolittle WF. Archaea and the prokaryote-to-eukaryote transition. Microbiol Mol Biol Rev 1997; 61:456-502.
61. Smith MW, Feng DF, Doolittle RF. Confidence in evolutionary trees from biological sequence data. Nature 1992; 364:440-442.
62. Syvanen M. Horizontal gene trasfer: Evidence and possible consequences. Ann Rev Genet 1994; 28:237-261.
63. Ochman H, Lawrence JG, Groisman EA. Lateral gene trasfer and the nature of bacterial innovation. Nature 2000; 405:299-304.
64. Gogarten JP, KibaK H, Dittrich P et al. Evolution of the vacuolar H+-ATPase: Implications for the origin of Eukaryotes. Proc Natl Acad Sci USA 1989; 86:6661-6665.
65. Iwabe N, Kuma K, Hasegawa S et al. Evolutionary relationship of archaebacteria. Eubacteria and eukaryotes inferred from phylogenetic trees from duplicated genes. Proc Natl Acad Sci USA 1989; 86:9355-9359.
66. Forterre P, Benanchenlou-Lafha N, Labedan B. Universal tree of life. Nature 1993; 362:795.
67. Snel B, Bork P, Huynen MA. Genome phylogeny based on gene content. Nature Genetics 1999; 21:1035-1041.

68. Tekia F, Lazcano A, Dujon B. The genomic tree as revealed from whole proteome comparisons. Genome Res 1999; 9:550-557.
69. Fitz-Gibbon ST, House CH. Whole genome-based phylogenetic analysis of free-living microorganisms. Nucleic Acid Res 1999; 27:4218-422.
70. Lin J, Gerstein M. Whole-genome trees based on the occurrence of folds and orthologs: Implications for comparing genomes on different levels. Genome Res 2000; 10:808-810.
71. Amils R, Ramírez L, Sanz JL et al. Phylogeny of antibiotic action. In: Hill WE, Dahlberg A, Garret RA et al, eds. The Ribosome: Structure, Function and Evolution. Washington: AMS, 1990:645-654.
72. Sánchez ME, Amils R. The use of functional inhibitors in the study of ribosomal evolution. Microbiol 1995; SEM, 11:251-262.
73. Briones C, Koroutchev K, Amils R. Functional phylogeny: the use of the sensitivity of ribosomes to protein synthesis inhibitors as a tool to study the evolution of organisms. Orig Life Evol Biosph 1998; 28: 571-582.
74. Cammarano P, Teichner A, Londei P et al. Insensitivity of archaebacterial ribosomes to protein síntesis inhibitors. Evolutionary implications. EMBO J 1995; 4:811-816.
75. Sanz JL, Huber G, Huber H et al. Using protein synthesis inhibitors to establish the phylogenetic relationships of the Sulfolobales order. J Mol Evol 1994; 39: 528-532.
76. Amils R, Sanz JL. Inhibitors of protein synthesis as phylogenetic markers. In: Hardesty B, Kramer G, eds. Structure, Function, Genetics of Ribosomes. New York: Springer-Verlag, 1986:605-620.
77. Sanz JL, Marín I, Ureña D et al. Functional analysis of seven ribosomal systems from extremely halophilic archaea. Can J Microbiol 1992; 39:311-317.
78. Marín I, Abad J, Ureña D et al. High ionic strength interference of ribosomal inhibition produced by aminoglycoside antibiotics. Biochem 1995; 34:16519-16523.
79. Oliver JL, Sanz JL, Amils R et al. Inferring the phylogeny of Archaebacteria: the use of ribosomal sensitivity to protein-synthesis inhibitors. J Mol Evol 1987; 24:281-288.
80. Amils R, Ramírez L, Sanz JL et al. The use of functional analysis of the ribosome as a tool to determine archaebacterial phylogeny. Can J Microbiol 1989; 35:141-147.
81. Briones C, Amils R. The evolution of function: A new method to asses the phylogenetic value of ribosomal sensitivity to antibiotics. Intl Microbiol 1998; 1:301-306.
82. Pace NR. New perpective on the natural microbial world: molecular microbial ecology. ASM News 1996; 62:463-470.
83. Margulis L. A review: Genetic and evolutionary consequences of symbiosis. Exp Parasitol 1976; 39:277-349.
84. Manly BFJ. Multivariate Statistical Methods: A Primer. London: Chapman and Hall, 1986.
85. Kamekura M, Dyall-Smith ML. (1995) Taxonomy of the family Halobacteriaceae and the description of two new genera Halorubrobacterium and Natrialba. J Gen Appl Microbiol 41:330-350.
86. Londei P, Altamura S, Huber R et al. Ribosomes of the extreme thermophilic eubacterium Thermotoga maritima are uniquely insensitive to the miscoding inducing action of aminoglycodide antibiotics. J Bacteriol 1988; 170:4353-4360.
87. Rodrígez-Fonseca C, Amils R, Garret R. Fine structure of the peptydil transferase center on 23S-like rRNAs deduced from chemical probing of antibiotic-ribosome complexes. J Mol Biol 1995; 247:224-235.
88. Rodríguez-Fonseca C, Phan H, Long KS et al. Puromycin-rRNA interaction sites at the peptidyl transferase center. RNA 2000; 6:744-754.
89. Sánchez E, Teixidó J, Guerrero R et al. Hypersensitivity of Rhodobacter sphaeroides ribosomes to protein synthesis inhibitors. Can J Microbiol 1994; 40:699-704.
90. Schlünzen F, Zarivach R, Harms J et al. Structural basis for the interaction of antibiotics with the peptydil transferase centre in eubacteria. Nature 2001; 413:814-821.

Chapter 8

Aminoacyl-tRNA Synthetases as Clues to Establishment of the Genetic Code

Lluís Ribas de Pouplana and Paul Schimmel

Introduction

Aminoacyl-tRNA synthetases catalyze the specific aminoacylation of tRNAs with their cognate amino acids, thus establishing the rules of the genetic code.[1,2] The enzymes are universally distributed, and their sequences and structures reveal that the majority of them were established by the time of the divergence of archaeal and bacterial organisms, over 3000 millions years ago.[3-5]

Throughout all living organisms tRNAs are specifically recognized by their cognate aminoacyl-tRNA synthetases, which also recognize the amino acid that corresponds to the respective anticodons, as determined by the genetic code.[6,7] The aminoacylated tRNAs are bound to the amino acid through an ester bond formed between the carboxyl of the amino acid and either the 2′- or 3′-hydroxyl groups at the 3′-terminal ribose of the tRNA. This reaction takes place within a single active site domain and typically proceeds in two steps.[1] First, the amino acid (AA) is activated with ATP to form an enzyme-bound aminoacyl-adenylate (AA-AMP) with release of pyrophosphate (PPi). Next, the amino acid is transferred to the 3′-end of the tRNA to generate aminoacyl-tRNA (AA-tRNA) and AMP.

$$E + AA + ATP \Leftrightarrow E(AA\text{-}AMP) + PPi \quad (1)$$

$$E(AA\text{-}AMP) + tRNA \Leftrightarrow AA\text{-}tRNA + AMP + E \quad (2)$$

In this chapter the relationship between the evolution of aminoacyl-tRNA synthetases and the formation of the genetic code is explored. Although the origin of the code is subject to debate, most agree that its present configuration is a result of gradual growth from a simple to a complex coding system.[8-19] In particular, the 'RNA world' theory postulates that the initial code was based on RNAs that performed all functions that today are carried out by a combination of proteins, RNA and DNA.[15]

The unique capacity of RNA to store genetic information and simultaneously catalyze biochemical reactions provides a logical solution to the chicken-and-egg paradox: DNA makes proteins, and proteins make DNA. Selection studies showed that ribozymes can catalyze the aminoacylation of tRNAs as well as the formation of peptide bonds.[20-28] Consistent with an origin of the genetic code nested within the RNA world are the central roles of RNA in extant transcription and translation cellular machineries.[29]

The transition from the hypothetical RNA world could have been driven by the gradual replacement of ribozymes by catalytic proteins.[15] These molecular substitutions would have been positively selected because proteins make more efficient and stable catalysts than RNAs. However, the heritable synthesis of complex proteins must have required the establishment of

The Genetic Code and the Origin of Life, edited by Lluís Ribas de Pouplana.

a system for gene transcription and translation before the replacement of ribozymes took place. Thus, a primitive machinery for protein synthesis would have evolved within the RNA world, eventually growing in efficiency and complexity until the polypeptides that it generated managed to functionally replace the pre-existing RNAs.

Sudden replacement of these ribozymes by proteins acting on the same substrates is unlikely. More probable is the replacement of the original ribozyme by a polypeptide already physically involved in the substrate interaction. In addition, ribonucleoproteins (RNPs) may have formed as catalytic entities where the catalytic center resided in the RNA but the protein added stability. Enzymatically inactive proteins associated with catalytic RNAs can be seen in contemporary systems such as RNAseP (which processes tRNA precursors) and the ribosome.[21,29-31]

Possibly, all ancestral protein families that currently interact with RNA might be the result of this substitution process, thus requiring an early co-evolution between interacting molecules. The evolution of polypeptides in complexes with central RNA components of the genetic code machinery might thus be correlated with the process of establishment and evolution of the code. For instance, the evolution of ribosomal proteins is probably related to the evolution of rRNAs during the development of the ribosome.

The central role played by aminoacyl-tRNA synthetases in the extant translation of the genetic code begs the question of how these enzymes were incorporated into the mostly RNA-based apparatus that is central to the RNA world theory. A possible answer is that tRNA synthetases were late additions to this mechanism, appearing after the code had been established, and gradually replacing a pre-existing set of aminoacylating ribozymes that operated on essentially the same set of tRNA molecules that are utilized today. If aminoacyl-tRNA synthetases evolved after final settlement of the code, then there is little reason to expect evolutionary relationships between the code and these enzymes.

Alternatively, aminoacyl-tRNA synthetases could have co-developed with the code. Thus, the establishment of the code itself may have shaped their evolutionary history. In this scenario, features of extant tRNA synthetases should correlate with those displayed by the tRNAs and the codons in the genetic code. An extensive database of biochemical properties, sequences, and crystal structures provides the information needed to understand the relationship of tRNA synthetases to the present day genetic code.

The Two Classes of Aminoacyl-tRNA Synthetases

Aminoacyl-tRNA synthetases are classified in two distinct structural families, known as class I and class II.[32-36] Of the twenty aminoacyl-tRNA synthetases, ten are found in each class.[35] Throughout evolution, the assignment of a given enzyme to a particular class is fixed. The exception is LysRS, which is generally a class II enzyme, but is occasionally found as a class I enzyme.[37] No evidence supports a common ancestor to the two classes. All the enzymes in each class evolved from a unique single-domain protein that formed the conserved active site characteristic of the class (Fig. 1).[38,39] However, the evolution of the individual members in the two ARS classes was not independent. Within each class, detailed structural classifications have revealed the existence of three distinct subclasses.[40] The amino acids recognized by the enzymes in each subclass are chemically related. Each subclass within class I has a matching subclass in class II that recognizes similar amino acids and contains a similar number of enzymes (Fig. 1).

All the members of class I ARS share an active site domain that forms a Rossmann nucleotide-binding fold. All the members of class II have an active site domain that contains an anti-parallel β-sheet flanked by two long α-helices.[40-42] These two folds are so fundamentally different that the two classes almost certainly evolved from two distinct ancestors. Genetic dissections showed that aminoacyl-tRNA synthetases developed from ancestral catalytic cores through the addition of large domains and insertions.[43,44] Crystallographic studies expanded further on this scheme and its details.[40-42] The active site domain always recognizes the acceptor stem end of the tRNA, where the amino acid is attached. Frequently, additional domains are involved in binding other regions of the tRNA, like the anticodon stem and loop, and in providing editing functions that improve the fidelity of the code.

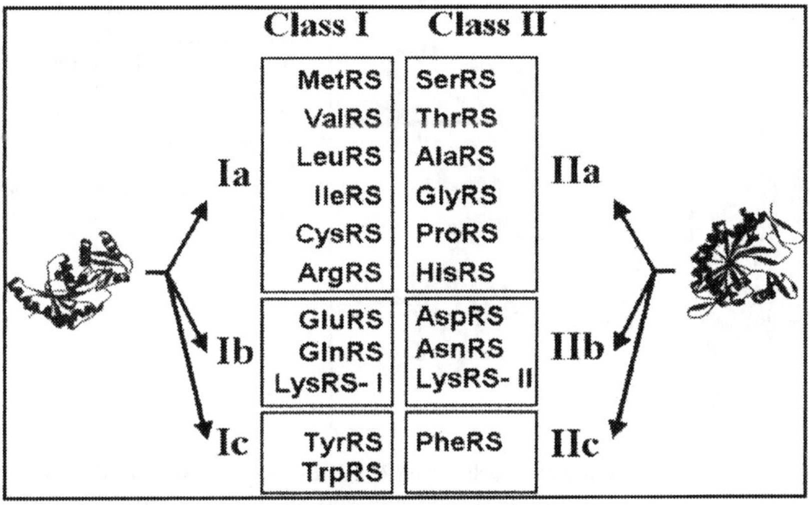

Figure 1. The two ARS classes and subclasses. Each class of enzymes evolved from a common ancestor, and the two ancestral proteins have completely different folds. Lysyl-tRNA synthetase (LysRS) varies from subclass Ia to IIa, depending on the organism (see text).

Class I ARS are subdivided into three subclasses, Ia, Ib, and Ic. Subclass Ia enzymes are specific for the amino acids leucine, isoleucine, valine, methionine, cysteine, and arginine. Subclass Ib enzymes recognize the amino acids glutamate, glutamine and lysine. Subclass Ic ARS recognize the aromatic residues tyrosine and tryptophan (Fig. 1).[40]

Generally, the active site domains of class I ARS bind the tRNA from the minor groove side of the acceptor stem (Fig. 2). This interaction relates acceptor stem sequences/structures to specific amino acids (Indeed, substrates based on just acceptor stems can be acylated by many of the synthetases in a specific fashion. These observations have led to the concept of an operational RNA code, imbedded in the acceptor stems of tRNAs.[38,45] This operational code may have preceded the genetic code.) In addition, most class I enzymes recognize the anticodon stem-loop structure of their cognate tRNAs. Interactions with the anticodon-stem loop are achieved with idiosyncratic domains that vary from enzyme to enzyme.

Several subclass Ia enzymes possess an editing activity to prevent misacylation of their cognate tRNAs. Valyl-, leucyl-, and isoleucyl-tRNA synthetases activate cognate amino acids that are difficult to discriminate from stereo-chemically similar ones.[46] In these enzymes, the hydrolysis of non-cognate aminoacyl adenylates or misacylated tRNAs is catalyzed by an independent domain.[47] This editing domain is inserted into the catalytic domain for aminoacylation, thereby creating a separate active site.[47,48]

Class II enzymes are subdivided into three subclasses, IIa, IIb, and IIc. Subclass IIa enzymes are specific for serine, threonine, glycine, alanine, proline and histidine. Subclass IIb enzymes recognize aspartate, asparagine, and lysine. Subclass IIc ARS recognize the aromatic residue phenylalanine (Fig. 1).[49]

As with class I enzymes, the active site domains of class II ARS bind to the tRNA acceptor stem but, in this case, class II enzymes approach the tRNA molecule from its major groove side (Fig. 2). Many class II enzymes also bind the anticodon stem-loop of cognate tRNAs.[49] As with class I ARS, recognition of the anticodon is achieved with additional domains that are idiosyncratic to each enzyme. In addition, glycyl-, alanyl-, prolyl-, and threonyl-tRNA synthetases (GlyRS, AlaRS, ProRS, and ThrRS) contain editing activities similar in nature to those present in class I enzymes.[50-55] In the cases of AlaRS, ThrRS, and ProRS, these activities have been localized to domains outside the active site.[51,53,55] Those editing domains are completely different than those found in class I enzymes.[54]

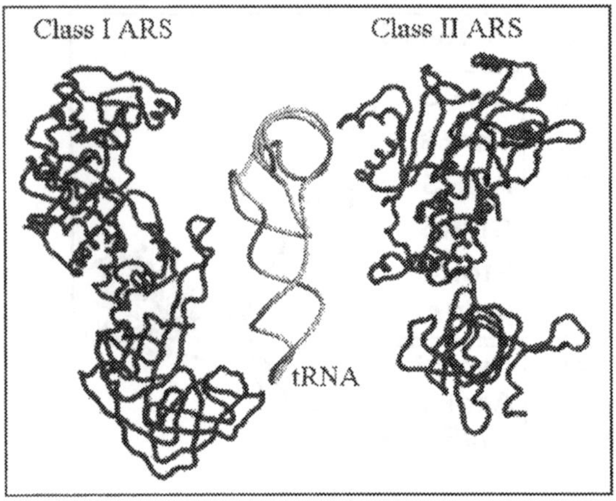

Figure 2. Structures of a class I and class II ARS with their tRNAs. The structures of glutaminyl-tRNA synthetase (class I) and aspartyl-tRNA synthetase (class II) are shown in their approximate binding orientation with respect to the tRNA molecule; [81,82] Most class I enzymes bind the tRNA from the minor groove side, while class II enzymes do it from the major groove side.

Thus, each class of enzyme approaches the tRNA from opposite sides of the acceptor stem (Fig. 2). This structural observation explains early work showing that most class I ARS attach amino acid to the 2′ OH group of the terminal ribose of tRNA, while most class II ARS attach amino acid to the 3′ OH.[56,57] Significant exceptions are synthetases that bind aromatic residues. Tyrosyl- and tryptophanyl-tRNA synthetases (TyrRS and TrpRS) are class I enzymes, but they bind the tRNA from the major groove side and indistinctly catalyze the attachment of the amino acid to the 2′ or 3′ OH of the tRNA.[56,57] On the other hand, phenylalanyl-tRNA synthetase (PheRS), a class II enzyme, binds the tRNA on the minor groove side and catalyzes the attachment of the amino acid to the 2′ OH.[56,57]

Thus, despite their independent origins, class I and class II synthetases evolved symmetrically to generate two families that display striking similarities and complementarities. This observation suggests that their evolution was driven by common constraints, which forced the symmetrical nature of the two classes. Recently, the available crystal structures of tRNA-ARS complexes have been used to generate a simple explanation for the distinguishing features of the two ARS classes.[58,59] This explanation links the evolution of the two ARS families to the development of the genetic code. Support is found in the structure of the genetic code itself and, as a result, produces a general framework within which the growth in complexity of tRNAs and codon families can be examined.[59]

Evolution of Aminoacyl-tRNA Synthetases from Phylogenetic Studies

Phylogenetic studies of synthetases can be divided between global analyses that study the overall phylogenetic relationships of all ARS and specific studies aimed at evolution of a single enzyme or of closely related ARS clusters.

That at least some synthetases were closely related and could be grouped together was established in early investigations of sequence alignments that also used the available crystal structures. These enzymes are now known as part of the class I enzymes.[32,34]

Shortly after the second class of ARS was identified,[35,36] cladistic methods were used in an attempt to infer the evolutionary picture for each class from a few representative sequences for each enzyme.[4,5] These studies took advantage of the consensus sequence motifs that identify each class to align the sequences. Despite the difficulty of obtaining a correct sequence alignment for all enzymes, these early studies succeeded (with few exceptions) in identifying the subclasses of closely related enzymes that compose each class. Importantly, these initial studies demonstrated the suitability of aminoacyl-tRNA synthetases to explore general aspects of the evolution of life. For example, Brown and colleagues showed that the universal tree of life deduced from 16S RNA sequences could be reproduced with sequences of class I aminoacyl-tRNA synthetases.[5] Phylogenetic methods were also used to study more specific aspects of ARS evolution. For example, the sequences of valyl- and alanyl-tRNA synthetases were used as phylogenetic markers to investigate early events in the evolution of the eukaryotic cell.[60,61]

The sequence of the genome of *Methanococcus jannaschii* posed several puzzling questions regarding ARS that have also been addressed with these methods. In *M. jannaschii*, no gene coding for a clear homologue of lysyl-tRNA synthetase (LysRS) could be detected.[62] Biochemical analysis of the proteome of this organism revealed that the aminoacylation of $tRNA^{Lys}$ in *M. jannaschii* is carried out by a class I enzyme (canonical LysRSs are class II).[37] This situation raised the possibility that two independent aminoacylation systems had evolved in different organisms. Alternatively, $tRNA^{Lys}$ could be a universal molecule, and the two types of LysRS would represent simply different solutions to the aminoacylation of the same tRNA.

Cladistic studies of sequences of LysRS sequences showed that these two enzymes were ancient, rooted well within their respective classes.[63] Thus, the two forms of LysRS existed (and possibly coexisted in the same organisms) early in evolution. Interestingly, phylogenetic methods demonstrated that, when in the context of every other tRNA sequences, all $tRNA^{Lys}$ form a strong monophyletic group.[63] Thus the final establishment of a given LysRS within certain species took place in the context of a pre-existing $tRNA^{Lys}$. The identity of this $tRNA^{Lys}$ preceded the emergence of its cognate synthetase.[63]

Phylogenetic methods have also been used to explore the evolutionary history of the most recent ARSs, glutaminyl- and asparaginyl-tRNA synthetases (GlnRS [subclass Ib], AsnRS [subclass IIb]), and cysteinyl-tRNA synthetase (CysRS [subclass Ia]).[64-67] In contrast to most synthetases, these three enzymes are not universal. For example, in most archaeal organisms and several bacteria, the initial aminoacylation of $tRNA^{Asn}$ and $tRNA^{Gln}$ is catalyzed by aspartyl- and glutamyl-tRNA synthetases (AspRS and GluRS), respectively.[68] The transformation of Asp-$tRNA^{Asn}$ and Glu-$tRNA^{Gln}$ into Asn-$tRNA^{Asn}$ and Gln-$tRNA^{Gln}$ is catalyzed a posteriori by a transamidase enzyme. Many species, including all studied eukaryotes, posses an AsnRS and a GlnRS that can catalyze the direct aminoacylations of $tRNA^{Asn}$ and $tRNA^{Gln}$. Both phylogenetic analyses and crystallographic studies showed that AsnRS is closely homologous to AspRS, while GlnRS is similarly related to GluRS.[64-67] Moreover, AsnRS and GlnRS clearly evolved from genes coding initially for AspRS and GluRS, after the separation of the archaeal and bacterial branches. In the case of GlnRS, the duplication and divergence of a gene coding for GluRS probably took place in the eukaryotic branch.[65] Later, the new GlnRS-coding gene was incorporated into bacterial species through lateral gene transfer.

The case of cysteinyl-tRNA synthetase (CysRS) bears some resemblance to that of LysRS. A gene coding for a canonical CysRS (subclass Ia) is missing in the *M. jannaschii* genome, as well as in other archaeal organisms.[62] Surprisingly, biochemical studies revealed that, in *M. jannaschii*, aminoacylation of $tRNA^{Cys}$ with cysteine could be catalyzed by prolyl-tRNA synthetase (ProRS, subclass IIa).[69] A dual proline-cysteine-tRNA synthetase (ProCysRS) could be the ancestral enzyme, or CysRS could have been replaced in some organisms by a gain in function of ProRS.

Recent phylogenetic analyses of sequences of ProRSs confirmed that these enzymes form two related but distinct clades.[67,70,71] Structure-based alignments showed that the clade of ProRS sequences that does not contain ProCysRS enzymes tends to collapse into a polytomy, a trend that is not seen in the clade that holds dual-activity enzymes.[71] Possibly, the emergence of a canonical class Ia CysRS allowed for a simplification and improvement of the activity of the pre-existing ProCysRS, which later lost the capacity to aminoacylate $tRNA^{Cys}$. These ProRSs were rapidly adopted through lateral transfer, thus explaining the polytomous nature of this group of sequences.[71] Other factors that might have influenced the evolution of ProCysRS, and the emergence of CysRS are the evolution of an editing activity in ProRS, and the appearance of other enzymes capable of aminocylating $tRNA^{Cys}$. Illustrating this last point, an example of a second alternative to the missing *M. jannaschii* CysRS has been recently reported.[72]

In contrast to most ARS, CysRS, AsnRS, and GlnRS evolved into their extant forms after the first split of the archaeal and bacterial branches.[65,66,69] A similar scenario has been proposed for tyrosyl- and tryptophanyl-tRNA synthetases (TyrRS and TrpRS), two homologous enzymes that constitute subclass Ic.[73] A first analysis of bacterial and eukaryal TyrRS and TrpRS sequences suggested the possibility that these two enzymes had independently emerged from a single ancestor,[73] in the bacterial and the eukaryal/archaeal branches of the 16S RNA tree. This initial hypothesis was questioned when analyses performed with a larger sequence dataset best agreed with the hypothesis that all TyrRS and all TrpRS sequences formed two separate monophyletic groups.[74] Further work found support for both hypotheses,[75,76] indicating that the final answer to the evolutionary origin of these two proteins will require the determination of the crystal structures of the prokaryotic and eukaryotic forms of both enzymes.

The realization that some aminoacyl-tRNA synthetases might have evolved after the first split of the universal phylogenetic tree has important implications for understanding the relationship between these enzymes and the genetic code. If there is a close connection between aminoacyl-tRNA synthetases and the development of the genetic code, then the amino acids recognized by these late synthetases may represent the last additions to the code. It should be noted that cysteine, tyrosine, and tryptophan have been postulated to be among the last residues to be incorporated to protein synthesis.[12]

Pairs of Subclasses and Their Significance

One explanation for the two classes of aaRS is that the two groups evolved from an ancestral complex where a single tRNA molecule was recognized simultaneously by a class I and a class II ancestor.[58] The extant subclasses would have originated from two proteins that were paired in a single complex with one tRNA. This scenario can explain several features displayed by extant ARS. For example, the equivalence in sizes of the two classes, and their subclasses, would result from coupled evolution. Thus, each event of gene duplication and divergence that generated a new tRNA species was followed by the duplication and divergence of the genes coding for the class I- and class II-type active site domains. This process would result in an equivalent numbers of class I and class II ARS. Similarly, the association of a class I and a class II ARS active site with a given tRNA can explain why the synthetases resulting from the evolution of this initial complex recognize sterically similar residues (see below).

For this hypothesis, formation of a complex between a single tRNA and two ARS must be sterically possible. In their extant forms, the association of two ARS on a single tRNA would be prevented by steric clashes (Fig. 2). However, the ARS ancestors were small proteins that contained only the active site domain.[39] To investigate the possibility that two ancestral ARS active site domains formed a complex with a single tRNA molecule, the interactions were modeled using available crystallographic data.[58]

All available crystal structures of ARS-tRNA complexes were edited to obtain the coordinates of each tRNA bound only to the respective active site domain.[48,54,77-83] The available structures cover at least one representative from each subclass (Because of close homology between enzymes of the same subclass, the mode of binding to the acceptor stem is thought to be the same for each subclass member.) The structures for all possible subclass Ia,b,c—subclass IIa,b,c pairs bound to tRNA were individually generated. The resulting structures were inspected for steric compatibility of the two bound active site domains. Not all superimpositions generated sterically compatible models. Several pairs, like that of AspRS (subclass IIb) and IleRS (subclass Ia), generated severe steric clashes because large parts of the respective active sites occupied the same three-dimensional space.[58]

Several superimpositions generated compatible pairs where two synthetases cover the tRNA acceptor stem without major steric clashes. Remarkably, these pairs link together specific ARS subclasses. In particular, the only combinations that accommodated all enzymes followed exactly a pairing of subclasses. Thus, subclass Ia enzymes (IleRS or ValRS) pair best with subclass IIa enzymes (SerRS or ThrRS). Similarly, a subclass Ib enzyme (GlnRS) forms a compatible pair with a subclass IIb enzyme (AspRS). Finally, the TyrRS complex (subclass Ic) can only form a compatible pair with PheRS (subclass IIc) (Fig. 3).[58]

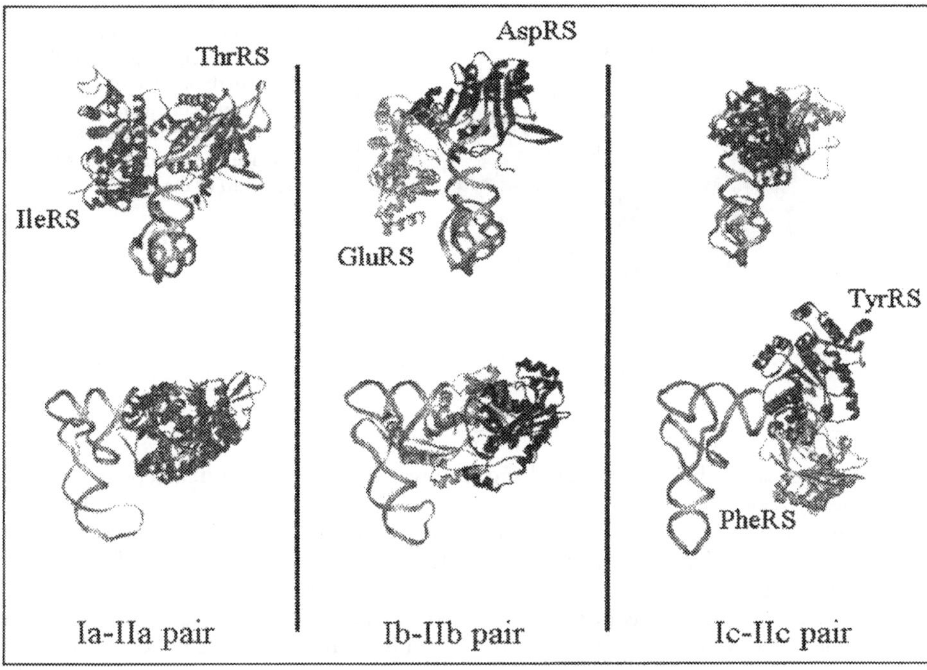

Figure 3. Pairs of ARS active sites. Depiction of the set of synthetase-pairs proposed as ancestors for the extant set of enzymes.[2] The active site domains of the proteins are depicted docked to a single tRNA molecule, and shown from two different angles.

Large translational and rotational differences between the different pairs (with respect to the axis of the tRNA acceptor stem) are an important feature of these complexes. The differences are particularly evident in the Ic-IIc pair (TyrRS and PheRS), which binds the tRNA acceptor stem at a 90° angle with respect to the other pairs. Thus, ancestral ARS pairs have large variations in their orientations around the tRNA acceptor stem (Fig. 3).[58]

This analysis supports the idea that the two extant classes of synthetases can be interpreted as a consequence of an early interaction of specific synthetase pairs in complex with tRNA (Fig. 3). These pairs may have formed to cover and protect the acceptor stem, in an environment (e.g., high temperature) where the structure of RNA was susceptible to chemical degradation or denaturation, or where the ester link between the tRNA molecule and its attached amino acid was particularly labile.

Further Support for the ARS-Pair Theory from the Editing Domains

All members of each ARS subclass emerged from a common ancestor that recognized one specific amino acid or had a loose specificity for a family of related side chains. Discrimination among similar residues was a necessary requirement for the development of an error-free protein synthesis machinery, and was achieved through two main strategies. First, differentiating between most side chains was accomplished by enzymes that developed specific active sites. Secondly, discrimination between closely similar residues was achieved by the evolution of editing domains. These editing domains proofread aminoacyl-AMP and aminoacyl-tRNA, and hydrolyze them if the amino acid is not the cognate one.

Editing has been most studied among subclass Ia enzymes, namely valyl-, leucyl-, and isoleucyl-tRNA synthetases whose editing domains share the same fold and are homologous.[48,78,84] Molecular discrimination by IleRS of isoleucine and valine, or by ValRS of valine

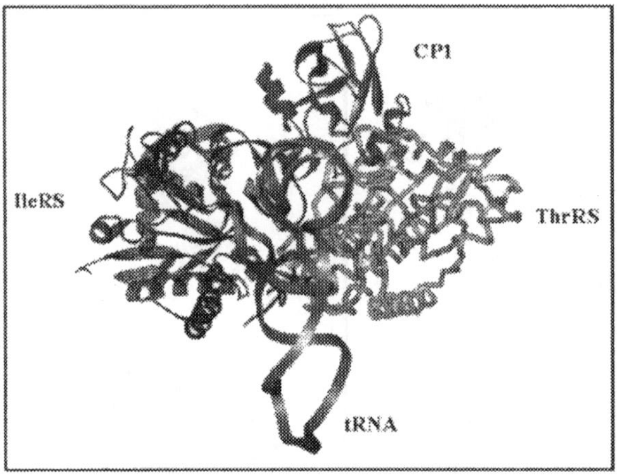

Figure 4. CP1 and Ia-IIa pair. The three-dimensional position of the CP1 domain of isoleucyl-tRNA synthetase is shown in the context of a hypothetical Ia-IIa synthetase pair.[48,77] The CP1 structure occupies the space left by the two active site domains, over the acceptor stem of the tRNA.

and threonine, cannot be easily achieved and requires editing to prevent unacceptable levels of tRNA misacylation.[85-87]

Because the three editing domains are homologous and are fused to the active site of each synthetase at the same point, they had to be added to the common ancestor of these three enzymes before posterior duplications and divergence produced the three related synthetases. Thus, the editing domain was in principle added to an ancestral synthetase during an early evolutionary time, when ARS pairs were perhaps still bound to a single tRNA. As a consequence, the spatial positioning of the editing domains of these enzymes would have to be compatible with the formation of a synthetase co-complex.

Incorporation of the editing domain into the models of Ia-IIa ARS pairs showed that the three domains are perfectly compatible, forming a molecular ring that completely surrounds the acceptor stem of the tRNA (Fig. 4) (It should be noted that the editing domain of class IIa ThrRSs produces a steric clash with its Ia partner. This domain is not universally distributed and is a relatively modern addition to ThrRS.)[54] Thus, the position of an ancient editing domain is sterically compatible with the proposed structures of the ancestral ARS pairs. A more recent editing domain does not display this compatibility, suggesting that the positioning of older domains is not a constraint imposed by the editing activity.

Evolution of the Genetic Code According to Aminoacyl-tRNA Synthetases

ARS-pairing suggests a mechanism for evolution of the genetic code. The establishment of the final code was likely achieved through duplication and mutation of tRNA genes.[8-19] If the evolution of the ARS classes was indeed coupled with duplications of tRNA genes, then the ARS pairs predicted by our theory should be correlated with the final population of tRNAs and their anticodons. A simple comparison of the distribution of codons in the code to the composition of the ARS subclasses can test this prediction. If subclasses of ARS evolved with formation of tRNAs, then paired ARS subclasses should recognize groups of tRNAs that have similar anticodons.[59]

The genetic code table shows that amino acids that are substrates of enzymes from paired subclasses have codons that cluster together (Fig. 5). The clearest case is that of subclasses Ib and IIb. The enzymes that constitute these two subclasses recognize charged amino acids—Asp, Glu, and Lys—and two derivatives, Asn and Gln. In turn, the tRNAs used by these enzymes occupy a common region of the code, and are distinguished by using the same second base (A) for their codons. Similarly, the pairing of subclasses Ic and IIc brings together enzymes that recognize aromatic residues tyrosine, tryptophan, and phenylalanine. Their cognate tRNAs share the same first base (T) of the code.[59]

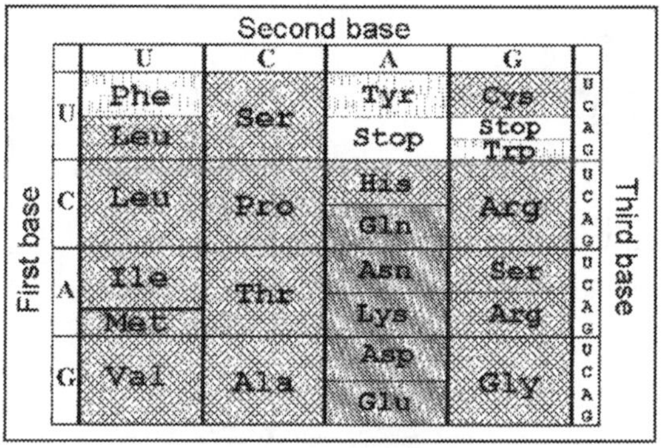

Figure 5. Genetic code and codon distribution. The distribution of codons in the genetic code table correlates well with the pairs of synthetases proposed in the previous figure. The codons that correspond to the amino acids recognized by the proposed paired synthetases are shadowed with the same pattern. A clear aggregation of codons related by our pairing scheme is apparent.

In the largest subclasses—Ia and IIa—a symmetry based on second-base composition is seen. Four of the six class Ia enzymes charge amino acids that are encoded by a U at the second position, and differentiate themselves mainly through changes of the first base (Fig. 5). Strikingly, four of the six IIa enzymes are for amino acids that are encoded using a fixed base (C) at the second position and differentiating through variations of the first base. Thus, codon assignments of 16 amino acids may have been constrained by selective pressures that operated through the subclass pairings of the associated synthetases.[59] Experimental demonstration of the possibility for aaRS pairings on tRNA acceptor stem would reinforce this conclusion.

A Model for the Emergence of Extant ARS and Establishment of the Genetic Code

Starting from the initial pairing complex, we propose that one of the two members of the pair replaced the aminoacylation activity of a pre-existing ribozyme. This replacement was preceded by the acquisition of substrate specificity (ATP, or the amino acid) by the protein. The result was a set of at least three tRNAs (corresponding to the three subclass pairs) being aminoacylated by one of the components of the bound protein pair (Fig. 6).

The evolution of a second amino acid specificity may have been driven by the appearance of a second sidechain in the milieu that was similar to the one being recognized previously. Discrimination errors started to appear, and systems that could efficiently discriminate between two similar sidechains were selected. This discrimination was achieved at first with the other member of the protein pair, thus establishing the symmetry in side chain specificity among the two classes. In some cases this discrimination resulted in the assignment to the new amino acid of part of the codon set recognized by the bound tRNA. The new assignment required duplication of the tRNA molecule. This process drove the growth of the genetic code (Fig. 6).

Members of the ARS complex were also duplicated, to give the ancestors of each ARS subclass (Fig. 4). Editing domains compatible with the synthetases pairs might have been developed at this point, to distinguish between very similar side chains (like valine and threonine.)

The introduction of two new sets of codons allowed each ternary complex (one tRNA with two synthetases) to evolve in two independent directions. Each independent complex mutated and diverged under the pressure to optimize the recognition of its specific amino acid side chain, and to improve discrimination against all related tRNAs. New domains were added to the active site domain to achieve better substrate specificity (Fig. 6). The addition of new domains that could recognize parts of the tRNA beyond the acceptor stem might have caused steric clashes that separated the ARS pairs.

A class I and a class II LysRS aminoacylation with lysine may have been incorporated by both active sites of the ancestral ternary complex. This incorporation gave rise to two distinct structures that recognized the same tRNA and had the same biochemical activity. The two

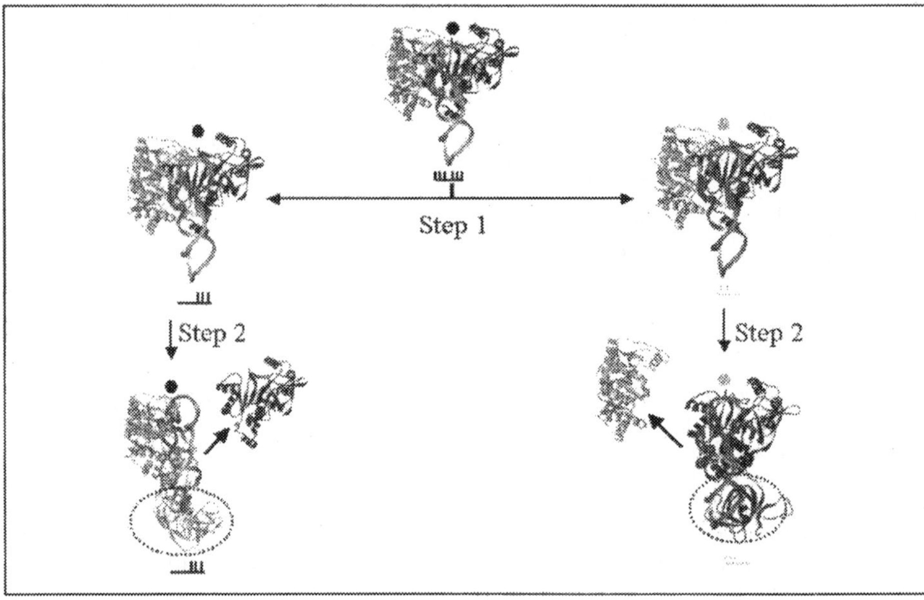

Figure 6. Evolution scheme of synthetases, tRNAs and codons. The proposed evolutionary scheme for the tRNA synthetases starts with an ancestral tRNA bound by two ARS active sites. The identity of this ancestral tRNA is unknown, but it depends on codon-anticodon pairings. The incorporation of a new amino acid to the code (step 1) requires the split of the codons, and the duplication and divergence of the genes coding for the three components of the tRNA-ARS complex. The pressure to maintain translational fidelity forces the appearance of anticodon-binding domains (step 2), and the separation of the initial ARS active site complex, to produce the extant synthetases.

paired active sites eventually separated, thus evolving into the class I and class II LysRS that can be found in extant species. In the case of the dual activity ProCysRS enzyme discussed earlier, we propose that, if cysteine was a late addition to the code, then aminoacylation with cysteine was incorporated first into the class II component of an ARS pair that already had a ProRS activity. (CysRS activity may also have been tentatively recruited by adaptation of a non-class I, non-class II architecture.)[72] This incorporation required the development of the discrimination mechanisms used by extant dual-specificity ProCysRS. Once a class I CysRS evolved, and was adopted by most organisms, the dual-specificity ProCysRS enzyme lost its capacity to aminoacylate $tRNA^{Cys}$.

The ARS-Pairs in the Context of Theories About the Origin of the Genetic Code

Two theories, namely the 'co-evolution theory' and the 'stereochemical theory', have historically been at the center of the ongoing debate about the origin of the genetic code.[12,18,19,88] More recently, the concept of a "second genetic code" being the precursor of the extant code has been advanced.[38,45,89] The existence of evidence supporting the co-evolution and the stereochemical theories can mean that both evolutionary mechanisms were involved in shaping of code.[18,19] The co-evolution theory, first postulated by Wong, proposes that the assignment of codons followed the increase in complexity of amino acid biosynthetic pathways. According to this proposal, codons assigned to each emerging amino acid originated from a pool that previously coded for metabolic precursors of the new side chain.[12] For instance, codons for methionine and isoleucine would be derived from a pool of triplets originally coding for threonine, a metabolic precursor to both methionine and isoleucine. Some aspects of this theory are related to the ARS-pairing theory, in that incorporation of new side chains would follow rules of steric similarity.

The initial proposal by Wong was based on a statistical analysis of the distribution of codons, which was seen as supporting a significant level of similarity between codons for amino acids that are linked by common biosynthetic pathways. Both the statistical basis of the theory and the physiological relevance of the biosynthetic pathways proposed by Wong have been subject to intense debate.[90-92] It seems likely that several of the relationships initially proposed by Wong are not statistically supported, or do not correspond to likely metabolic reactions. However, several of the original relationships remain suggestive of a link between the distribution of codons and the development of amino acid biosynthetic pathways in primitive organisms.

Another explanation offered for the origin of the genetic code is known as the 'stereochemical theory'.[93-95] It proposes that the amino acid-codon relationships are the result of a direct interaction between side chains and RNA structures, which would contain a large proportion of specific trinucleotide sequences that, later, became the codon sequences of the genetic code. Experimental support for this theory has been offered by the selection of RNA molecules that specifically recognize certain amino acids.[96-101] The selected RNAs contain, in their amino acid binding site, a high proportion of trinucleotide sequences that match the codon sequences for their amino acid ligand. The frequency of these trinucleotide sequences was found to be significantly higher than in other regions of the selected RNA molecules. However, other numerical analyses of the same sequences cast doubt over these conclusions.[102] A second weakness of this theory is its inability to explain the evolution of the codon-anticodon interactions of extant tRNA molecules and messenger RNAs.

The advent of the tRNA molecule is at the root of this debate. If the code was established mainly by codon-side chain interactions, then the modern tRNA molecule would be a late addition to the system. It physically separated out the initial codon-side chain pairings. However, the discovery of the potential ancestral pairings of ARS active sites suggests that the assignment of certain side chains to the code was contemporaneous with the evolution of new tRNA identities.

Perhaps the side chains that constituted the first code were assigned via the mechanism proposed by the stereochemical theory, and were the basis for the evolution of a simple translation machinery based on a smaller code. Once that limited system was in place, duplication and divergence of the tRNA-ARS complexes became the primary mechanism for new codon assignment, following steric considerations that would be determined, in part, by the development of the amino acid biosynthetic pathways.

Interestingly, a statistical analysis of tRNAs revealed a non-random distribution of sequences that forms specific tRNA pairs based on the complementarity of their anticodon sequences.[103,104] An evolutionary model for the emergence of the tRNA molecule from minihelix structures has been postulated to account for this distribution.[89,104] This idea also suggests that the anticodon-containing domain of tRNA arose from a stem-loop RNA structure (minihelix) that was a substrate for aminoacylation and was the progenitor of the extant tRNA molecule.[89] The operational RNA code (for amino acids) imbedded in tRNA acceptor stem is seen as the precursor to the genetic code.[13,34] Of the thirteen cases of complementary pairs of tRNAs that were considered in the study of complementary tRNA molecules, eight correspond to tRNAs that would be bound by an ancestral Ia-IIa ARS pair.[104] This indicates that the evolutionary model proposed for ARS might be in concordance with independent analyses of the evolution of other components of the translation apparatus. In addition, a further analysis supports the idea that at least the second base of the anticodon has its counterpart in the acceptor stem.[100]

Thus, the model of an ancestral complex between one tRNA molecule and two ARS active site domains is powerful enough to provide answers to specific characteristics of the two ARS classes, and offers a logical framework to understand the evolution of an emerging genetic code. The predictions that are derived from this model are consistent with the structure of the code, but final confirmation will require the experimental demonstration of the potential of two ARS active sites domains to recognize a tRNA acceptor stem simultaneously.

References

1. Schimmel PR, Soll D. Aminoacyl-tRNA synthetases: General features and recognition of transfer RNAs. Ann Rev Biochem 1979; 48:601-48.
2. Ribas de Pouplana L, Schimmel P. Operational RNA code for amino acids in relation to genetic code in evolution. J Biol Chem 2001; 276:6881-4.
3. Nagel GM, Doolittle RF. Evolution and relatedness in two aminoacyl-tRNA synthetase families. Proc Natl Acad Sci USA 1991; 88:8121-5.
4. Nagel GM, Doolittle RF. Phylogenetic analysis of the aminoacyl-tRNA synthetases. J Mol Evol 1995; 40:487-98.
5. Brown JR, Doolittle WF. Root of the universal tree of life based on ancient aminoacyl-tRNA synthetase gene duplications. Proc Natl Acad Sci USA 1995; 92:2441-5.
6. Giege R, Sissler M, Florentz C. Universal rules and idiosyncratic features in tRNA identity. Nuc Ac Res 1998; 26:5017-35.
7. Beuning PJ, Musier-Forsyth K. Transfer RNA recognition by aminoacyl-tRNA synthetases. Biopolymers 1999; 52:1-28.
8. Rich A. In: Kasha M, Pullman B, eds. Horizons in Biochemistry. New York: Academic Press, 1962:103-126.
9. Woese CR, Dugre DH, Kondo M et al. On the fundamental nature and evolution of the genetic code. CSH Symp Quant Biol 1966; 31:723-736.
10. Crick FH. The origin of the genetic code. J Mol Biol 1968; 38:367-79.
11. Orgel LE. Evolution of the genetic apparatus. J Mol Biol 1968; 38:381-93.
12. Wong JT. A co-evolution theory of the genetic code. Proc Natl Acad Sci USA 1975; 72:1909-12.
13. De Duve C. Blueprint for a cell: The nature and origin of life. Burlington, North Carolina: Neil Patterson, 1991.
14. Maizels N, Weiner AM. In: Gesteland RF, Atkins JF. The RNA world. New York: Cold Spring Harbor Laboratory Press, 1993.
15. Joyce GF, Orgel LE. In: Gesteland RF, Atkins JF. The RNA world. New York: Cold Spring Harbor Laboratory Press, 1993:1-26.
16. Szathmary E, Smith JM. The major evolutionary transitions. Nature 1995; 374:227-32.
17. Di Giulio, M. Reflections on the origin of the genetic code: A hypothesis. J Theor Biol 1998; 191:191-6.
18. Knight RD, Freeland SJ, Landweber LF. Selection, history and chemistry: the three faces of the genetic code. Trends Biochem Sci 1999; 24:241-7.
19. Szathmary E. The origin of the genetic code: amino acids as cofactors in an RNA world. Trends Genet 1999; 15:223-9.
20. Ellington AD, Szostak JW. In vitro selection of RNA molecules that bind specific ligands. Nature 1990; 346:818-22.
21. Guerrier-Takada C, Gardiner K, Marsh T et al. The RNA moiety of ribonuclease P is the catalytic subunit of the enzyme. Cell 1983; 35:849-57.
22. Illangasekare M, Sanchez G, Nickles T et al. Aminoacyl-RNA synthesis catalyzed by an RNA. Science 1995; 267:643-7.
23. Illangasekare M, Yarus M. A tiny RNA that catalyzes both aminoacyl-RNA and peptidyl-RNA synthesis. RNA 1999; 5:1482-9.
24. Jeffares DC, Poole AM, Penny D. Relics from the RNA world. J Molr Evol 1998; 46:18-36.
25. Saito H, Kourouklis D, Suga H. An in vitro evolved precursor tRNA with aminoacylation activity. EMBO J 2001; 20:1797-806.
26. Lee N, Suga H. A minihelix-loop RNA acts as a trans-aminoacylation catalyst. RNA 2001; 7:1043-51.
27. Saito H, Suga H. A ribozyme exclusively aminoacylates the 3'-hydroxyl group of the tRNA terminal adenosine. J Am Chem Soc 2001; 123:7178-9.
28. Saito H, Watanabe K, Suga H. Concurrent molecular recognition of the amino acid and tRNA by a ribozyme. RNA 2001; 7:1867-78.
29. Noller HF. In: Gesteland RF, Atkins JF. The RNA world. New York: Cold Spring Harbor Laboratory Press, 1993:137-156.
30. Noller HF. tRNA-rRNA interactions and peptidyl transferase. FASEB J 1993; 7:87-9.
31. Cate JH, Yusupov MM, Yusupova GZ et al. X-ray crystal structures of 70S ribosome functional complexes. Science 1999; 285:2095-104.
32. Webster T, Tsai H, Kula M et al. Specific sequence homology and three-dimensional structure of an aminoacyl transfer RNA synthetase. Science 1984; 226:1315-7.
33. Hountondji C, Dessen P, Blanquet S. Sequence similarities among the family of aminoacyl-tRNA synthetases. Biochimie 1986; 68:1071-8.

34. Ludmerer SW, Schimmel P. Gene for yeast glutamine tRNA synthetase encodes a large amino-terminal extension and provides a strong confirmation of the signature sequence for a group of the aminoacyl-tRNA synthetases. J Biol Chem 1987; 262:10801-6.
35. Eriani G, Delarue M, Poch O et al. Partition of tRNA synthetases into two classes based on mutually exclusive sets of sequence motifs. Nature 1990; 347:203-6.
36. Cusack S, Berthet-Colominas C, Hartlein M et al. A second class of synthetase structure revealed by X-ray analysis of Escherichia coli seryl-tRNA synthetase at 2.5 A [see comments]. Nature 1990; 347:249-55.
37. Ibba M et al. A euryarchaeal lysyl-tRNA synthetase: resemblance to class I synthetases. Science 1997; 278:1119-22.
38. Schimmel P, Giege R, Moras D et al. An operational RNA code for amino acids and possible relationship to genetic code. Proc Natl Acad Sci USA 1993; 90:8763-8.
39. Schimmel P, Ribas de Pouplana L. Transfer RNA: From minihelix to genetic code. Cell 1995; 81:983-6.
40. Cusack S. Aminoacyl-tRNA synthetases. Cur Op Struct Biol 1997; 7:881-9.
41. Moras D. Structural and functional relationships between aminoacyl-tRNA synthetases. Trends Biochem Sci 1992; 17:159-64.
42. Carter CW Jr. Cognition, mechanism, and evolutionary relationships in aminoacyl-tRNA synthetases. Annu Rev Biochem 1993; 62:715-48.
43. Jasin M, Regan L, Schimmel P. Modular arrangement of functional domains along the sequence of an aminoacyl tRNA synthetase. Nature 1983; 306:441-7.
44. Shiba K, Schimmel P. Functional assembly of a randomly cleaved protein. Proc Natl Acad Sci USA 1992; 89:1880-4.
45. de Duve, C. Transfer RNAs: The second genetic code. Nature 1988; 333:117-118.
46. Schimmel P, Schmidt E. Making connections: RNA-dependent amino acid recognition. Trends Biochem Sci 1995; 20:1-2.
47. Lin L, Hale SP, Schimmel P. Aminoacylation error correction [letter]. Nature 1996; 384:33-4.
48. Nureki O et al. Enzyme structure with two catalytic sites for double-sieve selection of substrate [see comments]. Science 1998; 280:578-82.
49. Cusack S. Sequence, structure and evolutionary relationships between class 2 aminoacyl-tRNA synthetases: an update. Biochimie 1993; 75:1077-81.
50. Tsui WC, Fersht AR. Probing the principles of amino acid selection using the alanyl-tRNA synthetase from Escherichia coli. Nuc Ac Res 1981; 9:4627-37.
51. Beuning PJ, Musier-Forsyth K. Species-specific differences in amino acid editing by class II prolyl-tRNA synthetase. J Biol Chem 2001; 276:30779-85.
52. Beuning PJ, Musier-Forsyth K. Hydrolytic editing by a class II aminoacyl-tRNA synthetase. Proc Natl Acad Sci USA 2000; 97:8916-20.
53. Dock-Bregeon A et al. Transfer RNA-mediated editing in threonyl-tRNA synthetase. The class II solution to the double discrimination problem. Cell 2000; 103:877-84.
54. Sankaranarayanan R et al. The structure of threonyl-tRNA synthetase-tRNA(Thr) complex enlightens its repressor activity and reveals an essential zinc ion in the active site. Cell 1999; 97:371-81.
55. Beebe K, Ribas de Pouplana L, Schimmel P. Characterization of the editing activity of alanyl-tRNA synthetase. EMBO J; In press.
56. Sprinzl M, Cramer F. Site of aminoacylation of tRNAs from Escherichia coli with respect to the 2'- or 3'-hydroxyl group of the terminal adenosine. Proc Natl Acad Sci USA 1975; 72:3049-53.
57. Fraser TH, Rich A. Amino acids are not all initially attached to the same position on transfer RNA molecules. Proc Natl Acad Sci USA 1975; 72:3044-8.
58. Ribas de Pouplana L, Schimmel P. Two classes of tRNA synthetases suggested by sterically compatible dockings on tRNA acceptor stem. Cell 2001; 104.
59. Ribas de Pouplana L, Schimmel P. Aminoacyl-tRNA synthetases: potential markers of genetic code development. Trends BiochemSci 2001; 26:591-6.
60. Hashimoto T, Sanchez LB, Shirakura T et al. Secondary absence of mitochondria in Giardia lamblia and Trichomonas vaginalis revealed by valyl-tRNA synthetase phylogeny. Proc Natl Acad Sci USA 1998; 95:6860-5.
61. Chihade JW, Brown JR, Schimmel PR et al. Origin of mitochondria in relation to evolutionary history of eukaryotic alanyl-tRNA synthetase. Proc Natl Acad Sci USA 2000; 97:12153-7.
62. Bult CJ et al. Complete genome sequence of the methanogenic archaeon, Methanococcus jannaschii. Science 1996; 273:1058-73.
63. Ribas de Pouplana L, Turner RJ, Steer BA et al. Genetic code origins: tRNAs older than their synthetases? Proc Natl Acad Sci USA 1998; 95:11295-11300.

64. Brown JR, Doolittle WF. Gene descent, duplication, and horizontal transfer in the evolution of glutamyl- and glutaminyl-tRNA synthetases. J Mol Evol 1999; 49:485-95.
65. Lamour V et al. Evolution of the Glx-tRNA synthetase family: The glutaminyl enzyme as a case of horizontal gene transfer. Proc Natl Acad Sci USA 1994; 91:8670-4.
66. Becker HD et al. Thermus thermophilus contains an eubacterial and an archaebacterial aspartyl-tRNA synthetase. Biochemistry 2000; 39:3216-30.
67. Woese CR, Olsen GJ, Ibba M et al. Aminoacyl-tRNA synthetases, the genetic code, and the evolutionary process. Microbiol Mol Biol Rev 2000; 64:202-36.
68. Ibba M, Becker HD, Stathopoulos C et al. The adaptor hypothesis revisited. Trends Biochem Sci 2000; 25:311-6.
69. Stathopoulos C et al. One polypeptide with two aminoacyl-tRNA synthetase activities. Science 2000; 287:479-82.
70. Burke B, Lipman RS, Shiba K et al. Divergent adaptation of tRNA recognition by Methanococcus jannaschii prolyl-tRNA synthetase. J Biol Chem 2001; 276:20286-91.
71. Ribas de Pouplana L, Brown JR, Schimmel P. Structure-based phylogeny of class IIa tRNA synthetases in relation to an unusual biochemistry. J Mol Evol 2001; 53:261-8.
72. Fabrega C et al. An aminoacyl tRNA synthetase whose sequence fits into neither of the two known classes. Nature 2001; 411:110-4.
73. Ribas de Pouplana L, Frugier M, Quinn CL et al. Evidence that two present-day components needed for the genetic code appeared after nucleated cells separated from eubacteria. Proc Natl Acad Sci USA 1996; 93:166-70.
74. Brown JR, Robb FT, Weiss R et al. Evidence for the early divergence of tryptophanyl- and tyrosyl-tRNA synthetases. J Mol Evol 1997; 45:9-16.
75. Diaz-lazcoz Y et al. Evolution of genes, evolution of species: The case of aminoacyl-tRNA synthetases. Mol Biol Evol 1998; 15:1548-1561.
76. Wolf YI, Aravind L, Grishin NV et al. Evolution of aminoacyl-tRNA synthetases—analysis of unique domain architectures and phylogenetic trees reveals a complex history of horizontal gene transfer events. Genome Res 1999; 9:689-710.
77. Biou V, Yaremchuk A, Tukalo M et al. The 2.9 A crystal structure of T. thermophilus seryl-tRNA synthetase complexed with tRNA(Ser). Science 1994; 263:1404-10.
78. Fukai S et al. Structural basis for double-sieve discrimination of l-valine from l-isoleucine and l-threonine by the complex of tRNA(Val) and valyl-tRNA synthetase. Cell 2000; 103:793-803.
79. Bedouelle H. Recognition of tRNA(Tyr) by tyrosyl-tRNA synthetase. Biochimie 1990; 72:589-98.
80. Silvian LF, Wang J, Steitz TA. Insights into editing from an ile-tRNA synthetase structure with tRNAile and mupirocin. Science 1999; 285:1074-7.
81. Rould MA, Perona JJ, Soll D et al. Structure of E. coli glutaminyl-tRNA synthetase complexed with tRNA(Gln) and ATP at 2.8 A resolution [see comments]. Science 1989; 246:1135-42.
82. Ruff M et al. Class II aminoacyl transfer RNA synthetases: crystal structure of yeast aspartyl-tRNA synthetase complexed with tRNA(Asp). Science 1991; 252:1682-9.
83. Goldgur Y et al. The crystal structure of phenylalanyl-tRNA synthetase from thermus thermophilus complexed with cognate tRNAPhe. Structure 1997; 5:59-68.
84. Cusack S, Yaremchuk A, Tukalo M. The 2 A crystal structure of leucyl-tRNA synthetase and its complex with a leucyl-adenylate analogue. EMBO J 2000; 19:2351-61.
85. Baldwin AN, Berg P. Purification and properties of isoleucyl ribonucleic acid synthetase from Escherichia coli. J Biol Chem 1966; 241:831-8.
86. Eldred EW, Schimmel PR. Investigation of the transfer of amino acid from a transfer ribonucleic acid synthetase-aminoacyl adenylate complex to transfer ribonucleic acid. Biochemistry 1972; 11:17-23.
87. Fersht AR, Dingwall C. Evidence for the double-sieve editing mechanism in protein synthesis. Steric exclusion of isoleucine by valyl-tRNA synthetases. Biochemistry 1979; 18:2627-31.
88. Pelc SR, Welton MG. Stereochemical relationship between coding triplets and amino-acids. Nature 1966; 209:868-70.
89. Musier-Forsyth K, Schimmel P. Atomic determinants for aminoacylation of RNA minihelices and relationship to genetic code. Acc Chem Res 1999; 32:368-375.
90. Di Giulio M, Medugno M. Physicochemical optimization in the genetic code origin as the number of codified amino acids increases. J Mol Evol 1999; 49:1-10.
91. Ronneberg TA, Landweber LF, Freeland SJ. Testing a biosynthetic theory of the genetic code: Fact or artifact? Proc Natl Acad Sci USA 2000; 97:13690-5.
92. Di Giulio M. A blind empiricism against the coevolution theory of the origin of the genetic code. J Mol Evol 2001; 53:724-32.

93. Yarus M. An RNA-amino acid complex and the origin of the genetic code. New Biol 1991; 3:183-9.
94. Szathmary E. Coding coenzyme handles: a hypothesis for the origin of the genetic code. Proc Natl Acad Sci USA 1993; 90:9916-20.
95. Yarus M. RNA-ligand chemistry: A testable source for the genetic code. RNA 2000; 6:475-84.
96. Connell GJ, Illangesekare M, Yarus M. Three small ribooligonucleotides with specific arginine sites. Biochemistry 1993; 32:5497-502.
97. Zinnen S, Yarus M. An RNA pocket for the planar aromatic side chains of phenylalanine and tryptophane. Nuc Ac Symp Ser 1995; 33:148-51.
98. Tao J, Frankel AD. Arginine-binding RNAs resembling TAR identified by in vitro selection. Biochemistry 1996; 35:2229-38.
99. Majerfeld I, Yarus M. Isoleucine: RNA sites with associated coding sequences. RNA 1998; 4:471-8.
100. Yarus M. Amino acids as RNA ligands: A direct-RNA-template theory for the code's origin. J Mol Evol 1998; 47:109-17.
101. Mannironi C, Scerch C, Fruscoloni P et al. Molecular recognition of amino acids by RNA aptamers: The evolution into an L-tyrosine binder of a dopamine-binding RNA motif. RNA 2000; 6:520-7.
102. Ellington AD, Khrapov M, Shaw CA. The scene of a frozen accident. RNA 2000; 6:485-98.
103. Rodin S, Ohno S, Rodin A. Transfer RNAs with complementary anticodons: could they reflect early evolution of discriminative genetic code adaptors? Proc Natl Acad Sci USA 1993; 90:4723-7.
104. Rodin S, Rodin A, Ohno S. The presence of codon-anticodon pairs in the acceptor stem of tRNAs. Proc Natl Acad Sci USA 1996; 93:4537-42.

Chapter 9

The Relation between Function, Structure and Evolution of Elongation Factors Tu

Mathias Sprinzl

Abstract

In each protein synthesis cycle, during which a new peptide bond is formed, EF-Tu forms a network of interactions with several different proteins and ribosomal RNAs. This fact explains the high similarity and the low evolution rate of bacterial elongation factors Tu and their eucaryotic orthologues. In order to sustain a constant rate of error-free translation, the thermodynamic parameters for interaction of aminoacyl-tRNAs with EF-Tu GTP have to be adjusted to provide the correct concentration of all different aminoacyl-tRNA EF-Tu GTP ternary complexes. As a result, the translation machinery has evolved on EF-Tu an almost universal binding site for all 20 different aminoacyl residues. The remaining differences in regard to amino acid side chain affinities are compensated by mutations in the tRNA sequence. Thus, whereas aminoacyl-tRNA synthetases evolved to discriminate between the 20 amino acids that have to be assigned to a corresponding anticodon of their specific tRNAs, the recognition of aminoacyl-tRNA by EF-Tu tRNA sequences was selected to roughly generate similar binding energies for all aminoacyl-tRNA EF-Tu GTP ternary complexes.

Introduction

Several elongation factors involved in protein synthesis are GTPases.[1] The GTPase activity is localized in the N-terminal G-domain of these proteins and is structurally and mechanistically similar to the GTPase domains of a large family of regulatory GTPases, including receptor-coupled heterotrimeric G-proteins and small GTPases of the ras family. Prokaryotic elongation factor Tu (EF-Tu), archeal elongation factor 1α and eukaryotic elongation factor 1α transport the aminoacyl-tRNA to the ribosomal decoding site (pre A site) and support the correct decoding of the A-site-located codon of the mRNA by anticodon of the aminoacyl-tRNA. Hydrolysis of GTP to GDP is coupled with this decoding process and promotes the dissociation of EF-Tu GDP from ribosomes. The GDP form of EF-Tu is activated by a nucleotide exchange factor, elongation factor Ts (EF-Ts), to the GTP form. The three-dimensional structures of several prokaryotic EF-Tus and their eukaryotic orthologues in different functional states have been determined. However, despite this detailed structural information and the recent progress in the determination of the ribosome structure, the molecular mechanism by which EF-Tu controls the GTPase and the decoding of the mRNA remains unclear.[2]

Functions of Elongation Factor Tu

Bacterial EF-Tu promotes the binding of aminoacyl-tRNAs to ribosomes. In eukaryots and archea the same function is fulfilled by elongation factor 1α. These proteins are GTPases that consecutively interact during their functional cycle with GDP, elongation factor Ts, GTP,

The Genetic Code and the Origin of Life, edited by Lluís Ribas de Pouplana.

aminoacyl-tRNA and ribosomes.[3-6] EF-Tu functions as a transporter of the aminoacyl-tRNA to a ribosomal decoding site (pre A-site) that is different from the ribosomal A-site, in which the aminoacyl-tRNA acceptor is located during the peptide transfer.[7-9] The second important role of EF-Tu is that of a timing device that determines the pace of translation and the correlation between the rate of codon recognition and the rate of peptide bond formation.[10-12] This function may be essential for maintaining the high precision of the interaction between the codon of the mRNA and the anticodon of aminoacyl-tRNA. Two concepts that contribute to the maintenance of high translational fidelity are discussed in this article:

i. The direct steric recognition of the codon/anticodon complex by ribosomes plays probably a major role in determining the precise codon reading. This reaction should be dependent entirely on the structure of tRNA anticodon and thus, be independent on the sequence and the structure of tRNA. This uniform affinity of all aminoacyl-tRNAs for the ribosomal A-site programmed by a particular codon is probably achieved by the mediator function of the EF-Tu GTP complex.
ii. Proofreading correlates the residual error frequency of the codon/anticodon interaction with the kinetic control of near cognate aminoacyl-tRNA EF-Tu GDP complex abortion from the ribosomal A-site, before an incorrect peptide bond is formed.[11-13]

The contribution of the direct steric recognition of the codon/anticodon complex and the proofreading to the maintenance of the overall fidelity of translation remains to be established.[2] As recently demonstrated, bacterial EF-Tu is also active as a chaperon to support correct folding of other protein.[14]

Mammalian elongation factors 1α have yet additional functions. Together with aminoacyl-tRNA synthetases their channel the tRNA from the ribosomal exit site, via an aminoacylation complex, to the ribosomal decoding site.[15-16] This channelling function is probably also the reason why EF 1α is associated with the multi-subunit exchange factor EF-1$\beta\gamma\delta$ during its whole functional cycle. EF 1α also interacts with the eukaryotic cytoskeleton.[17] A new function of mammalian EF1α in anchoring mRNA in cell protrusions emerged recently from the studies of interactions between elongation factor 1α and F-actin.[18-19] Unresolved is the question about the role of EF1α in the nucleocytoplasmic transport of tRNA.[20] EF1α and aminoacyl-tRNAs were found in the nucleus of mammalian cells and not only in the cytoplasm. The formation of the aminoacyl-tRNA EF1α GTP complex may be, therefore, important for the regulation of the transport of tRNAs from the nucleus to cytoplasm, and directly connected to the channelling of newly synthetised aminoacyl-tRNAs to the translating ribosomes.

It has also been proposed that the expression levels of EF-1α may affect the lifespan of *Drosophila melanogaster*.[21] However, in more recent investigations a simple correlation between the overexpression of EF-1α gene and longevity in *D. melanogaster* could not be confirmed.[22]

The Functional Cycle of EF-Tu

The functional cycle of EF-Tu resembles that of other regulatory GTPases e.g., heterotrimeric G-proteins and small GTPases of the p21 protein families.[23] The common feature of these regulatory cycles (Fig. 1) is the switch from the "inactive" GDP into the "active" GTP conformation that is promoted by a nucleotide exchange factor (NEF). Nucleotide exchange factors enhance the rate of GDP to GTP exchange on EF-Tu by catalysis of the GDP dissociation under condition of high cellular GTP concentration. The nucleotide exchange factors are usually the receptors of the extracellular signal that is transmitted. The GTP form of the G-protein interacts with the "effector", a molecule executing the signal that is transmitted. The activated "effector" finally transduces the signal to the cellular target.

GTPase activating proteins (GAP) regulate the hydrolysis of GTP to GDP by which the signal transduction is switched off. The unidirectional cycle of the GTPase is thus completed by GTP hydrolysis, bringing the protein back to its "inactive" GDP ground state. Similarly to the scheme of the functional cycle, the tertiary structures and the sequences of the nucleotide binding domains (G-domains) of all regulatory GTPases are closely related (Fig. 2). On the

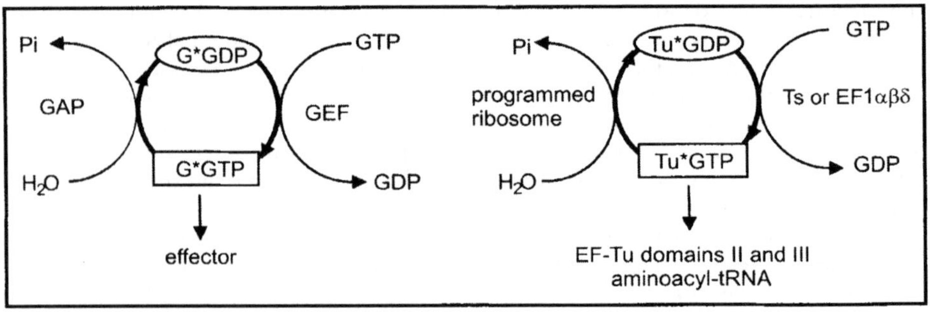

Figure 1. Generalized functional cycle of the regulatory GTPases (G-proteins) (left). The cycle is driven by a GTPase, and GDP to GTP exchange that is catalysed by the guanosine nucleotite nucleotide exchange factor (GEF). The GTPase is stimulated by a GTPase activating protein (GAP). In the active GTP-conformation the G-protein interacts with the target to which the signal is transmitted (effector). For comparison, the functional cycle of EF-Tu is shown (right). Here the function of GEF is accomplished by the elongation factor Ts (EF-Ts) and the role of GAP is fulfilled by the programmed ribosome with an A-site-bound aminoacyl-tRNA.

GTPase		P-loop	Switch I	Switch II	Nucleoside	
EF-Tu	(66)	G H D H G K	E R G I T I	D P G H A D Y	N K D	S A
EF-G	(26)	G H D G K T T	E R G I T I	D T P G H V D F	N K D	S A
IF2	(14)	G H V D H G K T	E G G I T Q	D T P G H A F	N K D	S A
RF3	(5)	S H P D A G K T T	E R G I S	D T P G H D F	N K D	
Translation factors		H D G K T	E G I	D P G H	N K D	
p21	**(36)**	G G K	Y P T	D T A G Q Y	N K D	S A
Gsα	**(72)**	G H G K T	L R R T	D V G G Q R E	F L N K D	S A

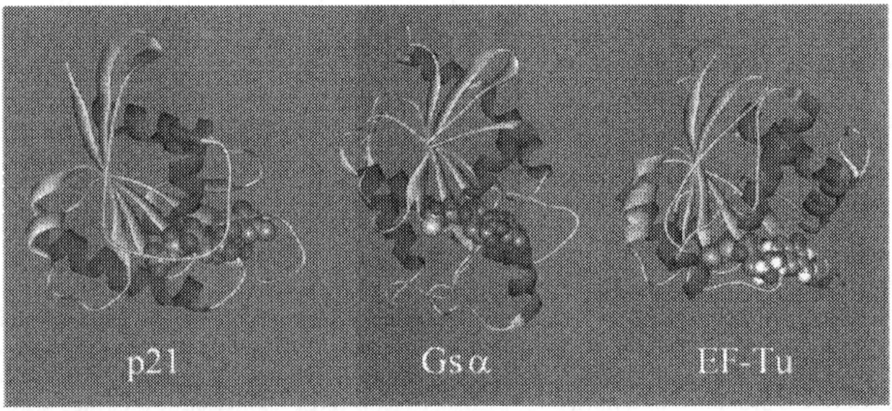

Figure 2. Different levels of sequence identity in the G-Domains of regulatory GTPases: Top) The sequence similarities. The capital letters in the boxes indicate, in one letter code, the consensus sequences of bacterial translation factors EF-Tu, EF-G, IF2 and RF3 in the nucleotide binding and switch regions. The residues common to all translation factors are then compared to consensus sequences in this region of eucaryotic p21 and Gsα proteins. The numbers in brackets correspond to the number of sequences. Bottom) The structural similarity; schematic diagrams of the tertiary structures of the G-domains of human p21 ras,[60] bovine Gsα[61] and *Thermus thermophilus* EF-Tu[24-25] in their GTP forms. The switch II regions (helix B) are pointed at by arrows. The nucleotide is shown in space-filling form.

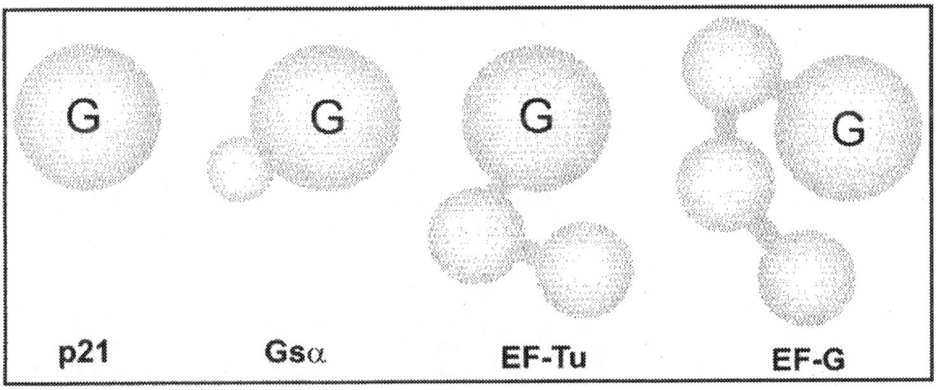

Figure 3. Schematic diagram of the domain arrangement in the structures of p21, Gsα, EF-Tu and EF-G.

other hand the molecular mechanisms of GTPase activation, nucleotide exchange, GTP hydrolysis and interaction with the particular effectors may be different for different types of regulatory GTPases. To fulfil the diverse regulatory functions, the basic structure represented by the GTP binding domain (G-domain) contains additional structural elements in form of binding sites and domains that are specific for each regulatory function (Fig. 3).

In the case of EF-Tu the function of the nucleotide exchange factor is executed by the elongation factor Ts (EF-Ts) and the function of the GTPase activating protein (GAP) is fulfilled by the codon-programmed ribosomes after binding of the aminoacyl-tRNA (Fig. 1). It is most probable that a ribosomal component, a ribosomal RNA or a ribosomal protein are involved in this GTPase activation. A signal transmission from the ribosomal decoding site to the EF-Tu GTPase centre is postulated for this process. The precise mechanism of GTPase activation of EF-Tu on the ribosome is, however, not yet understood.[2]

The Structure of EF-Tu

The tertiary structures of several bacterial, archaebacterial and eukaryotic elongation factors Tu (EF-1α) were determined as complexes with GTP,[24-26] GDP,[26] nucleotide exchange factors,[27] or as a ternary complexes with aminoacyl-tRNAs and GDPNP.[28,29] These structures provided information for understanding EF-Tu function, and underline the close structural relation between elongation factors Tu from different organisms.

EF-Tu is composed of three domains (Fig. 4). The G-domain that harbours the nucleotide-binding site consists of five β-sheets and six α-helices. Domains II and III are composed exclusively of antiparallel β-sheets forming two β-barrels. Domains II and III move as a rigid unit over a distance of about 40 Å when the GTP conformation changes to the GDP conformation (Fig. 4). This large conformational change takes place as a result of GTP hydrolysis that converts EF-Tu GTP to EF-Tu GDP, during the binding of GTP to the nucleotide-free EF-Tu EF-Ts complex. The conformation of EF-Tu in the ternary aminoacyl-tRNA EF-Tu GTP complex is essentially like in the GTP form. In the EF-Tu EF-Ts complex the EF-Tu adopts a EF-Tu GDP-like conformation.

There are two parts in the G-domain of EF-Tu that undergo a significant conformational change upon transition from GDP to GTP conformation. The "effector" loop (the name is related to the homology of EF-Tu with other GTPases) contains a short antiparallel β-sheet in the GDP form, whereas a short α-helix is formed in this region in the GTP form.[26] The switch II region is located around α-helix B (Figs. 2 and 3). This short helix changes significantly its position during the transition between the two forms of EF-Tu and provides the main new contacts between domain I and domains II and III that are necessary to stabilize the conformation essential for the interaction with aminoacyl-tRNA.

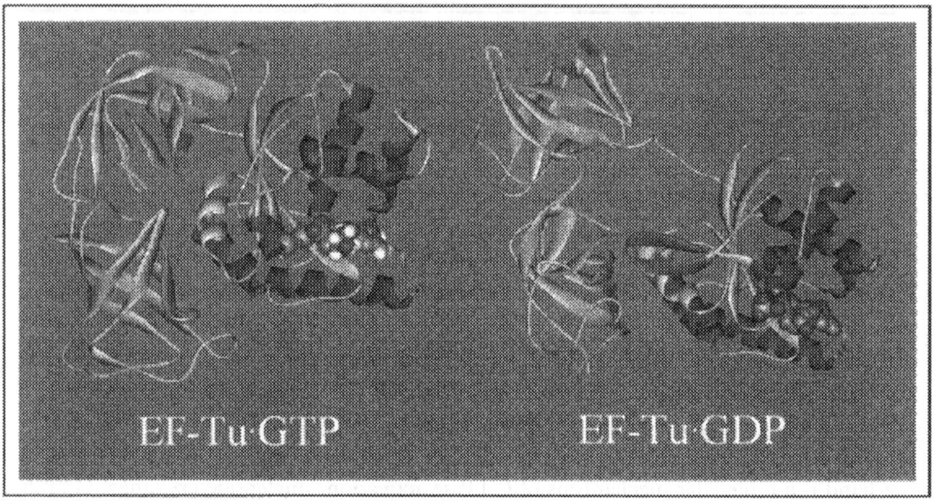

Figure 4. Comparison of the structures of *Escherichia coli* EF-Tu GDP,[26] and *Thermus thermophilus* EF-Tu GTP.[24] The nucleotide is shown in space-filling form.

X-ray structure analyses were performed also on EF-1α EF-1β complex from yeast,[30] EF-α from the archaebacterium *Sulfobolus sulphataricus*[31] and from yeast mitochondria.[32] These structures confirmed the high structural homology between EF-Tu and EF-1α. But, the structural homology goes beyond different EF-Tu species. The nucleotide binding domain of EF-Tu has the same architecture as the nucleotide binding domains of other GTPases as translation factors, p21 proteins and heterotrimeric G-proteins (Fig. 3). In all these structures the nucleotide interacts with loops of the G-domain that contain several consensus amino acid residues. These loops are the P-loop, the guanine recognition loop, and the two switch regions.

Mechanism of GTPase Activation

One of the most intriguing questions concerning the function of regulatory GTPases is how the signal induced by interaction with GAP is transmitted to the GTPase center to trigger the hydrolysis of GTP to GDP. In the case of EF-Tu, the rate of intrinsic (in the absence of ribosomes) GTPase activity is stimulated only fivefold by its binding to ribosomes in the absence of aminoacyl-tRNA, whereas the codon/anticodon interaction-dependent binding of ternary aminoacyl-tRNA EF-Tu GTP complex to ribosomes leads to a 100,000-fold increase of the intrinsic GTPase rate.[33] Thus cooperative interaction of programmed ribosomes and the aminoacyl-tRNA EF-Tu GTP complex and the correct recognition of the particular codon are essential for the GTPase stimulation by EF-Tu. Evidence is available which suggests that the effector region of EF-Tu participates in this process. Proteolytic cleavage of the effector loop does not affect the binding nor GTP or GDP, or the interaction with aminoacyl-tRNA[34,35] but abolishes the ability of ribosomes to stimulate the GTPase.[36] Thus, it seems likely that a cut in the effector region intercepts the transfer of information from the GTPase site of EF-Tu to the ribosomal decoding site.

A central feature of the GTPase mechanism in EF-Tu is the catalytic triad activating the water molecule placed in the vicinity of the γ-phosphate of GTP.[24,25] The mechanism, how the transition state of GTP hydrolysis is stabilized, is not yet understood in the case of EF-Tu. Despite the significant homologies in the GTP binding sites of different GTPases (Fig. 3) the mechanism of their GTPase reaction is not entirely uniform. It is remarkable that the amino acid residues, located in the effector loop region of EF-Tu, which seem to prevent the optimal formation of this catalytic triad, are conserved in all elongation factors Tu. This indicates the possible function of these residues in the control of GTPase.

The EF-Tu GDP and EF-Tu GTP structures in Figure 4 provide a picture of the dramatic change caused by the presence or absence of a single γ-phosphate group of GTP. This conformational change is propagated from the G-domain. The conformational switch of the helix B, identified by comparison of the p21ras GDP and p21ras GTP structures (Fig. 2) takes also place in EF-Tu. Thus the mechanism of these GTP-induced structural changes is similar in both EF-Tu and p21ras and probably universal for all nucleotide binding domains of regulatory GTPases. Helix B with its variable GTP/GDP-dependent conformations (Fig. 4) is well suited to serve as a switch in the interaction with an effector protein. Since in EF-Tu GTP the domains II/III interact with helix B, it is most likely that these domains represent in EF-Tu an "integrated effector".[3] An analogous intermolecular interaction of the effectors with heterotrimeric G-proteins and the small GTPases of the p21 family, respectively, also use the corresponding B-helices.

Interaction of EF-Tu GTP with Aminoacyl-tRNA

The essential function of EF-Tu is to transport aminoacyl-tRNAs to the ribosomomal A-site. Consequently, EF-Tu GTP should bind efficiently only aminoacyl-tRNAs and not the uncharged tRNA. There are 50 to 60 tRNA genes in the bacterial genomes coding for tRNAs specific for 20 amino acids (www.trna.uni-bayreuth.de). Depending on the particular codon, a given amino acid is attached to one or several tRNAs, e.g., there is only one gene for $tRNA^{Phe}$, $tRNA^{Cys}$ and $tRNA^{Ile}$ but four different genes for $tRNA^{Arg}$ and six for $tRNA^{Leu}$. Considering the different chemical properties of the amino acid side chains of the particular aminoacyl-tRNA, and the differences in the sequences and posttranscriptional modifications of the tRNAs, the question arises as to how is such heterogeneity compensated to provide the approximately equal affinity to EF-Tu GTP that is needed to maintain a constant rate of error-free translation.

The determination of the three-dimensional structure of Phe-$tRNA^{Phe}$ EF-Tu GTP and Cys-$tRNA^{Cys}$ EF-Tu GTP ternary complexes by X- ray crystallography provided an insight into the important structural elements that determine the specificity of interaction between aminoacyl-tRNA and EF-Tu GTP. As demonstrated in Figure 5, two types of essential interactions take place in these ternary complexes;

i. An interaction of the aminoacyl residue in the deep cleft formed in the interface between the G-domain and domain II in the GTP conformation of EF-Tu, and
ii. Interactions of basic amino acids (lysine and arginine) with the coaxial helix formed by the aminoacyl- and T-stems of the tRNA.

The results of biochemical experiments in which the aminoacyl-tRNA EF-Tu GTP ternary complexes were studied as a function of tRNA and EF-Tu modification are in good agreement with these structural results.

The K_D-values for different aminoacyl-tRNAs are in the range between 2 x 10^{-9} and 6,5 x 10^{-8} M.[37,38] The interaction of uncharged tRNA with EF-Tu GDP is measurable only at milimolar concentrations of the components.[39] Both the aminoacyl residue on the aminoacyl-tRNA and the GTP conformation of EF-Tu are therefore required for the formation of a stable complex. The dissociation constants of both aminoacyl-tRNA EF-Tu GDP and not aminoacylated tRNA EF-Tu GTP complexes are at least five orders of magnitude higher than that of the aminoacyl-tRNA EF-Tu GTP ternary complex. The thermodynamic contribution of individual ionic interactions between the polynucleotide chain and the basic amino acid side chains contribute less to binding energies than the binding of the aminoacyl-adenosine, as determined by site-directed mutagenesis of EF-Tu and modifications of aminoacyl-tRNA.[40-43] This is a surprising observation because EF-Tu cannot possess a specific binding site for the amino acid side chains, given the structural variability of the amino acid side chains.

Instead, these side chains are placed in a binding pocket that provide sufficient space, but are bare of specific interactions. Still, the affinity of the aminoacyl-adenosine residue to EF-Tu GTP is relatively high. This was demonstrated by the study of the interaction of an aminoacyl-adenosine analogue, anthraniolyl-adenosine, with EF-Tu GTP.[44] This compound, which consist of a single nucleotide (75 nucleotide residues of the aminoacyl-tRNA are miss-

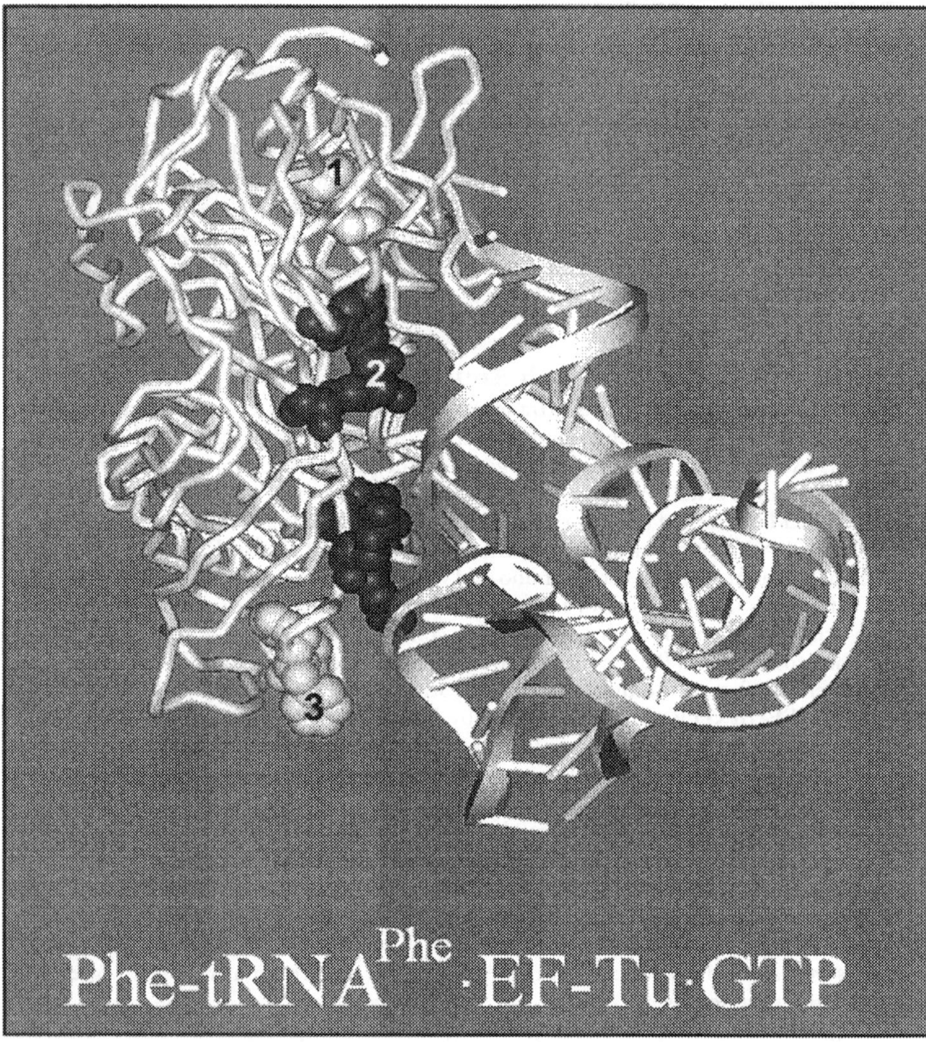

Figure 5. Structure of Phe-tRNA EF-Tu GTP[28] ternary complex depicting the amino acid binding pocket (labelled as 1), basic amino acid side chains in the interface with the "aminoacyl domain" of the tRNA (labelled as 2), and the invariant aromatic residues in the vicinity of the tRNA "elbow" (labelled as 3). The EF-Tu chain is shown as a white tube, the tRNA chain is shown as a ribbon.

ing) is still bound to the aminoacyl-adenosine binding site of EF-Tu GTP with a dissociation constant, K_d, in the micromolar range.

Why is the interaction of aminoacyl-adenosine without the possibility of a direct recognition of the amino acid side chain so efficient? The structures of Phe-tRNA EF-Tu GTP and Cys-tRNA EF-Tu GTP, which were determined by X-ray crystallography[29,45] provide a reasonable answer to this question. The aminoacylation status of the tRNA is not recognized by EF-Tu GTP by interactions with the aminoacyl side chains. Unlike in uncharged tRNA, the 3′-terminal adenine base in the aminoacyl- tRNA EF-Tu GTP ternary complex is not stacked to the neighbouring cytidine residue of the invariant CCA 3′-end, but is instead placed in a protein binding pocket, and stabilized by hydrophobic interactions. The ribose residue of the terminal aminoacyl-adenosine is in an unusual 2′-endo conformation that determines also the location of the aminoacyl residue in an empty channel in the protein structure. In addition, it allows the interaction of the 2′-OH group with an invariant glutamic acid residue. Therefore, a conformational switch of the terminal ribose from 2′-exo conformation in the uncharged tRNA to 2′- endo-conformation in the aminoacyl-tRNA seems to be the general recognition principle that is used by EF-Tu for all aminoacyl-tRNAs.[46] As seen in Figure 5, the binding site for the aminoacyl residue is connected by the CCA terminus of tRNA with the aminoacyl- and T-stem coaxial helix that shares several ionic interactions with EF-Tu GTP.[53-54] It has been demonstrated that variations in the tRNA sequences and modifications of the nucleosides in this area modulate the affinity of aminoacyl-tRNAs to EF-Tu GTP. This has important physiological role of exclusion of initiator tRNAs, misacylated tRNAs, or the tRNAs responsible for the incorporation of selenocysteine into proteins, from interaction with EF-Tu.

Evolution of EF-Tu and tRNA

Due to the central function in protein biosynthesis and its functionally essential interactions with a number of other macromolecular partners, elongation factor Tu is a slowly evolving protein. For this reason, sequences of EF-Tu are often used to root the universal tree of life.[47] However, some EF-Tu species have evolved to fulfil special functions in metabolism, like recognition of strongly truncated tRNAs in mammalian mitochondria;[48] or recognition of selenocysteinyl-tRNA.[49] The high evolutionary stability of EF-Tu and EF1α rises an interesting question about the evolution of EF-Tu into an universal protein that recognizes all chemically different aminoacyl-tRNA structures with approximately the same efficiency. The affinities of all different aminoacyl-tRNAs to EF-Tu GTP must be similar in order to provide roughly equal concentrations of all aminoacyl-tRNA EF-Tu GTP ternary complexes for translation.

Let us suppose that the space formed by the 64 triplet codons of the genetic code was gradually filled by aminoacyl-tRNAs. The distribution of the amino acids between different triplets of the genetic code and the evolutionary splitting of aminoacyl-tRNA synthetases into two classes[50-52] support the hypothesis of the gradual appearance of amino acids during the evolution of live. This means that EF-Tu GTP was also gradually confronted with a rising number of different amino acids and aminoacyl-tRNAs. As a consequence, the existence of aminoacyl-tRNAs charged with 20 different aminoacyl residues must have led to an evolution of EF-Tu possessing a binding pocket that can accommodate the side chains of all proteinogenic amino acids. Another possibility is that the differences in binding energies due to the amino acid side chains were compensated in evolution by mutations in the sequence of tRNA. Available experimental evidence indicates that both principles are used in living cells:

1. The structures of aminoacyl-tRNA EF-Tu GTP ternary complexes reveal that the binding pocket of EF-Tu for the aminoacyl residue is not designed for specific interactions with a particular side chain, and can accommodate all twenty amino acids. The discrimination between the aminoacyl-tRNA and uncharged tRNA is based on a conformational switch of the terminal adenine and ribose residue. The direct interaction of the amino acid side chain, although probably existent, has only minor significance to the binding energy.
2. The residual thermodynamic differences, due to different amino acid side chains, can be compensated by alterations in the tRNA sequence and posttranscriptional tRNA modification.

This thermodynamic compensation by sequence variations was demonstrated by the determination of the dissociation constants of correctly aminoacylated and uncharged tRNA transcripts.[53-54] This systematic study revealed that the amino acids that bind strongly into the EF-Tu binding site are associated with weak-interacting polynucleotide chains and, conversely, amino acids that are weak binders are attached to efficiently interacting tRNA chains. The modulation of the affinity of EF-Tu for aminoacyl-tRNA by sequence variations in the aminoacyl- and T-stems could be the mechanism to evolve tRNA species that fulfil a specific function as initiation of translation,[55-56] incorporation of selenocysteine,[57] and tRNA-dependent synthesis of amino acids and related metabolites.[58-59]

Acknowledgement

The work from author's laboratory was supported by Fonds der Chemischen Industrie and the Deutsche Forschungsgemeinschaft, Sp 243/-2.

References

1. Czworkowski J, Moore PB. The Elongation phase of protein synthesis. Progr Nucl Acids Res Mol Biol 1996; 54:293-333.
2. Ramakrishnan V. Ribosome structure and the mechanism of translation. Cell 2002; 108:557-572.
3. Sprinzl M. Elongation factor Tu: A regulatory GTPase with an integrated effector. Trends Biochem Sci 1994; 19:245-250.
4. Krab IM, Parmeggiani A. EF-Tu, a GTPase odyssey. Biochim Biophys Acta 1998; 1443:1-22.
5. Krab IM, Parmeggiani A. Mechanisms of EF-Tu, a pioneer GTPase. Prog Nucleic Acid Res Mol Biol 2002; 71:513-551.
6. Moazed D, Noller HF. Intermediate states in the movement of transfer RNA in the ribosome. Nature 1989; 342:142-148.
7. Stark H, Rodnina MV, Wieden HJ et al. Ribosome interactions of aminoacyl-tRNA and elongation factor Tu in the codon-recognition complex. Nat Struct Biol 2002; 9:849-854.
8. Valle M, Sengupta J, Swami NK et al. Cryo-EM reveals an active role for aminoacyl-tRNA in the accommodation process. EMBO J 2002; 21:3557-3567.
9. Noller HF, Yusupov MM, Yusupova GZ et al. Translocation of tRNA during protein synthesis. FEBS Lett 2002; 514:11-16.
10. Thompson RC. EFTu provides an internal kinetic standard for translational accuracy. Trends Biochem Sci 1988; 13:91-93.
11. Pape T, Wintermeyer W, Rodnina M. Induced fit in initial selection and proofreading of aminoacyl-tRNA on the ribosome. EMBO J 1999; 18:3800-3807.
12. Ogle JM, Brodersen DE, Clemons WMJ et al. Recognition of cognate transfer RNA by the 30S ribosomal subunit. Science 2001; 292:897-902.
13. Hopfield JJ. Origin of the genetic code: A testable hypothesis based on tRNA structure, sequence, and kinetic proofreading. Proc Natl Acad Sci USA 1978; 75:4334-4338.
14. Kudlicki W, Coffman A, Kramer G et al. Renaturation of rhodanese by translational elongation factor (EF) Tu. Protein refolding by EF-Tu flexing. J Biol Chem 1997; 272:32206-32210.
15. Negrutskii BS, Shalak VF, Kerjan P et al. Functional interaction of mammalian valyl-tRNA synthetase with elongation factor EF-1α in the complex with EF-1H. J Biol Chem 1999; 274:4545-4550.
16. Stapulionis R, Deutscher MP. A channeled tRNA cycle during mammalian protein synthesis. Proc Natl Acad Sci USA 1995; 92:7158-7161.
17. Yang F, Demma M, Warren V et al. Identification of an actin-binding protein from Dictyostelium as elongation factor 1α. Nature 1990; 347:494-496.
18. Liu G, Grant WM, Persky D et al. Interactions of elongation factor 1α with F-actin and beta-actin mRNA: Implications for anchoring mRNA in cell protrusions. Mol Biol Cell 2002; 13:579-592.
19. Edmonds BT, Wyckoff J, Yeung YG et al. Elongation factor-1α is an overexpressed actin binding protein in metastatic rat mammary adenocarcinoma. J Cell Sci 1996; 109:2705-2714.
20. Lund E, Dahlberg JE. Proofreading and aminoacylation of tRNAs before export from the nucleus. Science 1998; 282:2082-2085.
21. Shepherd JC, Walldorf U, Hug P et al. Fruit flies with additional expression of the elongation factor EF-1α live longer. Proc Natl Acad Sci USA 1989; 86:7520-7521.
22. Shikama N, Ackermann R, Brack C. Protein synthesis elongation factor EF-1α expression and longevity in Drosophila melanogaster. Proc Natl Acad Sci USA 1994; 91:4199-4203.
23. Bourne HR, Sanders DA, McCormick F. The GTPase superfamily: Conserved structure and molecular mechanism. Nature 1991; 349:117-127.
24. Berchtold H, Reshetnikova L, Reiser CO et al. Crystal structure of active elongation factor Tu reveals major domain rearrangements [published erratum appears in Nature 1993 Sep 23; 365(6444):368]. Nature 1993; 365:126-132.
25. Kjeldgaard M, Nissen P, Thirup S et al. The crystal structure of elongation factor EF-Tu from Thermus aquaticus in the GTP conformation. Structure 1993; 1:35-50.
26. Abel K, Yoder MD, Hilgenfeld R et al. An alpha to beta conformational switch in EF-Tu. Structure 1996; 4:1153-1159.
27. Kawashima T, Berthet-Colominas C, Wulff M et al. The structure of the Escherichia coli EF-Tu. EF-Ts complex at 2.5. [A resolution published erratum appears in Nature 1996 May 9; 381(6578):172]. Nature 1996; 379:511-518.

28. Nissen P, Kjeldgaard M, Thirup S et al. Crystal structure of the ternary complex of Phe-tRNAPhe, EF-Tu, and a GTP analog. Science 1995; 270:1464-1472.
29. Nissen P, Thirup S, Kjeldgaard M et al. The crystal structure of Cys-tRNACys-EF-Tu-GDPNP reveals general and specific features in the ternary complex and in tRNA. Structure Fold Des 1999; 7:143-156.
30. Andersen GR, Pedersen L, Valente L et al. Structural basis for nucleotide exchange and competition with tRNA in the yeast elongation factor complex eEF1A: eEF1Balpha. Mol Cell 2000; 6:1261-1266.
31. Vitagliano L, Masullo M, Sica F et al. The crystal structure of Sulfolobus solfataricus elongation factor 1alpha in complex with GDP reveals novel features in nucleotide binding and exchange. EMBO J 2001; 20:5305-5311.
32. Andersen GR, Thirup S, Spremulli LL et al. High resolution crystal structure of bovine mitochondrial EF-Tu in complex with GDP. J Mol Biol 2000; 297:421-436.
33. Pape T, Wintermeyer W, Rodnina MV. Complete kinetic mechanism of elongation factor Tu-dependent binding of aminoacyl-tRNA to the A site of the E. coli ribosome. EMBO J 1998; 17:7490-7497.
34. Wittinghofer A, Frank R, Leberman R. Composition and properties of trypsin-cleaved elongation factor Tu. Eur J Biochem 1980; 108:423-431.
35. Ott G, Jonak J, Abrahams IP et al. The influence of different modifications of elongation factor Tu from Escherichia coli on ternary complex formation investigated by fluorescence spectroscopy. Nucleic Acids Res 1990; 18:437-441.
36. Peter ME, Schirmer NK, Reiser CO et al. Mapping the effector region in Thermus thermophilus elongation factor Tu. Biochemistry 1990; 29:2876-2884.
37. Louie A, Ribeiro NS, Reid BR et al. Relative affinities of all Escherichia coli aminoacyl-tRNAs for elongation factor Tu-GTP. J Biol Chem 1984; 259:5010-5016.
38. Ott G, Schiesswohl M, Kiesewetter S et al. Ternary complexes of Escherichia coli aminoacyl-tRNAs with the elongation factor Tu and GTP: Thermodynamic and structural studies. Biochim Biophys Acta 1990; 1050:222-225.
39. Shulman RG, Hilbers CW, Miller DL. Nuclear magnetic resonance studies of protein-RNA interactions. J Mol Biol 1974; 90:601-607.
40. Sprinzl M, Graeser E. Role of the 5'-terminal phosphate of tRNA for its function during protein biosynthesis elongation cycle. Nucleic Acids Res 1980; 8:4737-4744.
41. Nawrot B, Sprinzl M. Aminoacyl-tRNA analogues; synthesis, purification and properties of 3'-anthraniloyl oligoribonucleotides. Nucleosides Nucleotides 1998; 17:815-829.
42. Rattenborg T, Nautrup PG, Clark BF et al. Contribution of Arg288 of Escherichia coli elongation factor Tu to translational functionality. Eur J Biochem 1997; 249:408-414.
43. Wiborg O, Andersen C, Knudsen CR et al. Mapping Escherichia coli elongation factor Tu residues involved in binding of aminoacyl-tRNA. J Biol Chem 1996; 271:20406-20411.
44. Limmer S, Vogtherr M, Nawrot B et al. Specific recognition of a minimal model of aminoacylated tRNA by the elongation factor Tu of bacterial protein biosynthesis. Angew Chem Int Ed Engl 1997; 36:2485-2489.
45. Nissen P, Kjeldgaard M, Thirup S et al. The ternary complex of aminoacylated tRNA and EF-Tu-GTP. Recognition of a bond and a fold. Biochimie 1996; 78:921-933.
46. Schlosser A, Nawrot B, Grillenbeck N et al. Fluorescence-monitored conformational change on the 3'-end of tRNA upon aminoacylation. J Biomol Struct Dyn 2001; 19:285-291.
47. Gaucher EA, Miyamoto MM, Benner SA. Function-structure analysis of proteins using covarion-based evolutionary approaches: Elongation factors. Proc Natl Acad Sci USA 2001; 98:548-552.
48. Ohtsuki T, Sato A, Watanabe Y et al. unique serine-specific elongation factor Tu found in nematode mitochondria. Nat Struct Biol 2002; 9:669-673.
49. Forchhammer K, Rucknagel KP, Bock A. Purification and biochemical characterization of SELB, a translation factor involved in selenoprotein synthesis. J Biol Chem 1990; 265:9346-9350.
50. Eriani G, Delarue M, Poch O et al. Partition of tRNA synthetases into two classes based on mutually exclusive sets of sequence motifs. Nature 1990; 347:203-206.
51. Sprinzl M, Cramer F. Site of aminoacylation of tRNAs from Escherichia coli with respect to the 2'- or 3'-hydroxyl group of the terminal adenosine. Proc Natl Acad Sci USA 1975; 72:3049-3053.
52. Fraser TH, Rich A. Amino acids are not all initially attached to the same position on transfer RNA molecules. Proc Natl Acad Sci USA 1975; 72:3044-3048.
53. LaRiviere FJ, Wolfson AD, Uhlenbeck OC. Uniform binding of aminoacyl-tRNAs to elongation factor Tu by thermodynamic compensation. Science 2001; 294:165-168.

54. Asahara H, Uhlenbeck OC. The tRNA specificity of Thermus thermophilus EF-Tu. Proc Natl Acad Sci USA 2002; 99:3499-3504.
55. Kiesewetter S, Ott G, Sprinzl M. The role of modified purine 64 in initiator/elongator discrimination of tRNA(iMet) from yeast and wheat germ. Nucleic Acids Res 1990; 18:4677-4682.
56. Drabkin HJ, Estrella M, RajBhandary UL. Initiator-elongator discrimination in vertebrate tRNAs for protein synthesis. Mol Cell Biol 1998; 18:1459-1466.
57. Rudinger J, Hillenbrandt R, Sprinzl M et al. Antideterminants present in minihelix(Sec) hinder its recognition by prokaryotic elongation factor Tu. EMBO J 1996; 15:650-657.
58. Stanzel M, Schon A, Sprinzl M. Discrimination against misacylated tRNA by chloroplast elongation factor Tu. Eur J Biochem 1994; 219:435-439.
59. Becker HD, Kern D. Thermus thermophilus: A link in evolution of the tRNA-dependent amino acid amidation pathways. Proc Natl Acad Sci USA 1998; 95:12832-12837.
60. Pai EF, Kabsch W, Krengel U et al. Structure of the guanine-nucleotide-binding domain of the Ha-ras oncogene product p21 in the triphosphate conformation. Nature 1989; 341:209-214.
61. Sunahara RK, Tesmer JJG, Gilman AG et al. Crystal structure of the adenylyl cyclase activator $G_s\alpha$. Science 1997; 278:1943-1947.

Chapter 10

Origin and Evolution of DNA and DNA Replication Machineries

Patrick Forterre, Jonathan Filée and Hannu Myllykallio

Summary

The transition from the RNA to the DNA world was a major event in the history of life. The invention of DNA required the appearance of enzymatic activities for both synthesis of DNA precursors, retro-transcription of RNA templates and replication of single- and double-stranded DNA molecules. Recent data from comparative genomics, structural biology and traditional biochemistry have revealed that several of these enzymatic activities have been invented independently more than once, indicating that the transition from RNA to DNA genomes was more complex than previously thought. The distribution of the different protein families corresponding to these activities in the three domains of life (*Archaea, Eukarya,* and *Bacteria*) is puzzling. In many cases, *Archaea* and *Eukarya* contain the same version of these proteins, whereas *Bacteria* contain another version. However, in other cases, such as thymidylate synthases or type II DNA topoisomerases, the phylogenetic distributions of these proteins do not follow this simple pattern. Several hypotheses have been proposed to explain these observations, including independent invention of DNA and DNA replication proteins, ancient gene transfer and gene loss, and/or nonorthologous replacement. We review all of them here, with more emphasis on recent proposals suggesting that viruses have played a major role in the origin and evolution of the DNA replication proteins and possibly of DNA itself.

Introduction

All cellular organisms have double-stranded DNA genomes. The origin of DNA and DNA replication mechanisms is thus a critical question for our understanding of early life evolution. For some time, it was believed by some molecular biologist that life originated with the appearance of the first DNA molecule![1] Watson and Crick even suggested that DNA was possibly replicated without proteins, wondering "*whether a special enzyme would be required to carry out the polymerization or whether the existing single helical chain could act effectively as an enzyme*".[2] Such extreme conception was in line with the idea that DNA was the aperiodic crystal predicted by Schroedinger in his influential book "*What's life*".[3] Times have changed, and several decades of experimental work have convinced us that DNA synthesis and replication actually require a plethora of proteins.[4] We are reasonably sure now that DNA and DNA replication mechanisms appeared late in early life history, and that DNA originated from RNA in an RNA/protein world. The origin and evolution of DNA replication mechanisms thus occurred at a critical period of life evolution that encompasses the late RNA world and the emergence of the Last Universal Cellular Ancestor (LUCA) to the present three domains of life (*Eukarya, Bacteria* and *Archaea.*).[5-7] It is an exciting time to learn through comparative genomics and

The Genetic Code and the Origin of Life, edited by Lluís Ribas de Pouplana.

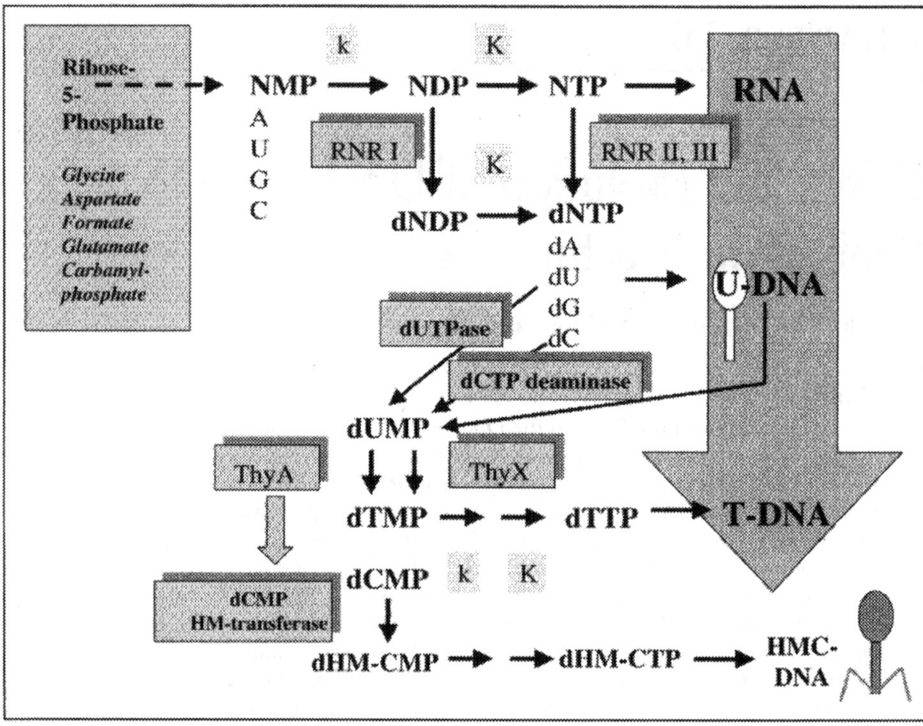

Figure 1. Metabolic pathways for RNA and DNA precursors biosynthesis: a palimpsest from the RNA to DNA world transition? The biosynthetic pathways for purine and pyrimidine nucleotides both start with ribose 5-monophosphate. The formation of the four bases requires several amino-acids, formate and carbamyl-phosphate. Nucleotide monophosphates (NMP) are converted into RNA precursors (NTP) by NMP kinases (k) and NDP kinases (K). These reactions probably are relics of the RNA-protein world. DNA precursors are produced from NDP and/or NTP by ribonucleotide reductases (RNR), except for dTTP, which results from methylation of dUMP. dTMP is produced from dUMP by Thymidylate synthases (ThyA or ThyX) and converted into dTTP by the same kinases that convert NMP into NTP. dUMP can be produced either by dUTPAse or by dCTP deaminase. In the U-DNA world, it could have been also produced by degradation of U-DNA. The mode of dTMP production clearly suggests that U-DNA was an evolutionary intermediate between RNA and T-DNA. Some viruses contain U-DNA, whereas others contain HMC-DNA (HMC= hydroxymethyl-cytosine). Transformation of C into HMC occurs at the level of dCMP, and conversion of dCMP into dHMCMP is catalyzed by a dCMP hydroxy-methyl transferase (dCMP HM transferase), which is homologue to ThyA (See refs. 11, 14, and 19 for more details).

molecular biology about the details of modern mechanisms for precursor DNA synthesis and DNA replication, in order to trace their histories.

Origin of DNA

DNA can be considered as a modified form of RNA, since the "normal" ribose sugar in RNA is reduced into deoxyribose in DNA, whereas the "simple" base uracil is methylated into thymidine. In modern cells, the DNA precursors (the four deoxyribonucleoties, dNTPs) are produced by reduction of ribonucleotides di- or triphosphate by ribonucleotide reductases (Fig. 1). The synthesis of DNA building blocks from RNA precursors is a major argument in favor of RNA preceding DNA in evolution. The direct prebiotic origin of is theoretically plausible (from acetaldehyde and glyceraldehyde-5-phosphate) but highly unlikely, considering that evolution, as stated by F. Jacob, works like a tinkerer, not an engineer.[8,9]

The first step in the emergence of DNA has been most likely the formation of U-DNA (DNA containing uracil), since ribonucleotide reductases produce dUTP (or dUDP) from UTP (or UDP) and not dTTP from TTP (the latter does not exist in the cell) (Fig. 1). Some modern viruses indeed have a U-DNA genome,[10] possibly reflecting this first transition step between the RNA and DNA worlds. The selection of the letter T occurred probably in a second step, dTTP being produced in modern cells by the modification of dUMP into dTMP by thymidylate synthases (followed by phosphorylation).[11] Interestingly, the same kinase can phosphorylate both dUMP and dTMP.[11] In modern cells, dUMP is produced from dUTP by dUTPases, or from dCMP by dCMP deaminases (Fig. 1).[11] This is another indication that T-DNA originated after U-DNA. In ancient U-DNA cells, dUMP might have been also produced by degradation of U-DNA (Fig. 1).

The origin of DNA also required the appearance of enzymes able to incorporate dNTPs using first RNA templates (reverse transcriptases) and later on DNA templates (DNA polymerases). In all living organisms (cells and viruses), all these enzymes work in the 5' to 3' direction. This directionality is dictated by the cellular metabolism that produces only dNTP 5' triphosphates and no 3' triphosphates. Indeed, both purine and pyrimidine biosyntheses are built up on ribose 5 monophosphate as a common precursor. The sense of DNA synthesis itself is therefore a relic of the RNA world metabolism. Modern DNA polymerases of the A and B families, reverse transcriptases, cellular RNA polymerases and viral replicative RNA polymerases are structurally related and thus probably homologous (for references, see a recent review on viral RNA-dependent RNA polymerases.)[12] This suggests that reverse transcriptase and DNA polymerases of the A and B families originated from an ancestral RNA polymerase that has also descendants among viral-like RNA replicases. However, there are several other DNA polymerase families (C, D, X, Y) whose origin is obscure (we will go back to this point below).

If DNA actually appeared in the RNA world, it was *a priori* possible to imagine that formation of the four dNTPs from the four rNTPs was initially performed by ribozymes. Most scientists, who consider that the reduction of ribose cannot be accomplished by an RNA enzyme, now reject this hypothesis.[9,13-19] The removal of the 2' oxygen in the ribose involves indeed a complex chemistry for reduction that requires the formation of stable radicals in ribonucleotide reductases. Such radicals would have destroyed the RNA backbone of a ribozyme by attacking the labile phosphodiester bond of RNA. Accordingly, DNA could have only originated after the invention of modern complex proteins, in an already elaborated protein/DNA world. This suggests that RNA polymerases were indeed available at that time to evolve into DNA polymerases (as well as kinases to phosphorylate dUMP).

Three classes of ribonucleotide reductases (I, II and III) have been discovered so far (for a review, see refs. 9, 16-19) (Fig. 1). Although they correspond to three distinct protein families, with different cofactors and mechanisms of action, these mechanisms are articulated around a common theme (radical based chemistry). In all cases, the critical step is the conversion of a cysteine residue into a catalytically essential thiol radical in the active center.[18] Recent structural and mechanistic analyses of several RNR at atomic resolution have suggested that all ribonucleotide reductases originated from a common ancestral enzyme, favoring the idea that U-DNA was invented only once.[17,18] It has been suggested that either class III (strictly anaerobic) or class II (anaerobic but oxygen tolerant) represent the ancestral form, and that new versions appeared in relation to different lifestyles by recruiting new mechanisms for radical activation (class III in strict anaerobes and class I in aerobes).[9,18]

The origin of U-DNA in a protein/RNA world logically implies that the second step in the synthesis of DNA precursors, the formation of the letter T, was catalyzed by ancestral thymidylate synthase. For a long time, it was believed that modern thymidylate synthases were all homologues of *E. coli* ThyA protein, indicating that the letter T was invented only once. However, comparative genomics has revealed recently that ThyA is absent in many archaeal and bacterial genomes, leading to the discovery of a new thymidylate synthase family (ThyX).[19] ThyX and ThyA share neither sequence nor structural similarity between each other and have different mechanisms of action,[19,20] indicating that thymidylate synthase activity was invented twice independently (Fig. 1). T-DNA might have appeared either in two different U-DNA cells, or the invention of a second thymidylate synthase might have occurred in a cell already containing a T-DNA genome. The first possibility would indicate that T-DNA itself has been invented twice, thus suggesting a strong selection pressure to select for uracil modification. In the second case, one should imagine that the new enzyme (either ThyA or ThyX) brought a selective advantage over the previous one in the organism where it appeared first.

A major question is why was DNA selected to replace RNA? The traditional explanation is that DNA replaced RNA as genetic material because it is more stable and can be repaired more faithfully.[4] Indeed, removal of the 2' oxygen of the ribose in DNA has clearly stabilized the molecule, since this reactive oxygen can attack the phophodiester bond (this explains why RNA is so prone to strand breakage). In addition, the replacement of uracil by thymine has made possible to correct the deleterious effect of spontaneous cytosine deamination, since a misplaced uracil cannot be recognized in RNA, whereas it can be pint-pointed as an alien base in DNA and efficiently removed by repair systems. Replacement of RNA by DNA as genetic material has thus opened the way to the formation of large genomes, a prerequisite for the evolution of modern cells.

The above scenario nicely explains why, through Darwinian competition, cell populations with DNA genomes finally eliminated cells with RNA genomes. However, this does not explain why the first organisms with a modified RNA (DNA-U), and later on with T-DNA, were successfully selected against the wild type organisms of that time? Indeed, the possibility to have a large genome or to repair cytosine deamination could not have been realized in that individual. In both cases, efficient DNA repair (to remove uracil from DNA) and replication proteins able to replicate large DNA genomes should have evolved first in order for the cell to take advantage of the presence of DNA.[15] To explain the origin of DNA, it is thus necessary to consider an advantage that could have been directly selected in the organism in which the transition occurred.

In order to solve this problem, it has recently been proposed that U-DNA first appeared in a virus, making this first U-DNA organism resistant to the RNAses of its host (Fig. 2).[6,7] Indeed, ribose reduction led to a drastic modification in the structure of the double helix (from the A to the B form) that explains why RNAses are usually inactive on DNA and DNAses inactive on RNA. Similarly, thymidylate synthase could have appeared later on in a virus with U-DNA, to makes its genome resistant to cellular U-DNAses (Fig. 2). The same process would have lead to modifications observed in modern DNA viruses (further base methylation in many viral genomes or hydroxymethylation of cytosines in T-even bacteriophages). These modifications are clearly designed to protect viral DNA against host DNAses. Interestingly, thymidylate synthase of the ThyA family are homologous to the T-even bacteriophages DNA modification enzyme dCMP hydroxymethyl-transferases.[21] Hydroxymethyl (HMC)-dCTP is directly incorporated into HMC-DNA by the viral polymerase (Fig. 1).[11] Restriction-modifications systems could be descendant of such viral mechanisms for genome protection; some of them being stolen later on by cells themselves.

If DNA replication and repair mechanisms also originated in viruses, it is easy to imagine that enzymes to correct cytosine deamination are of viral origin, and were later on transferred to cells, a prerequisite to understand the selective advantage of DNA cells over RNA cells in term of faithful replication (see a discussion of this problem in ref. 15). Several scenarios are possible for the transfer of a DNA genome from a virus to a cell: either a cell succeeded to capture several viral enzymes at once to change its genetic material from RNA to DNA, or a large DNA provirus, living in a carrier state inside an RNA cell, finally take over all functions of its host by retro-transcription, subsequently eliminating the labile RNA genomes.

The idea that viruses have played a critical role in the origin of DNA is in line with previous conception that retroviruses were relics of the RNA/DNA world transition.[22] In particular, production of DNA from RNA genome in Hepadnavirus could reflect the ancient pathway leading from RNA to DNA.[23] The invention of DNA by an RNA virus seems to be more likely than the invention of DNA by an RNA cell for protection against viral RNAses, because it has

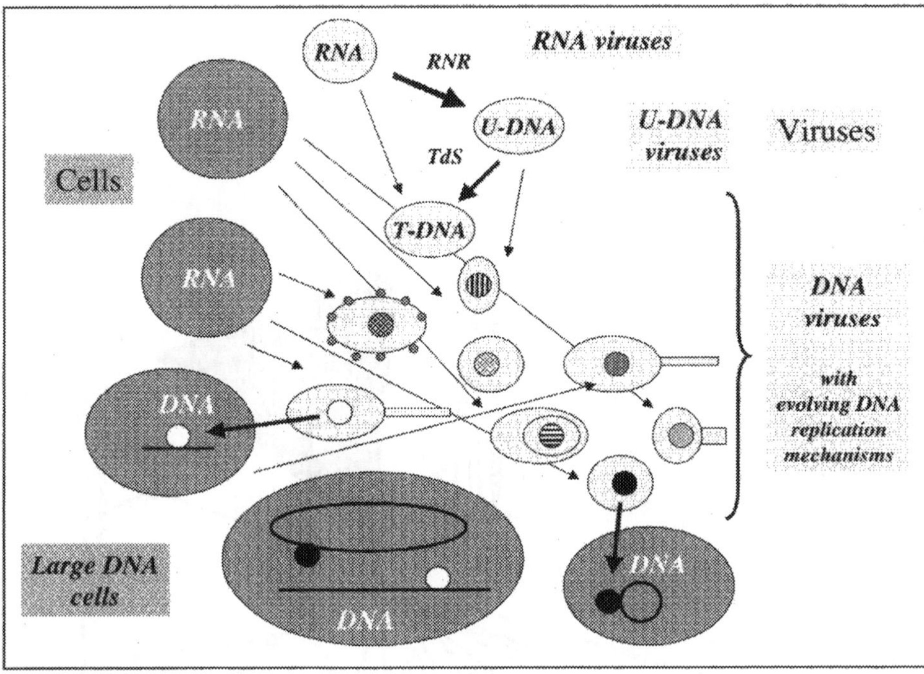

Figure 2. Evolution of DNA replication mechanisms in the viral world? This figure illustrates a coevolution scenario of cells and viruses in the transition from the RNA to the DNA world. Large gray circles or ovals indicate cells, whereas small light grey circles ovals (some with tails) indicate viruses. In this scenario, different replication mechanisms (inner circles with different colors) originated among various viral lineages after the invention of U-DNA and T-DNA by viruses (RNR= ribonucleotide reductase, TdS= Thymidylate synthase).[7] These mechanisms evolved through the independent recruitment of cellular or viral enzymes involved in RNA replication or transcription (polymerases, helicases, nucleotide binding proteins) to produce enzymes involved in DNA replication (thin arrows). Two different DNA replication mechanisms (black and white circles) were finally transferred independently to cells (thick arows). These two transfers can have occurred either before or after the LUCA. In the first case, the two systems might have been present in LUCA via cell fusion or successive transfers. One system could have also replaced the other in a particular cell lineage (for these different possibilities, see Fig. 6).

been probably easier for a virus, than for a cell, to change at once the chemical nature of its genome. This is exemplified by the fact that viruses have managed to multiply with very different types of genetic material (ssRNA, dsRNA, ssDNA, dsDNA, modified DNA) whereas, apart for localized methylation, all types of cells have the same kind of dsDNA genomes.

The hypothesis of a viral origin for DNA could explain why many DNA viruses encode their own ribonucleotide reductase and/or thymidylate synthase. This is usually interpreted as the recruitment of cellular enzymes by viruses, but, if DNA appeared in viruses, the opposite could be true as well. Many viral ribonucleotide reductases and thymidylate synthases branch far off from ribonucleotide reductases and thymidylate synthases of their hosts in phylogenetic trees, suggesting that the viral versions of these enzymes are indeed as ancient as their cellular versions (Fig. 3). Unfortunately, the direction of ancient transfer of these enzymes (either from cells to viruses or from viruses to cells) is difficult to determine, considering possible artifacts of long branch attraction that can be produced by differences in evolutionary rates between cellular and viral enzymes, i.e., viral sequences can be artificially separated from cellular ones because the latter have evolved more slowly and thus have conserved more common ancestral positions.

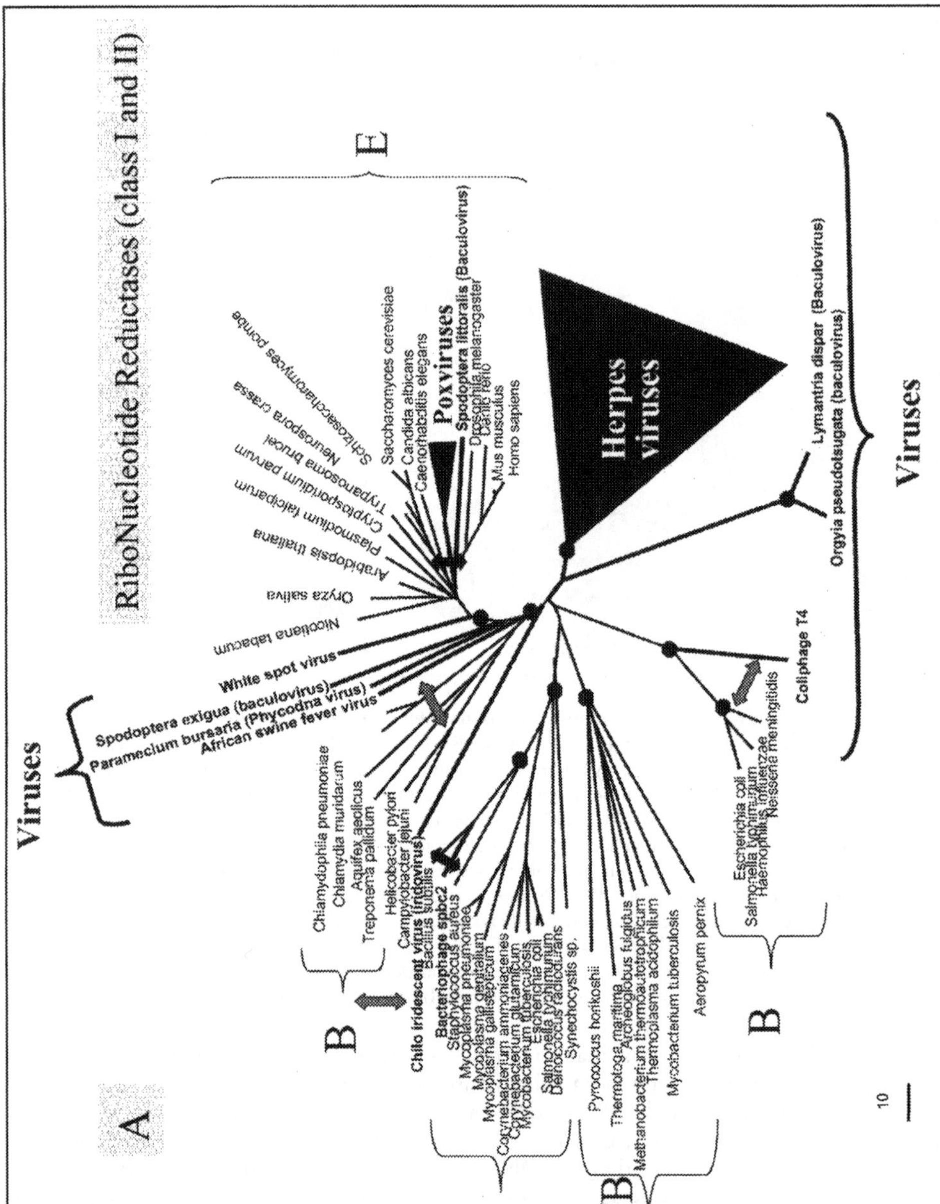

Figure 3A. Phylogenetic trees of the ribonucleotide reductases of Class I and II (A); type II DNA topoisomerase of the A family (B –left),[60] thymidylate synthases of the ThyA family (B-right) and DNA polymerases of the B family (RNA-primed) (from ref. 5) (C). Viral sequences are indicated in bold. A=Archaea, B= Bacteria, E= Eukarya. Grey double arrows indicate possible ancient transfer between cells and viruses. Black double arrows indicate recent transfers between cells and viruses. These trees were made with the program PROTML, using the quick add search, with the JTT model of amino acid substitution and retention of the 1,000 top-ranking trees. Boostrap values higher than 90% are indicated by filled circle. The scale bar represents the number of substitution per 100 sites for unit branch length.

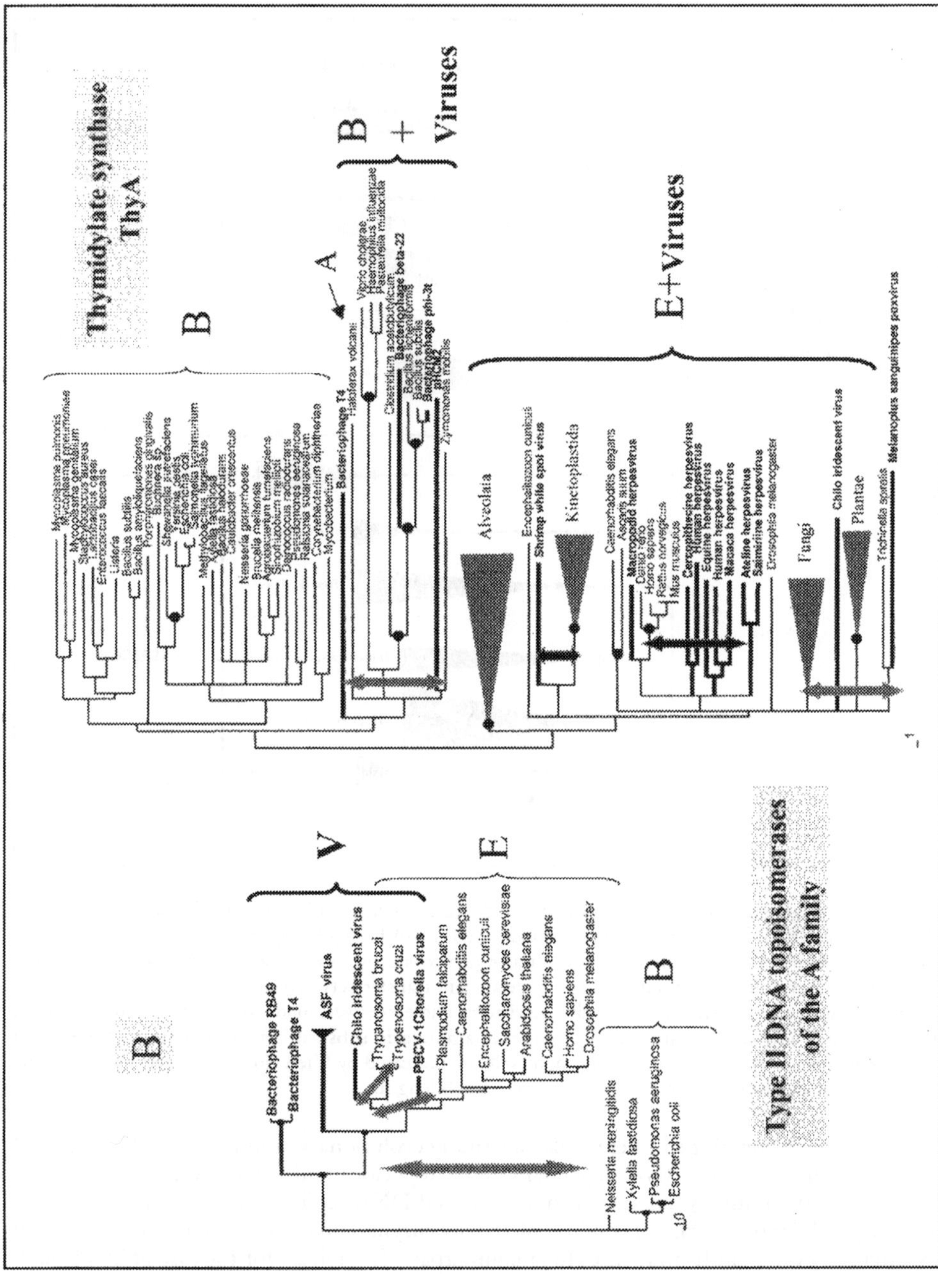

Figure 3B. Phylogenetic trees of the ribonucleotide reductases of Class I and II (A); type II DNA topoisomerase of the A family (B –left),[60] thymidylate synthases of the ThyA family (B-right) and DNA polymerases of the B family (RNA-primed) (from ref. 5) (C). Viral sequences are indicated in bold. A= Archaea, B= Bacteria, E= Eukarya. Grey double arrows indicate possible ancient transfer between cells and viruses. Black double arrows indicate recent transfers between cells and viruses. These trees were made with the program PROTML, using the quick add search, with the JTT model of amino acid substitution and retention of the 1,000 top-ranking trees. Boostrap values higher than 90% are indicated by filled circle. The scale bar represents the number of substitution per 100 sites for unit branch length.

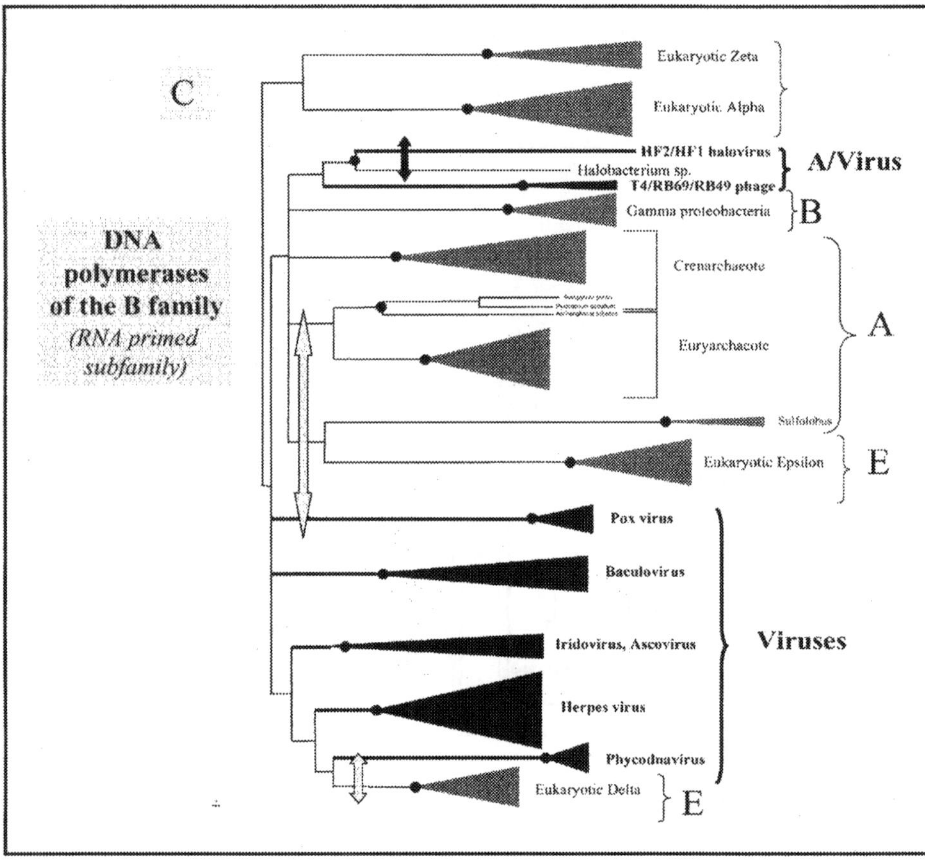

Figure 3C. Phylogenetic trees of the ribonucleotide reductases of Class I and II (A); type II DNA topoisomerase of the A family (B –left),[60] thymidylate synthases of the ThyA family (B-right) and DNA polymerases of the B family (RNA-primed) (from ref. 5) (C). Viral sequences are indicated in bold. A= Archaea, B= Bacteria, E= Eukarya. Grey double arrows indicate possible ancient transfer between cells and viruses. Black double arrows indicate recent transfers between cells and viruses. These trees were made with the program PROTML, using the quick add search, with the JTT model of amino acid substitution and retention of the 1,000 top-ranking trees. Boostrap values higher than 90% are indicated by filled circle. The scale bar represents the number of substitution per 100 sites for unit branch length.

As previously mentioned, there is also a striking evolutionary connection at the structural level between most viral RNA dependent RNA replicases and some modern DNA polymerases.[12,24] Interestingly, an ancient origin of viral DNA replication mechanisms (possibly predating cellular ones) (Fig. 2) would explain why enzymes involved in viral DNA replication are often very different from their cellular counterparts (see ref. 25 for the case of DNA polymerases) (see below for further discussion of this point).

These speculations on the origin of DNA fit well with hypotheses on viral origin that consider no longer viruses as fragments of genetic materials recently escaped from their hosts, but as ancient players in life evolution, possibly predating the divergence between the three domains of life.[26,27] The idea that viruses originated before LUCA has been recently supported by the discovery of structural and/or functional similarities between viruses infecting different cellular domains of life, such as those detected between some archaeal viruses (Lipothrixvirus and Rudivirus) and several large eukaryal DNA viruses (Poxviruses, ASFV, Chlorella viruses),[28] between Adenoviruses (eukaryal virus) and bacterial Tectiviruses,[29] or between eukaryal Flavivirus and bacterial Cystoviruses.[30]

Origin and Evolution of DNA Replication Mechanism

Viral DNA Replication Mechanisms

In contrast to cellular genomes, which are all made of double-stranded DNA, viral DNA genomes are very diverse; some viruses have circular or linear double-stranded DNA genomes, while others have circular single-stranded DNA genomes.[11] Single-stranded DNA genomes are replicated via rolling circle replication with a double-stranded DNA intermediate, whereas double-stranded viral DNA genomes are replicated either via classical theta or Y-shaped replication (for circular and linear genomes, respectively), by rolling circle, or by linear strand displacement (11, for recent reviews on eukaryal viral DNA replication, see ref. 31). In addition, replication can be symmetric, with both strands replicated simultaneously, but also asymmetric (the two strand are replicated not simultaneously but one after the other) or semi-asymmetric (the initiation of DNA replication on one strand being delayed until the first one is already partly replicated) (Fig. 1). Some viral replication mechanisms are also used by plasmids (rolling circle) and some plasmids encode DNA replication proteins homologous to viral ones (see below), suggesting that plasmids originated from ancient viruses that have lost their capsid genes.[26]

The initiation of viral DNA replication needs a specific viral encoded initiator protein that can be a site-specific endonuclease (rolling-circle replication) or a protein that trigger double-stranded unwinding. Interestingly, plasmid and viral endonucleases involved in rolling-circle replication are evolutionary related.[32] The minimal recruitment for DNA chain elongation is a DNA polymerase. In contrast to RNA polymerases, all DNA polymerases (viral or cellular) need a 3'OH primer to initiate strand synthesis. This primer can be a tRNA (for reverse transcriptases), or a short RNA, either produced by a classical RNA polymerase (also involved in transcription) or a DNA primase. This use of RNA to initiate DNA synthesis is also often considered as a relic of the RNA world.

Some primases have a strong DNA polymerase activity, suggesting that primases testify for the transition between RNA and DNA polymerases.[33] The definition of a DNA polymerase is thus becoming less straightforward, as also demonstrated by the recent characterization of DNA polymerases of the Y family that are involved in DNA repair and synthesize very short patches of DNA (much like a primase)[25,34] and by the discovery of structural similarities between eukaryal primase and DNA polymerases of the family X.[35]

As a consequence of the ancient metabolic pathway producing only 5' nucleotides, the strand moving in the 3' to 5' direction in symmetric or semi asymmetric replication has to be replicated backward in the form of short DNA pieces (Okazaki fragments) (Fig. 3). These fragments are primed by DNA primase and later on assembled by a DNA ligase, after removal of the RNA primer by RnaseH or various exonuclease activities, sometimes associated to DNA polymerases. In some cases of asymmetric replication (Adenovirus, bacteriophage ϕ29, mitochondrial linear plasmids), the DNA polymerases use a protein priming system to produce a free 3'OH for the DNA polymerase. All polymerases using this system belong to a subfamily of the DNA polymerase B family.[25]

Some DNA polymerases can perform strand displacement that is required for asymmetric DNA replication, while others, in order to improve the efficiency of this process associate with DNA helicases and/or single-stranded DNA binding proteins (ssb) to unwind the two DNA strands. The processivity of many viral DNA polymerases is further enhanced by specific processivity factors. In the case of T4, these include ring-shaped DNA clamps, and hand-shaped clamp-loader complexes that can open and close the ring-shaped DNA clamp around the DNA molecule.

In symmetric replication, the syntheses of the leading and lagging strands are coupled via an interaction between the primase and the helicase (Fig. 4). In some bacteriophages (T7, P4) and

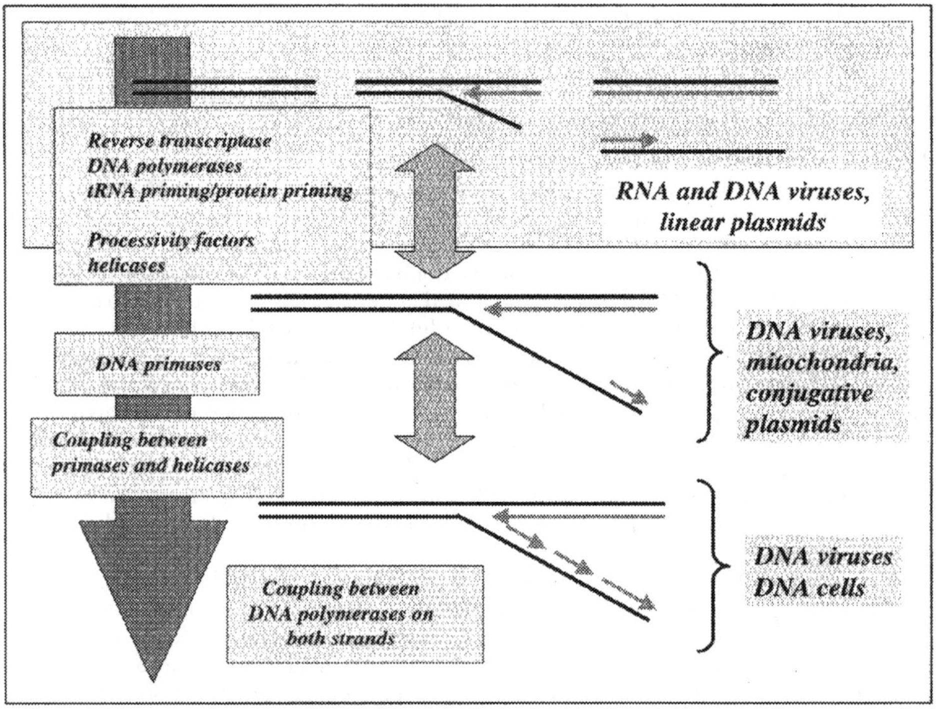

Figure 4. Evolution of DNA replication mechanisms from the simple asymmetric mode to the symmetric mode (or vice versa). In the fully asymmetric mode (top) that occured in RNA and DNA viruses, one strand is replicated entirely before the initiation of replication of the displaced strand. The minimal requirement of this mechanism is a DNA polymerase and a priming system. Strand displacement can be made more efficient by the recruitment of processivity factors and a helicase. In the semi-asymmetric mode (middle) a DNA primase initiates replication on the displaced strand before termination of the replication of the first (leading) strand. In the symmetric mode (bottom) coupling between primase and helicase allows the displaced strand (now the lagging strand) to be replicated together with the leading strand.

eukaryal viruses (Herpes), this coupling is achieved by the fusion of the helicase and the primase activities into a single polypeptide.[36,37] This is clearly a case of convergent evolution, since bacteriophages and Herpes primases belong to different protein families.[38]

The two DNA polymerases that replicate the lagging and the leading strands can be also physically linked. As a consequence, the lagging strand loops upon itself, and the two strands are replicated at once very rapidly, limiting the presence of single-stranded DNA to the fork vicinity. This is in striking contrast with asymmetric replication that requires complete denaturation of the two strands before replication of the lagging strand (Fig. 4).

Some DNA viruses replicate their genome using only replication proteins encoded by their host (with the exception of initiator proteins). However, many large DNA viruses encode also several proteins involved in the elongation step of DNA replication. Some of them (e.g., T4-phages) have reached a high level of complexity in their DNA replication machinery, and consequently encode functional analogs for all proteins involved in cellular DNA replication (Fig. 5)[39]

Considering that replication of double-strand RNA viruses is completely asymmetric, it is likely that DNA replication first occurred via the asymmetric mode and evolved toward fully symmetric theta mode via the semi-asymmetric mode (Fig. 3). If viruses recruited their DNA replication mechanisms from the cells, as proposed in the "escaped theory" for viral origin, this

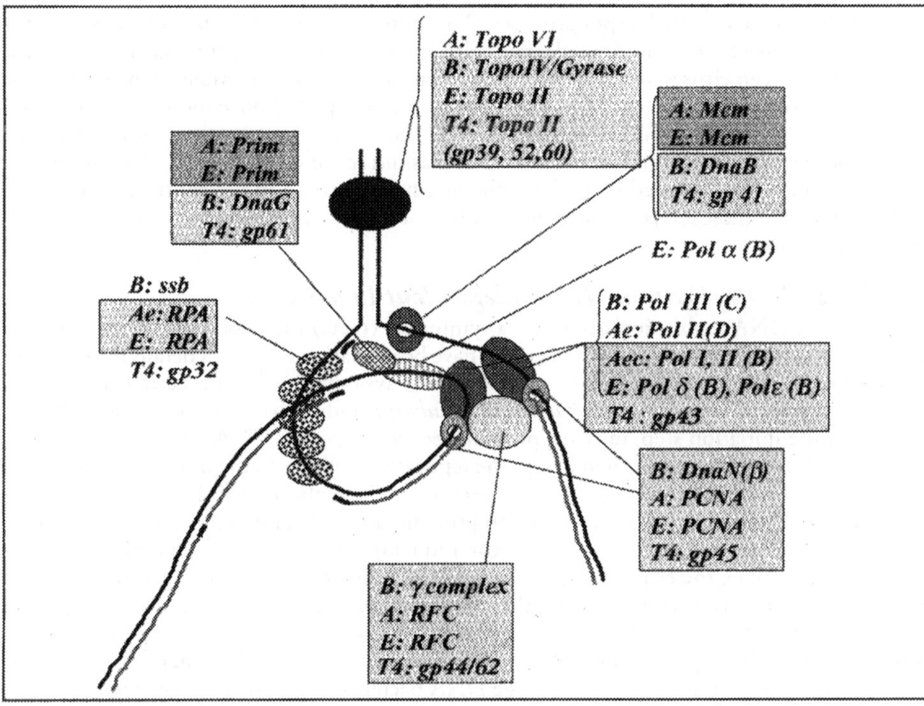

Figure 5. The universal replication fork for symmetric theta replication. Proteins with different activities are indicated with different colours and their usual names indicated for A= Archaea (Ae= euryarchaea, Ac= crenarchaea, B= Bacteria, E= Eukarya, and bacteriophage T4. Homologous proteins performing the same function are framed together. Letters in brackets indicate DNA polymerase families. The looping of the lagging strand, which allows concomitant replication of the leading and the lagging strand by a single replicase factory, is supported by experimental evidence for Bacteria and T4 phage. For an exhaustive analysis of the phylogenetic relationships between different cellular replication proteins see reference 48.

means either that viruses originated from early DNA cells that have not yet reached the stage of the symmetric mode of replication, or that this mode has been modified in many viruses to produce simpler systems. The latter possibility cannot be excluded, since there is some plasticity in the evolution of DNA replication mechanisms, and this evolution is not necessarily unidirectional (Fig. 4). For example, the replication of bacterial chromosome during conjugation can be changed from the symmetric theta mode to the asymmetric rolling-circle mode upon the integration of a conjugative plasmid.[11]

On the contrary, if DNA originated in viruses (7), one can even imagine that several DNA replication systems emerged and evolved independently from different lineages of RNA viruses. This hypothesis thus allows for a long period of DNA replication evolution purely in the viral world (Fig. 2). This would nicely explain the existence of different version of functionally analogs but nonhomologous DNA replication proteins. The diversity of viral replication proteins can be exemplified by those of Pox virus, Herpes viruses or T4, that are completely different from each others, and are no more related to the archaeal/bacterial systems (in term of protein similarities) than these systems are related between each others.[31,36,37,39] Recent sequencing of the 280 kbp bacteriophage phiKZ of *Pseudomonas aeruginosa* failed to identify virus-encoded DNA replication-associated proteins, suggesting that they may be strongly divergent from known homologous proteins.[40] Finally, it is noteworthy that several families of proteins involved in DNA replication also appears restricted to the virus world, such as helicase of the superfamily III,[41] the Herpes primases,[38] or protein-primed DNA polymerases of the B family.[25] Some linear mitochondrial plasmids also encode the latter enzyme, again suggesting a connection between viruses and plasmids. The recent discovery of a completely new family of DNA polymerase/primase encoded by the archaeal plasmid pRN2 once more emphasizes the potential of viruses and plasmids as source of novel DNA replication proteins.[78] It is difficult to understand the existence of all these viral and/or plasmid specific DNA replication proteins in the framework of the "escaped theory" for the origin of viruses. On the contrary, in the viral origin hypothesis, these enzymes have simply originated in viruses and were never been transferred to the cells.

Cellular DNA Replication: Two Independent Inventions

In all cells, DNA replication occurs by a symmetric (theta) mode of replication. The proteins involved and their mechanisms of action have been analyzed in much details during these last decades in several bacterial and eukaryal model systems.[11,31,43] The basic principles of DNA replication are very similar in *Bacteria* and *Eukarya*, and probably in *Archaea* as well (Fig. 5).[44,45] For the initiation step, initiator proteins recognize specific DNA sequences at replication origin(s). A loading factor then brings the replication helicase to the initiation complex to start the assembly of the replisome. The movement of the replication forks involves the concerted action of primases, DNA helicases, ssb proteins, and at least two processive DNA polymerases (with clamp and clamp loading factors) to couple replication of the leading and lagging strands, allowing the efficient replication of large cellular genomes. In turn, type II DNA topoisomerases became essential to solve the topological problems due to the unwinding of the double-helix in such large molecules, counteracting the production of positive superturns ahead of the forks and allowing separation of daughter molecules. This mechanism of DNA replication strikingly resembles those of some large DNA bacteriophages, such as T4 (Fig. 5).

Originally, the striking similarity between the enzymatic activities involved in bacterial and eukaryal DNA replication suggested that they originated from a common ancestral DNA replication mechanism already present in LUCA (in the nomenclature of the evolutionists, the bacterial and eukaryal DNA replication proteins were supposed to be orthologues, i.e., to have evolved in parallel to speciation from a common ancestor). In that case, bacterial, eukaryal and archaeal DNA replication proteins performing analogous function should be orthologous. However, comparative genomic analyses have shown that this is not the case (Fig. 5).[46-48] On the contrary, several critical DNA replication proteins identified in *Bacteria* by genetic and in vitro analyses have no homolog in *Archaea* or *Eukarya*, whereas others have only very distantly related homologues that are probably not orthologues. Similarly, most DNA replication proteins previously identified in *Eukarya* turned out to have readily detectable homologues only in *Archaea*.

The similarity between DNA replication proteins in *Archaea* and *Eukarya* is especially remarkable. It cannot be due to functional convergence since they have somewhat different modes of replication (unique origin and high-speed in *Archaea*, multiple origin and low speed in *Eukarya*),[49] whereas *Archaea* and *Bacteria* have dissimilar replication proteins but identical replication mode (unique origin, high speed, hot spot of recombination at the replication terminus, and major genomic recombination events occurring between bi-directional replication forks.)[49-50] The high level of similarities between the archaeal and eukaryal DNA replication proteins also cannot be explained by similar chromatin structure (as suggested by Cavalier-Smith),[51] since most archaeal proteins involved in DNA replication are similar in the two archaeal phyla the Crenarchaeota and the Euryarchaeota, whereas the presence of eukaryal-like histones is restricted to the Euryarchaeota.

Five alternative hypotheses have been proposed to explain the evolutionary gap between the bacterial and the eukaryal/archaeal replication systems (Fig. 6).

1. the replication proteins of *Bacteria* and *Archaea/Eukarya* are actually orthologues, but they have diverged to such an extent that their homology cannot be detected anymore at the sequence level.[46]

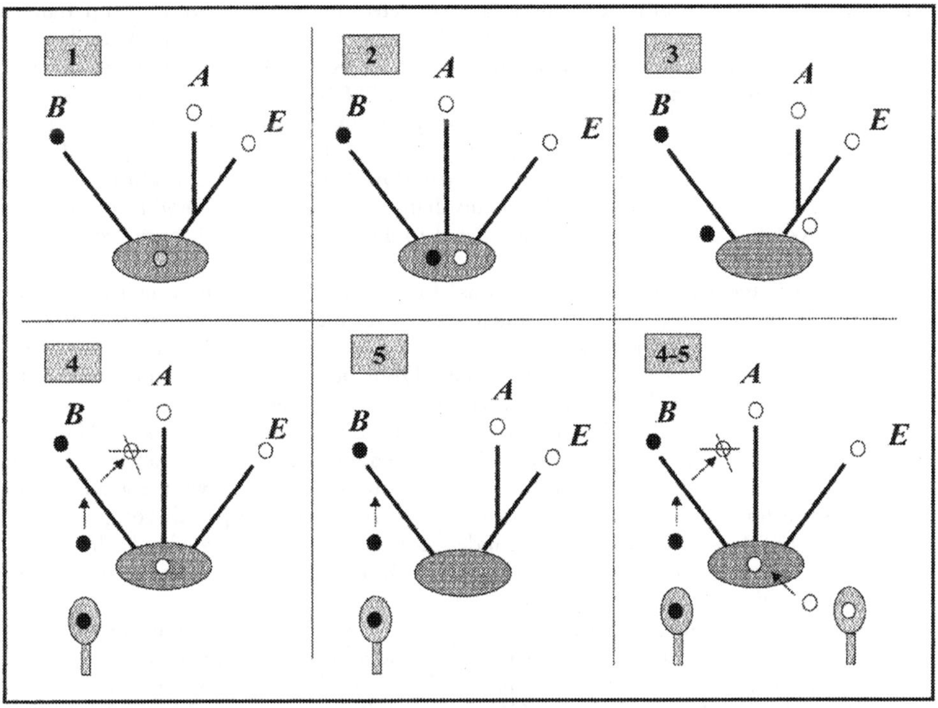

Figure 6. The different hypotheses for the origin and evolution of DNA and DNA replication mechanisms. A= Archaea, B= Bacteria, E= Eukarya. The universal trees of life are unrooted, except in the case of hypotheses 1, 3 and 5, which favor the bacterial rooting.[3-5] White circle: the archaeal/eukaryal DNA replication proteins: black circle: the bacterial DNA replication protein). The large gray circle represents LUCA. See the text for explanations.

2. two different replication systems were present in the LUCA; one was retained in *Bacteria*, the other in *Archaea/Eukarya*.[46]
3. LUCA had an RNA genome, and DNA and DNA replication were invented twice independently, once in *Bacteria* and once in the ancestral lineage common to *Archaea* and the Eukarya.[47-48]
4. The ancestral replication mechanism present in LUCA has been displaced either in *Bacteria* or in *Archaea/Eukarya* by a new one, corresponding to a nonorthologous displacement.[46,52] More specifically, it has been suggested that the bacterial replication system, or part of the eukaryal one, are of viral origin.[52-53]
5. Both bacterial, archaeal and eukaryal replication mechanisms are of viral origin and have been transferred to cells independently.[7]

The hypotheses 4 and 5 can be combined, if a first transfer from viruses to cells occurred before LUCA, and a second one displaced this ancestral cellular mechanism later on.

In addition several authors have proposed that the eukaryal nucleus itself originated from a large DNA virus (possibly an archaeal virus) that could be related to Poxviruses.[54-55]

The first hypothesis (the hidden orthology) can be clearly ruled out, since the bacterial and the archaeal/eukaryal versions of the two central players in the elongation step of DNA replication, the replicative polymerases and the primases, belong to different protein families.[25,35,48] In the case of primases, structural analyses have shown that the bacterial and the eukaryal/archaeal versions are completely unrelated, the latter being member of the DNA polymerase X family.[35] In the case of the replicase, the structure of the bacterial one (PolC/DnaE) has not yet been solved, but in-depth sequence analysis failed to detect any similarity with the superfamily of RNA polymerases, reverse transcriptase and DNA polymerases of the A and B families.[48]

In other cases (the replicative helicase, the single-stranded DNA binding proteins, the initiator proteins), comparative structural analyses and/or PSI-BLAST searches have shown that the bacterial and eukaryal/archaeal proteins belong to same superfamilies, since they share homologous domains. However, they are clearly not orthologues, since they belong to different families. For example, in the case the initiator protein (DnaA in *Bacteria*, Cdc6/Orc1 in *Archaea* and *Eukarya*) the bacterial and archaeal proteins share a common ATPase module of the same family (AAA+), but these modules are associated to different modules that are probably involved in DNA binding.[57]

The bacterial and archaeal/eukaryal versions of many DNA replication proteins have thus been certainly invented independently, probably by recruitment and modification of proteins previously involved in RNA replication and/or RNA gene regulation. However, a few DNA replication proteins (the clamp, the clamp loader, DNA ligase) could be orthologous in the three domains of life since they share sequence similarities that can be detected by elaborated PSI-BLAST analyses or structural similarity with unique fold and fold arrangement.[48] Furthermore, they are more similar to each other's, from one domain to another, than to any other proteins. We should thus explain why different replication systems that have emerged independently use some homologous accessory proteins. It is possible that these proteins originated late in the history of DNA replication and were independently recruited by evolving DNA replication systems. Alternatively, they might have predated DNA replication itself and were independently used by different emerging systems.

In order to better understand the evolution of the DNA replication apparatus, it would be necessary to determine with some confidence when and where the independent inventions of the bacterial and the eukaryal/archaeal versions of nonorthologous DNA replication mechanisms occurred (either before or after LUCA, either in cells or in viruses?). We will discuss now several specific points of the above hypotheses (except hypothesis 1 that we have ruled out) in an attempt to answer some of these questions.

The Genome of LUCA (DNA or RNA)

In hypotheses 2 and 4, LUCA had a DNA genome, whereas in 3, LUCA had an RNA genome (hypothesis 5 can be accommodated with both possibilities) (Fig. 6). The nature of the genome of LUCA is thus a major pending question. Obviously, LUCA had already a well-developed translation system (see other chapters), but the question of the status of transcription and replication in LUCA is by far more complex. The hypothesis of a primitive LUCA with an RNA genome was first formulated twenty years ago by Carl Woese.[58] This hypothesis was mainly based on the prejudice of a very simple LUCA (a progenote). It is remarkable that Carl Woese correctly predicted in 1977, based on this idea, that DNA replication mechanisms should not be homologous in prokaryotes and eukaryotes (if prokaryotes are for this purpose assimilated to *Bacteria*).

The idea of a simple LUCA without DNA was strongly disputed by Forterre following the discovery in *Archaea* of DNA polymerases (family B) and type II DNA topoisomerases (gyrase) that were homologous to bacterial and eukaryal enzymes.[59] However, it turned out that these were cases of lateral gene transfer.[25,60] More recently, several authors also argued for the presence of DNA in LUCA, as some proteins using DNA as substrate are probably orthologues in the three domains of life.[46,48] This is the case for the clamp, the clamp loader, and DNA ligase (as already mentioned), but also for DNA-dependent RNA polymerases, type I DNA topoisomerases of the A family, RecA-like recombinases, SMC proteins involved in chromosome condensation, and Mre11/Rad50 complex involved in homologous recombination. However, it is difficult to reject the RNA LUCA hypothesis simply based on this observation, because some of these proteins could have been already operational in the RNA world. For instance, cellular DNA-dependent RNA polymerases can also replicate the genome of RNA viruses,[61-62] type I DNA topoisomerases of the A family can act as RNA topoisomerase,[63] and the common ancestor of present-day DNA ligases could have been an RNA ligase involved in RNA processing.

The better argument in favor of a DNA-based LUCA could be actually the orthology of the clamp and clamp loader in the three domains, since RNA viruses apparently does not use clamp for their replication. However, the clamp and clamp loader of T-even bacteriophages being homologous to their cellular counterparts,[64] another possibility is that *Bacteria*, *Archaea*/*Eukarya* have recruited independently these homologous proteins from viruses related to T-even bacteriophages.

Perhaps a more direct strategy to decide whether LUCA had an RNA-based or a DNA-based genome could be to determine if ribonucleotide reductases and thymidylate synthases were already present in LUCA. Ribonucleotide reductases of class II and III are present in both *Bacteria* and *Archaea*. Bacterial and archaeal class II ribonucleotide reductases form two monophyletic groups, suggesting that class II was present in LUCA (Fig 3A, B). Similarly, ThyX is present in both *Archaea* and *Bacteria*,[19] suggesting that ThyX could have been present in LUCA. The overall distribution of ribonucleotide reductases and thymidylate synthases thus seems to favor a LUCA with a T-DNA genome, in agreement with the presence of the clamp and clamp loader in LUCA. However, the presence of an orthologous protein in *Bacteria* and *Archaea* can be also explained by the monophyly of prokaryotes if the root of the universal tree is in the eukaryal branch (reference 65 and see discussion below). Furthermore, many viral sequences are interspersed with cellular sequences both in the ribonucleotide reductase and thymidylate synthase tree (Fig 3A, B). Thus, one cannot exclude that these proteins have been transferred from viruses to the proto-archaeal and bacterial lineages shortly after their divergence from LUCA.

Finally, one should consider the possibility that LUCA had still an RNA genome but already contained retro-transcribed DNA. This hypothesis was proposed by Leipe and coworkers, in an attempt to reconcile the existence of two independent replication mechanisms with the possible presence of DNA in LUCA.[48] We will thus discus now specifically the problem of the origin of the two DNA replication mechanisms.

When and Where DNA Replication Mechanisms Originated

If the bacterial and the archaeal/eukaryal versions of DNA replication proteins were already present in LUCA (hypotheses 2), (Fig. 6), they might have appeared either successively in the same lineages ancestral to LUCA, or in different lineages (being later on mixed in LUCA or in one of its ancestor by cell fusion or gene transfer). In the first hypothesis, it is unclear how new DNA replication machinery could have been selected in any organism already containing a more evolved version? If the new version was more efficient by chance, why the old one survived? If the bacterial and the archaeal/eukaryal versions of DNA replication proteins appeared in different lineages, one can still imagine that they have evolved different properties explaining their coexistence in a single cell.

The hypothesis 3 implies that the two distinct sets of DNA replication proteins originated after LUCA, one in a proto-bacterium and one in a common lineage to *Archaea* and *Eukarya* (Fig. 6). This is in nice agreement with the classical rooting of the universal tree of life in the bacterial branch. However, this rooting is highly disputed.[66,67] It has been shown that phylogenetic data that support this rooting are not valid (which does not mean that this rooting is wrong) and other hypotheses have been proposed, such as an eukaryal rooting,[65] or a fusion between a proto-bacterium and a proto-archaeon to give *Eukarya*.[68]

Even if the bacterial rooting is correct, the hypothesis 3 did not explain the distribution of some DNA replication proteins, such as type II DNA topoisomerases or DNA polymerases, that can be also divided into nonhomologous families. The case of type II DNA topoisomerases is illuminating. Although these elaborated enzymes catalyze a complicated reaction (the crossing of a double helix by another DNA duplex) two versions have been invented independently.[59] This has been shown by the discovery in *Archaea* of an atypical topoisomerase (DNA topoisomerase VI, prototype of Topo IIB) whose modular organization and mechanism of action turned out to be distinct from classical type II DNA topoisomerases (Topo IIA.)[69] The existence of two families of nonhomologous type II DNA topoisomerases is *a priori* in line with the independent invention of two sets of nonhomologous DNA replication proteins. However, the phylogenomic distribution of Topo IIA and Topo IIB did not fit with those of other DNA replication proteins. Topo IIB is present in *Archaea* and plants, whereas Topo IIA is present in *Bacteria* and *Eukarya* (with few recent transfers from *Bacteria* to *Archaea*). The situation is even more complex in the case of DNA polymerases, since seven nonorthologous families have already been recognized (A, B, C, D, X, Y, the pRN2 plasmid polymerase). Bacterial replicases are of the C family, whereas archaeal replicases are either from the B or D families, and eukaryal polymerases from the B family. In a tree of family B, archaeal and eukaryal DNA polymerases do not form a clade, but the three eukaryal replicases (α,δ and ε) are interspersed with archaeal DNA polymerases and many groups of DNA polymerases from bacteriophages and animal viruses (Fig. 3C).[25] These atypical distributions can be reconciled with the different hypotheses that have been proposed for the universal tree of life only by introducing ancient gene transfers and gene losses, as well as nonorthologous displacements, suggesting a scenario for the origin of DNA replication proteins more complex than hypothesis 3.

The hypothesis 4 (Fig. 6) is based on the observation, from comparative genomics, that replacement of a protein belonging to a given family by a protein of similar function but belonging to another family (nonorthologous or even nonhomologous) has frequently occurred during genome evolution.[47] In particular, phylogenomic analyses of replication proteins also revealed that nonorthologous displacement occurred during the evolution of *Archaea*. The archaeal Topo II (family Topo IIB) has thus been "recently" displaced in *Archaea* of the order Thermoplasmatales by bacterial DNA gyrase (Topo IIA).[59] More ancient displacements have occurred between the two archaeal phyla. Thus, the eukaryal version of the single-strand binding (ssb) complex (RPA) that is present in Euryarchaea has been displaced in Crenarchaea by a novel form of ssb protein (or vice versa).[70] Similarly, one should refer to nonorthologous displacement to explain why Crenarchaea and Euryarchaea have probably different DNA replicases (belonging to DNA polymerase B and D families, respectively).[71] If nonorthologous displacement occurred during the diversification of *Archaea*, more drastic one (replacement of a nearly complete system by another, possibly carried by viral genomes) might well have occurred early on, during domain diversification, especially at a time when lateral gene transfer were more frequent and when primitive replication systems were probably even more plastic.

Interestingly, nonhomologous DNA polymerases of the B and D families interact at the replication forks with proteins that are homologous in Crenarchaea and Euryarchaea (Fig. 5). Similarly, a small DNA polymerase subunit present in *Archaea* and *Eukarya* can interact with catalytic subunits of either phylogenetically unrelated DNA polymerases B or D.[72] All these observations clearly indicate that nonorthologous displacement can affect interacting proteins of the replication complex at the forks. This could explain why the clamp and the clamp loader are still homologous in the bacterial and archaeal replication systems, if other elements have been displaced in the course of evolution.

Nonorthologous displacement can have also played an important role in modifying the relative rate of evolution of proteins that remained orthologues.[65] For example, if several ancestral replication proteins have been displaced in *Bacteria*, proteins that remain orthologues in the three domains (such as the clamp and the clamp loader) will become more divergent in *Bacteria* since they have coevolved now with different partners, without regards to the real phylogenetic relationships between the three domains.

A Viral Origin for Cellular DNA Replication Proteins?

If DNA and DNA replication proteins originated in viruses,[7] one can imagine that DNA replication mechanisms have been transferred from viruses to cells (Fig. 2). This possibly occurred either at different stages of viral evolution (giving birth to various types of cellular lineages with different DNA replication modes), or only at the symmetric stage. In the latter case, the first DNA cells would have had an immediate advantage compared to RNA cells still using the asymmetric mode of RNA replication. In particular, the symmetric mode allows the replication of large cellular DNA genomes without accumulating excessive amount of unstable single-stranded DNA. Hypotheses 4 and 5 are in agreement with these ideas (also hypothesis 4 can also be accommodated with the idea that viral DNA replication proteins are of cellular origin but diverged extensively from their ancestral versions during viral evolution (before being transferred back to some cell lineages.)

Many of the problems that beset previous hypotheses can be solved in the framework of the viral origin of DNA and DNA replication theory. It is no more necessary to explain why two distinct sets of nonhomologous DNA replication proteins with similar function coexisted in the same cell (hypothesis 2), neither to refer to the universal tree of life based on rRNA (hypothesis 3). The existence of puzzling and contradictory phylogenies for many DNA replication proteins is now readily explained by suggesting that different elements of the replisome have been recruited independently from different viruses. Finally, the origin of the proteins involved in the nonorthologous displacement postulated in hypothesis 4 is clearly identified. The implication of viruses in a massive nonorthologous displacement is appealing, since DNA replication proteins encoded by DNA viruses often form gene clusters that facilitate their transfer in a single step from a virus to its host.

Figure 6 illustrates two scenarios for the viral origin of cellular DNA replication proteins. In the first case (hypotheses 5), all DNA replication proteins originated from viruses, after the separation of the archaeal and bacterial lineages, in agreement with an RNA based LUCA, whereas in the other (hypothesis 4-5) a first transfer occurred before LUCA, and a second one occurred in the bacterial branch (post-LUCA). The second step corresponds to the nonorthologous displacement of hypothesis 4.

Many protein phylogenies support the idea of ancient transfers of replicative proteins between cells and viruses. As previously mentioned, it is striking that the various subtypes of eukaryal DNA polymerases in the B family (α, δ, ε) are not grouped together in phylogenetic trees but interspersed with archaeal DNA polymerases, bacteriophage T4 DNA polymerases, and various groups of viral DNA polymerases.[25] Furthermore, DNA polymerase δ branches off a group of viruses including Iridovirus in phylogenetic trees of DNA polymerases B (Fig. 3C).[25,53] The phylogenetic tree of type II DNA topoisomerases of the A family can also be explain by a viral origin. Indeed, both bacterial and eukaryal Topo IIA emerged independently from a group of viral sequences that include both bacteriophages and eukaryal viruses (Fig. 3D).[59]

These phylogenies clearly indicate that ancient transfers actually occurred between viruses and cells. Unfortunately, as in the case of thymidylate synthases and ribonucleotide reductases, the direction of these transfers (from cells to viruses or from viruses to cells) cannot be determined with complete confidence, since viral lineages have usually long branches that can be attracted by outgroup sequences, artificially separating viruses from cellular domains.

However, it is noteworthy that such a global replacement of cellular proteins by viral-encoded functional analogs actually occurred in the course of mitochondial evolution. Indeed, both the mitochondrial RNA polymerase that initiate replication of the H-strand, the mitochondrial DNA polymerase γ, and the mitochondrial primase that initiates on the L strand, are phylogenetically related to viral homologues of the T3/T7 superfamily (refs. 25, 73, and unpublished result from this laboratory), clearly indicating that nonorthologous displacement of cellular DNA replication proteins by viral ones is possible. It is remarkable that the present-day mitochondrial genome in yeast and mammals replicate via a semi-asymmetric mode, instead to replicate via the symmetric mode of the proteobacterial genomes, suggesting that a virus (or a plasmid) take over DNA replication in mitochondria both in term of proteins and replication mode.

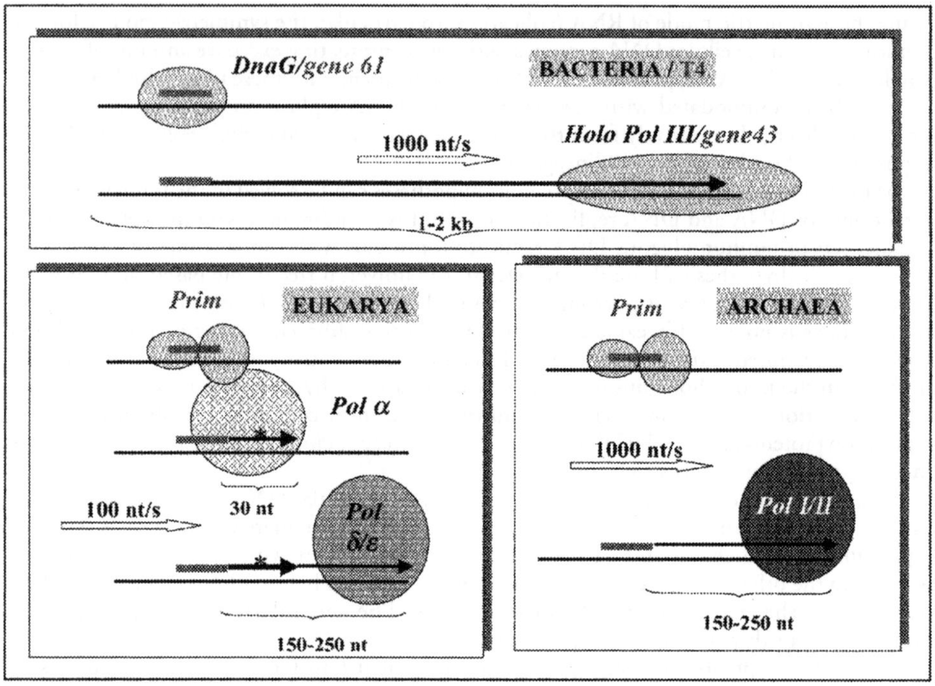

Figure 7. Synthesis of Okazaki fragments in the three domains of life and in T4. In Bacteria and T4 long Okazaki fragments are produced at high speed, by a single DNA polymerase using an RNA primer. In Eukarya short Okazaki fragments are produced at low speed by the successive actions of a two-subunit primase, an unfaithful DNA polymerase (DNA polymerase α) and DNA polymerase δ and/or ε. RNA primers are in bold line, Bold arrow indicates DNA fragment synthesized by DNA polymerase α. Putative mistakes are suggested made by DNA polymerase α are indicated by a star. In Archaea, *in silico* analyses suggest that an eukaryotic-like primase synthesizes an RNA primer which is elongated by a DNA polymerase of the B family (Pol I in Euryarchaea, Pol I or Pol II in Crenarchaea) or of the D family (Pol II in Euryarchaea). Experimental evidences suggest a rapid rate of DNA chain elongation,[49] and a similar size of the Okazaki fragment in Archaea and Eukarya.[75]

Evolution of Specific Mechanisms Associated to Cellular DNA Replication: Two Case Studies

Further progress in our understanding of the origin and evolution of DNA and DNA replication apparatus will certainly benefit from the sequencing of new genomes, especially from protists and viruses. However, it will be also critical to deepen our understanding of the "historical logic" hidden in various facets of the replication process itself. This will require more experimental data on a great variety of systems to get new insights from comparative molecular biology. We will finish this chapter by discussing two examples that illustrate this point.

The first one refers to the different sets of proteins performing the synthesis and processing of Okazaki fragments in *Bacteria*, and *Eukarya* (Fig. 7).[74] In *Bacteria*, DNA polymerase III directly used the RNA primers synthesized by the primase DnaG (a monomer) to produce at once full-length Okazaki fragment (about 1000 base pairs). A single protein, DNA Polymerase I, can both eliminate the RNA primer via its 5' to 3' exonuclease module, and fill the gap with

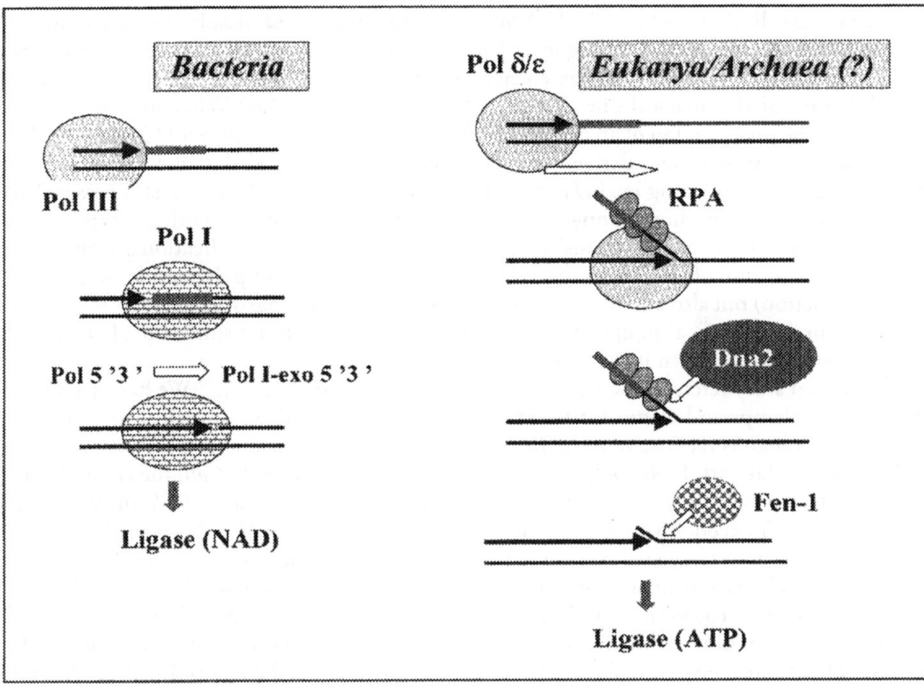

Figure 8. Processing of Okazaki fragments in the three domains of life (adapted from ref. 74). In Bacteria, the RNA primers of Okazaki fragments are removed by the concerted action of the 5' to 3' exonuclease and polymerase domains of DNA polymerase I. In Eukarya, the RNA primer and the DNA fragment containing putative mistakes synthesized by DNA polymerase α is displaced by DNA polymerases δ and/or ε. The displaced single-strand is cleaved by the successive actions of the endonuclease domain of the DNA 2 helicase and of the flap-endonuclease Fen-1. This fragment is covered first by the single-stranded binding protein RPA, which recruits Dna2. *In silico* analysis predict that processing of Okazaki fragments in Archaea occurs as in Eukarya.

its polymerase activity. The mechanism of Okazaki fragment synthesis and processing in *Eukarya* appears to be more complex and less "rational", even *bizarre*. The RNA primer synthesized by the eukaryal primase is first elongated in vivo by DNA polymerase α to produce a short RNA-DNA primer (about 30 base pairs) that is extended into a full length Okazaki fragment (about 100-150 base pairs) by DNA polymerases δ. The role of DNA polymerase α is puzzling, since DNA polymerase δ can extend an RNA primer to a full length Okazaki fragment in vitro.[54] Furthermore, DNA polymerase α lacks the editing 3' to 5' exonuclease activity required for faithful DNA synthesis. As a consequence, the processing of Okazaki fragments in *Eukarya* requires the removal of the RNA primer and, also, of the stretch of DNA containing possible errors that has been synthesized by DNA Polymerase α (Fig. 8). This involves the successive action of three proteins: RPA, Dna2 and FEN-1. DNA polymerase δ first displaces the RNA primer and the DNA portion synthesized by DNA Polymerase α. The displaced single-stranded DNA is then covered with RPA that both inhibits further progression by DNA polymerase δ and recruits the Dna2 protein. The displaced strand can then be cleaved by the endonuclease activity of Dna2, leaving a short single-stranded tail, which is finally degraded by the flap-endonuclease FEN-1.

Interestingly, RPA, Dna2 and FEN-1 are conserved in *Archaea*, despite the apparent absence of DNA Polymerase α ortholog in archaeal genomes! In the traditional view of evolution (from simple prokaryotes to complex eukaryotes), the eukaryal replication system is an improved version of the archaeal one. What is the significance of this? What improvement is gained from the introduction of an unfaithful DNA polymerase in the system? Could it be that the archaeal system is in fact derived from the eukaryotic one, and that the mechanism of Okazaki fragment processing in *Archaea* is a relic of the time when Pol α was still present? To answer this question, we should know more about the role of the Pol α and other actors in both the eukaryal and the archaeal systems! In general, the archaeal DNA replication system is not only a simpler version of the eukaryal one (with a smaller number of polypeptides to perform the same function) but also a more efficient one. For example, the rate of elongation is as fast in *Archaea* as in *Bacteria*,[49] although the sizes of Okazaki fragments are similar in *Archaea* and *Eukarya* (much shorter than in *Bacteria*, see Fig. 7).[75]

The second story refers to the structure of the bacterial clamp loader.[76] We have seen that, although the clamps and clamp loaders are homologues in the three domains, they interact with nonhomologous replicative proteins in *Bacteria* on one side and *Archaea/Eukarya* on the other (in particular with DNA polymerases of different families). In *E. Coli*, the clamp loader contains subunits called τ, γ, δ, δ' that are homologous to archaeal/eukaryal RFC proteins. The same gene (*dnaX*) encodes for the γ and τ subunits. The protein τ (71 kDa) is the full-length protein, whereas γ is a truncated version (47 kDa) due to a translational frameshift followed by a stop codon. The C-terminal amino-acid extension of 24 kDa in τ has been added to the clamp loader during bacterial evolution since it has no homolog in archaeal or eukaryal RFC proteins. This extension allows the dimeric clamp loader to connect the clamp loader to the helicase (DnaB) and the two replicases (Pol III) (Fig. 9). What is the reason for this? One can argue that it helps to structure the bacterial replisome, or that it compensates for the absence of one of the two types of bacterial replicase (PolC) that are found in other *Bacteria* (e.g., *Bacillus subtilis*).[77] On the other hand, in the hypothesis of nonorthologous displacement of ancestral replication proteins by DnaB and Pol III, one could imagine that this C-terminal extension is the trick found by these proteins to force the cellular clamp loader to interact with them, instead of interacting with the ancestral cellular machinery (much like the P protein of bacteriophage γ force the bacterial initiator protein DnaA to interact with it instead to DnaC). However, this scenario is challenged by the restricted distribution of this C-terminal extension in the bacterial domain. Clearly, we would like to know more about the different types of replisomes that are present in *Bacteria* and to understand how they are evolutionary related to figure out the signification of such oddities as the *dnaX* gene!

Conclusion and Future Prospects

Up to now, most scientists interested in the studying of DNA replication have not been apparently concerned by the problem of the origin and evolution of this central cellular mechanism. The problem of the origin of DNA is also largely ignored, with few exceptions. It is striking that recent hypotheses on the evolution of DNA replication have been proposed by evolutionists involved in comparative genomics, and not by people actively involved at the bench in the molecular study of DNA replication. The same is also true for transcription. In contrast, scientists working on translation have a long lasting interest in the origin and evolution of the genetic code and the translation apparatus. The central role of RNA in both the origin of life theory and the mechanism of protein synthesis can explain this trends, reinforced by the role that 16S/18S rRNA have played in evolutionary constructions. However, thanks to comparative genomics, scientists working on DNA replication on various model systems should now be encouraged to grasp a new cultural attitude and realize that their work is not only important to understand the functioning of modern cells, identify new drugs targets, or design

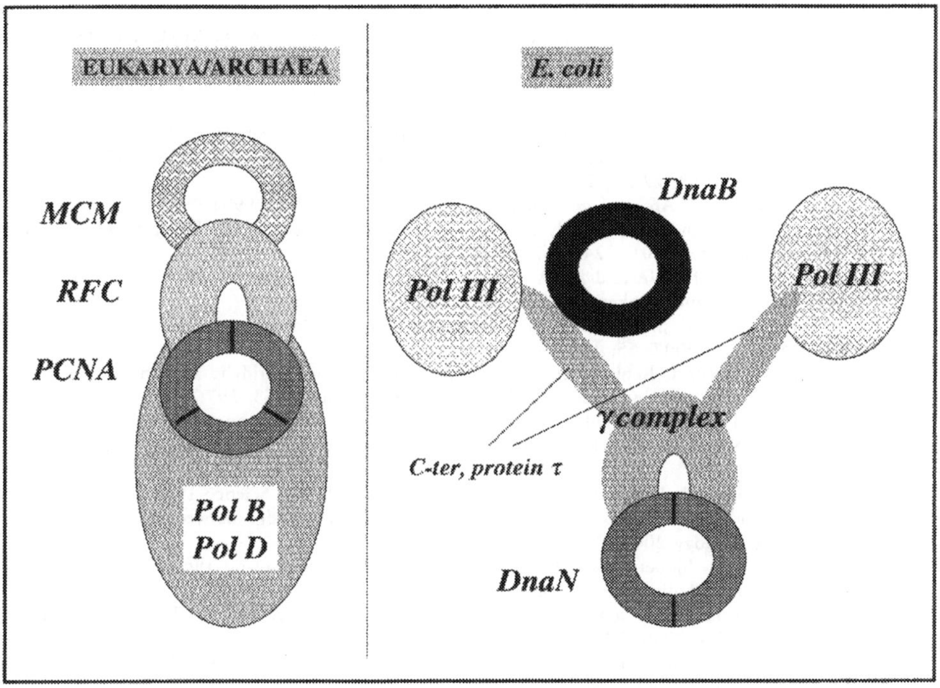

Figure 9. Interactions of the homologous clamp and clamp loader with nonhomologous components of the replisome in Archaea/Eukarya and in *E. coli* (adapted from ref. 76). In Eukarya and Archaea, RFC and PCNA (the respective homologues of bacterial γ complex and DnaN) interact with helicase (MCM) and polymerases of the B and/or D family). In *E. coli*, the dimeric γ complex interacts via a C-terminal extension of the protein τ with the helicase DnaB and two molecules of DNA polymerase III, two proteins that are not homologous to their functional counterparts in Archaea and Eukarya. The C-terminal extension of the protein τ is not present in all Bacteria.

new products for biotechnology, but that it is also critical for understanding the history of life. They should not only try to adapt their findings to current evolutionary theories, but also try to detect possibilities to check the validity of these theories in these findings. As we have seen in this chapter, there is no lack of alternative, and sometimes contradictory, hypotheses. We have emphasized the importance that viruses could have played in the story since their role is usually ignored or underestimated. In any case, viral replication systems should not be only considered as simple model system, giving possible clue to more complex cellular ones, but as mechanisms interesting to study on their own, as witnesses of critical aspects of early life evolution. The availability of many more replication protein sequences from viruses of the three domains of life and new methods to analyze viral protein phylogenies will possibly help to critically test some of the hypotheses we propose. Their comparison with other replication systems will certainly be productive at the end, if done with an evolutionary oriented mind.

Acknowledgments

Work on DNA replication in our laboratory was supported by grants from l'Association de Recherche contre le Cancer (ARC) and the Programme de Recherche Fondamentale en Microbiologie et Maladies infectieuses et parasitaires (PRFMMIP) of the Ministère de Education et de la Recherche.

References

1. Monod J. Chance and Necessity: An Essay on the Natural Philosophy of Modern Biology. New York: Knopf, 1971.
2. Watson JD, Crick FHC. The structure of DNA. Cold Spring Harbor Symp Quant Biol 1953; 18:123-113.
3. Shroedinger E. What is life, the physical aspect of the living cell. Cambridge 1944.
4. Lazcano A, Guerrero R, Margulis L et al. The evolutionary transition from RNA to DNA in early cells. J Mol Evol 1988; 27:283-290.
5. Olsen GJ, Woese CR. Archaeal genomics: an overview. Cell 1997; 89:991-994.
6. Forterre P. 2001, Genomic and early cellular evolution. The origin of the DNA world. CR Acad Sci Paris Life Sciences 2001; 324:1067-1076.
7. Forterre P. Origin of DNA and DNA genomes. Curr Opin in Microbiol 2002; 5:525-532.
8. Jacob F. Evolution and thinkering. Science 1997; 196:1161-1166.
9. Poole AM, Logan DT, Sjöberg B-M. The evolution of ribonucleotide reductase: much ado about oxygen. J Mol Evol 2002; 55: 180-196.
10. Takahashi I, Marmur J. Replacement of thymidylic acid by deoxyuridylic acid in the deoxyribonucleic acid of a transducing phage for Bacillus subtilis. Nature 1963; 197:794-5
11. Kornberg A, Baker T. DNA replication. New York: Freeman and Company, 1992.
12. Ahlquist P. RNA-dependent RNA polymerases, viruses, and RNA silencing. Science 2002; 296:1270-1273.
13. Freeland SJ, Knight R, Landweber LF. Do proteins predate DNA? Science 1999; 286:690-692.
14. Poole A, Penny D, Sjöberg B-M. Methyl-RNA: Evolutionary bridge between RNA and DNA? Chemistry and Biology 2000; 7:207-216.
15. Poole A, Penny D, Sjöberg B-M. Confounded cytosine! Tinkering and the evolution of DNA. Nature Rev Mol Cell Biol 2001; 2:147-151.
16. Stubbe JA. Ribonucleotide reductases: The link between an RNA and a DNA world? Current Opin Structural Biol 2000; 10:731-773.
17. Eklund H, Uhin U, Farnegardh M et al. Structutre and function of the radical enzyme ribonucleotide reductase. Prog Biophys Mol Biol 2001; 77:177-268.
18. Fontecave M, Mulliez E, Logan DT. Deoxyribonucleotide synthesis in anaerobic microorganisms: the class III ribonucleotide reductase. Prog Nucleic Acid Res and Mol Biol 2002; 72:95-128.
19. Myllykallio H, Lipowski G, Leduc D et al. An Alternative Flavin-Dependent Mechanism for Thymidylate Synthesis. Science 2002; 297:105-107.
20. Murzin AG. DNA building blocks reinvented. Science 2002; 297:61-62.
21. Song HK, Sohn SH, Suh SW. Crystal structure of deoxycytidylate hydroxymethylase from bacteriophage T4, a component of the deoxyribonucleoside triphosphate-synthesizing complex. EMBO J 1999; 18:1104-1113.
22. Lazcano A, Valverde V, Hernandez G et al. On the early emergence of reverse transcription : theeoretical basis and experimental evidence. J Mol Evol 1992; 35:524-536.
23. Wintersberger U, Wintersberger E. RNA makes DNA : A speculative view of the evolution of DNA replication mechanisms. Trends in Genet 1987; 3:198-202.
24. Ng KK, Cherney MM, Vazquez AL et al. Crystal structures of active and inactive conformations of a caliciviral RNA-dependent RNA polymerase. J Biol Chem 2002; 277:1381-1387.
25. Filee J, Forterre P, Sen-Lin T et al. Evolution of DNA polymerase families: evidences for multiple gene exchange between cellular and viral proteins. J Mol Evol 2002; 54:763-773.
26. Forterre P. New hypotheses about the origins of viruses, prokaryotes and eukaryotes. In: Trân Thanh Vân JK, Mounolou JC, Shneider J and Mc Kay C, eds. Gif-sur-Yvette, France: Editions Frontières, 1992:221-234.
27. Bamford DH, Burnett RM, Stuart DI. Evolution of viral structure. Theor Popul Biol 2002; 61:461-470.
28. Peng X, Blum H, Qhe Q et al. Sequence and replication of genomes of the archaeal Rudivirus SIRV1 and SIRV2: relationships to the archaeal Lipothrixvirus SIFV and some eukaryal viruses. Virology 2001; 291:226-234.
29. Benson SD, Bamford JKH, Bamford D et al. Viral evolution revealed by bacteriophage PRD1 and human adenovirus coat protein structures. Cell 1999; 98:825-833.
30. Butcher SJ, Grimes JM, Makeyev EV et al. A mechanism for initiating RNA-dependent RNA polymerization. Nature 2001; 410:235-240.
31. De Pamphilis ML. DNA replication in eukaryotic cells. Cold spring Harbor Laboratory Press 1996.

32. Ilyina TV, Koonin EV. Conserved sequence motifs in the initiator proteins for rolling circle DNA replication encoded by diverse replicons from eubacteria, eucaryotes and archaebacteria. Nucleic Acids Res 1992; 20:3279-3285.
33. Bocquier AA, Liu L, Cann IK et al. Archaeal primase: bridging the gap between RNA and DNA polymerases. Curr Biol 2001; 11:452-456.
34. Ohmori H, Friedberg EC, Fuchs RP et al. The Y-family of DNA polymerases. Mol Cell 2001; 8:7-8.
35. Kirk BW, Kuchta RD. Arg304 of human DNA primase is a key contributor to catalysis and NTP binding: primase and the family X polymerases share significant sequence homology. Biochemistry 1999; 38:7727-7736.
36. Kato M, Frick DN, Lee J et al. A complex of the bacteriophage T7 primase-helicase and DNA polymerase directs primer utilization. J Biol Chem 2001; 276:21809-1820.
37. Lehman IR, Boehmer PE. Replication of herpes simplex virus DNA. J Biol Chem 1999; 274:28059-18062.
38. Dracheva S, Koonin EV, Crute JJ. Identification of the primase active site of the herpes simplex virus type 1 helicase-primase. J Biol Chem 1995; 270:14148-14153.
39. Trakselis MA, Mayer MU, Ishmael FT et al. Dynamic protein interactions in the bacteriophage T4 replisome. Trends Biochem Sci. 2001; 26:566-572. Review.
40. Mesyanzhinov VV, Robben J, Grymonprez B et al. The genome of bacteriophage phiKZ of Pseudomonas aeruginosa. J Mol Biol 2002; 317:1-19.
41. Lakshminarayan MI, Aravind L, Koonin E. Common origin of four large families of large eukaryotic DNA viruses. J Virol 2001; 75:11720-11734.
42. Waga S, Stillman B. The 46, 48DNA replication fork in eukaryotic cells. Annu Rev Biochem 1998; 67:721-751.
43. Keck JL, Berger JM. DNA replication at high resolution. Chem Biol 2000: 7(3):R63-71
44. Bohlke K, Pisani FM, Rossi M et al. Archaeal DNA replication: spotlight on a rapidly moving field. Extremophiles 2002; 6:1-14.
45. Matsunaga F, Forterre P, Ishino Y et al. In vivo interactions of archaeal Cdc6/Orc1 and minichromosome maintenance proteins with the replication origin. Proc Natl Acad Sci USA 2001; 98:11152-11157.
46. Edgell DF, Doolittle WF. Archaea and the origin[s] of DNA replication proteins. Cell 1997; 89:995-998.
47. Mushegian AR, Koonin EV. A minimal gene set for cellular life derived by comparison of complete bacterial genomes. Proc Natl Acad Sci USA 1996; 93:10268-10273.
48. Leipe DD, Aravind L, Koonin EV. Did DNA replication evolve twice independently ? Nucleic Acids Res 1999; 27:3389-3401.
49. Myllykallio H, Lopez P, Lopez-Garcia P et al. Bacterial mode of replication with eukaryotic-like machinery in a hyperthermophilic archaeon. Science 2000; 288:2212-2215.
50. Zivanovic Y, Lopez P, Philippe H et al. Pyrococcus genome comparison evidences chromosome shuffling-driven evolution. Nucleic Acids Res 2002; 30:1902-1910.
51. Cavalier-Smith T. The neomuran origin of archaebacteria, the negibacterial root of the universal tree and bacterial megaclassification. Int J Syst Evol Microbiol 2002; 52:7-76.
52. Forterre P. Displacement of cellular proteins by functional analogues from plasmids or viruses could explain puzzling phylogenies of many DNA informational proteins. Mol Microbiol 1999; 33:457-465.
53. Villarreal LP, DeFilippis A. Hypothesis for DNA viruses as the origin of eukaryotic replication proteins. J Virol 2000; 74:7079-7084.
54. Takemura M. Poxviruses and the origin of the eukaryotic nucleus. J Mol Evol 2001; 52:419-425.
55. Bell PJL. Viral eukaryogenesis: Was the ancestor of the nucleus a complex DNA virus? J Mol Evol 200; 5:251-256.
56. Keck JL, Roche DD, Lynch AS et al. Structure of the RNA polymerase domain of E. coli primase. Structure of the RNA polymerase domain of E. coli primase. Science 2000; 287:2482-2486.
57. Erzberger JP, Pirruccello MM, Berger JM. The structure of bacterial DnaA: implications for general mechanisms underlying DNA replication initiation. EMBO J 2002; 21:4763-4773.
58. Woese CR, Fox GE. The concept of cellular evolution. J Mol Evol 1977; 1:1-6.
59. Forterre P, Benhachenou N, Confalonieri F et al. The nature of the last universal ancestor and the universal tree of life, still open questions. Biosystem 1993; 28:15-32-LUCA DNA.
60. Gadelle D, Filée J, Bulher C et al. Phylogenomics of type II DNA topoisomerases, Bioassay 2003; 25:232-242.

61. MacNaughton TB, Shi ST, Modahl LE et al. Rolling circle replication of hepatitis delta virus RNA is carried out by two different cellular RNA Polymerases. J Virol 2002; 76:3920-3927.
62. Moraleda G, Taylor J. Host RNA polymerase requirements for transcription of the human hepatitis delta virus genome. J Virol 2001; 75:10161-10169.
63. Wang H, Di Gate RJ, Seeman NC. An RNA topoisomerase. Proc Natl Acad Sci USA 1996; 93:9477-9482.
64. Davey MJ, Jeruzalmi D, Kuriyan J et al. Motors and switches: AAA+ machines within the replisome. Nat Rev Mol Cell Biol 2002; 3:826-835.
65. Brinckman H, Philippe H. Archaea sister group of Bacteria ? Indications from tree reconstruction artifacts in ancient phylogenies. Mol Biol Evol 1999; 16:817-825.
66. Forterre P, Philippe H. Where is the root of the universal tree of life? Bioessays 1999; 21:871-879.
67. Gribaldo S, Philippe H. Ancient phylogenetic relationships. Theor Pop Biol 2002; 61:391-408.
68. Lopez-Garcia P, Moreira D. Metabolic symbiosis at the origin of eukaryotes. Trends Biochem Sci 1999; 24:88-93.
69. Bergerat A, de Massy B, Gadelle D et al. An atypical topoisomerase II from Archaea with implications for meiotic recombination. Nature 1997; 386:414-417.
70. Wadsworth RI, White MF. Identification and properties of the crenarchaeal single-stranded DNA binding protein from Sulfolobus solfataricus. Nucleic Acids Res 2001; 29:914-920.
71. Myllykallio H, Forterre P. Mapping of a chromosome replication origin in an archaeon: Response. Trends Microbiol 2000; 8:537-539.
72. Makiniemi M, Pospiech H, Kilpelainen S et al. A novel family of DNA-polymerase- associated B subunits. Trends Biochem Sci 1999; 24:14-16.
73. Moreira D. Multiple independent horizontal transfers of informational genes from bacteria to plasmids and phages implications for the origin of bacterial replication machinery. Mol Microbiol 2000; 35:1-5.
74. MacNeill SA. DNA replication: partners in the Okazaki two-step. Current Biol 2001; 11:R842-R844.
75. Matsunaga F, Norais C, Forterre P et al. Identification of short 'eukaryotic' Okazaki fragments synthesized from a prokaryotic replication origin EMBO Report 2003; 4:154-158.
76. O'Donnell M, Jeruzalmi D, Kuriyan J. Clamp loader structure predicts the architecture of DNA polymerase III holoenzyme and RFC. Curr Biol 2000; 11:R935-R946.
77. Dervyn E, Suski C, Daniel R et al. Two essential DNA polymerases at the bacterial replication fork. Science 2001; 294:1716-1719.
78. Lipps G, Röther S, Hart C et al. A novel type of replicative enzyme barbouring ATPase, primase and DNA polymerase activity. EMBO J 2003; 22:2516-2525.

CHAPTER 11

Early Evolution of DNA Repair Mechanisms

Jocelyne DiRuggiero and Frank T. Robb

Abstract

DNA repair is critical for the maintenance of genome integrity and replication fidelity in all cells, and therefore was arguably of major importance in the Last Universal Cellular Ancestor (LUCA) as well. Archaea, and hyperthermophiles in particular, are well suited for studying early DNA repair mechanisms from two perspectives. First, these prokaryotes embody a mix of bacterial and eukaryal molecular features. Second, DNA in many archaea is subject to ongoing damage during normal growth under extreme conditions such as high temperature or low pH. Third, recent work suggests that the mutation rates of model hyperthermophiles are quite close to norms for bacteria, indicating that their replication/repair processes are operating with high fidelity at elevated temperatures. The Archaea also have minimal sets of genes involved in all of the major cellular information transfer processes, compared with Eukarya, which have highly paralogous and redundant sets of genes for DNA replication, repair and recombination.

Repair activities have been demonstrated for several hyperthermophiles including our studies with *Pyrococcus furiosus*, an archaeon growing optimally at 100°C. In addition, using comparative genomic analysis and the genome sequence of several hyperthermophilic archaea, homologs of conserved eukaryotic and bacterial DNA repair proteins have been identified. Although close to 100 microbial genome sequences have been analyzed, including 16 from the Archaea, so far many highly conserved repair genes are missing in some or in all of the archaeal genomes. This may be the result of low sequence conservation across the three domains of life, preventing identification using sequence similarity searches. It is possible, and proven in some instances, that Archaea have novel versions of repair proteins.

Here we argue that the commonality of mechanisms and protein sequences, shared between prokaryotes and eukaryotes for several modes of DNA repair, reflects diversification from a minimal set of genes. However, for several pathways, the close similarity between components of eukaryal and archaeal repair pathways suggests that those specific processes likely evolved independently in the bacterial and archaeal/eukaryal lineages.

Introduction: Life in Extremis

For cells to survive, they must repair DNA lesions continuously and accurately. In response to continuous hydrolytic and oxidative DNA damage, cells must have an effective network of repair systems that recognize, remove, and rebuild the injured sites. DNA damage such as abnormal nucleotides (modified, fragmented, cross-linked) single-strand (ss) and double-strand (ds) breaks, abasic sites, inter- and intra-strand crosslinks are produced endogenously by metabolic byproducts or exogenously by environmental conditions.[1,2] Once these lesions are in place, they must be reversed by enzymatic DNA repair systems to prevent lethal replication blocks, transcriptional stoppages, and mutations.[2] Evidence of the importance of DNA repair systems is clearly shown by the great diversity of DNA repair processes.

The Genetic Code and the Origin of Life, edited by Lluís Ribas de Pouplana.

The Archaea, one of the three domains of life proposed by Carl Woese,[3] are now known to have very significant molecular similarities to the Eukarya, particularly in functions of information processing such as transcription and DNA synthesis and modification. Since DNA repair belongs to this class, it is tempting to speculate that simple, and possibly ancestral versions of the eukaryotic DNA repair processes will be found in the Archaea. Our emphasis in this chapter will be on the Archaea, since they are to the base of the Tree of Life in most of its versions, and because they combine many of the otherwise distinctly bacterial or eukaryotic molecular characteristics.[4,5] Archaea are, with few exceptions, associated with some type of extreme environment — extreme heat or cold, high salt, extremes in pH and pressure.[5] In addition, because of the life styles they pursue in extreme environments, we and others have previously established that ongoing DNA damage in extremophiles necessitates continuously active DNA repair processes.[6-8] The hyperthermophiles generally have small genomes and relatively sparse growth requirements, and this has led many to propose that the LUCA may have been hyperthermophilic.[9-11] If this were true, the DNA repair mechanisms and processes common to hyperthermophiles could be our closest approach to the systems that were essential to the existence of the LUCA.

One of the fascinating features of hyperthermophiles is their ability to grow at temperatures exceeding the nominal denaturation temperatures of their DNA and RNA. Counter intuitively, the G+C content of genomic DNA in hyperthermophiles is spread over a wide range, and many hyperthermophiles have genomes with relatively low G+C content.[12] Part of the explanation might be in the fact that in all cases that have been studied, salts and compatible solutes are present at high concentrations in the cytoplasm of hyperthermophiles. For example, *Pyrococcus furiosus* has an intracellular potassium concentration of 700 mM. The counterions for K^+ are either Di-myo-inositol-1,1'-phosphate, or glutamate.[13] *Methanopyrus kandleri*, a strain that grows up to 110°C, and cannot grow below 90°C has a cytoplasmic concentration of up to 2.5 M cyclic diphosphoglycerate.[14] These solutes have been shown in many studies to be critical for achieving full thermostability in enzymes from hyperthermophiles,[13] and to lead to stabilization of DNA duplexes. High salt has also been shown to prevent DNA breakage in vitro, and covalently closed circular DNA is highly resistant to DNA breakage compared to nicked- or linear-DNA.[15]

Variation in temperature will alter the superhelical density of a DNA replicon by changing the pitch of the DNA helix.[15] Therefore, DNA topoisomerase activity has been studied in several hyperthermophiles. All hyperthermophilic genomes, from bacteria and archaea, sequenced so far contain at least one copy of a type I topoisomerase, called reverse gyrase.[16-19] The enzyme consists of a helicase domain and a topoisomerase I domain.[16-19] The apparent role of reverse gyrase is in maintaining the superhelical density of chromosomes and plasmids in neutral or positively supercoiled condition,[16-19] however, mechanisms of regulation of superhelical density are still unclear. Topoisomerase V from *M. kandleri* is apparently another dual functional enzyme with an important function to carry out backbone scission during base excision repair at high growth temperatures.[20]

DNA binding proteins in Euryarcheota include archaeal histones, which form tetrameric nucleosomes that maintain DNA in positive supercoils in high salt conditions,[21,22] stabilizing and compacting DNA in vitro. In the Crenarcheota, DNA binding proteins, including a novel chromatin forming protein named Alba,[23,24] contribute to DNA stability by raising the T_m of nucleoprotein complexes. The binding properties of Alba are modulated by acetylation in a manner analogous to eukaryotic histones.[23,24] Mechanisms of prevention of hydrolytic DNA breakage at high temperature,[6,25] as well as rapid and precise ds break repair[6,25] are dependent on nucleoprotein complexes, which must alternate between being tight enough to protect the DNA, yet able to release according to the demands of transcription and translation. Interestingly, the genome sequences of *Thermoplasma acidophilum*[26,27] and *T. volcanium*[26,27] do not contain either Crenarcheote or Euryarcheote variants of archaeal histones, but instead contain copies of the bacterial basic, DNA binding protein HU. Presumably this was the result of lateral gene transfer, however it raises the interesting question as to whether the *Thermoplasma* spp. represent upward or downward mobility in terms of their temperature ranges.

Most of the genes that are required for DNA replication have been identified in archaeal genomes recently.[4] In the Euryarcheota, Ishino et al[28] identified a dimeric replicative DNA polymerase with no obvious sequence similarity to either bacterial or eukaryotic polymerases. This polymerase is conserved in all of the sequenced genomes of the Euryarcheota.[29] Additional B-class DNA polymerases are also found in the Euryarcheota, functioning in lagging strand synthesis and DNA repair. In Crenarcheota, the B-polymerases are found, often in multiple paralogous genes, functioning in both repair and replication.[4] The close similarity between components of eukaryal and archaeal replication forks suggest that the archaeal process resembles that of eukaryotes, although in term of origin utilization and replication speed, the two prokaryotic domains share a similar replication mode.[30]

Experimental Evidence of DNA Repair Mechanisms

We previously reported that the hyperthermophile *P. furiosus* has exceptional resistance to ionizing radiation.[6] *P. furiosus* cells irradiated in late exponential phase showed no loss of viability up to 2,000 Gy. By comparison, a human exposed to 5 Gy (1 Gy = 100 Rad) of ionizing radiation would suffer almost certain death. We showed that the *P. furiosus* 2 Mb chromosome, fragmented into pieces from 500 to shorter than 30 kb, after 60-Co gamma-irradiation at a dose of 2,500 Gy, was fully reassembled upon incubation at 95°C. These results imply that *P. furiosus* is equipped with extremely efficient recombination repair mechanisms, as well as oxidative damage repair pathways. Extreme levels for resistance to gamma rays were also reported for *Thermoccoccus stetteri* and *Desulfurococcus amylolyticus*,[31] and more recently for *P. abyssi*.[32]

Exchange and recombination of genetic markers was demonstrated in the acidophilic hyperthermophile, *Sulfolbus acidocaldarius*, using auxotrophic mutants.[33,34] This microorganism's natural mechanism for conjugation and recombination (termed marker exchange) allowed for the first analysis of genetic properties of homologous recombination in hyperthermophilic archaea.[35] Simple mating was shown to resolve mutations separated by less than 28 base pairs, suggesting that this method could be use to precisely map large numbers of mutations to the *S. acidocaldarius* chromosome.[35]

Efficient photoreactivation following exposure to UV irradiation has been reported for *S. acidocaldarius*[36,37] and for several species of halophiles.[38,39] Kinetics of photoreactivation for several wavelengths of light suggests the presence of a broad action spectrum DNA photolysase in *S. acidocaldarius*.[36,37] Mc Cready[38,39] reported rapid repair of UV-induced cyclobutane dimers and 6-4 photoproducts in the light and in the dark, suggesting the presence of both photoreactivation and nucleotide excision repair mechanisms in two species of halophilic archaea. Evidence of nucleotide excision repair (NER) was also found in the mesophilic archaeon, *Methanobacterium thermoautotrophicum*.[40] Cell extracts of this methanogen were found to remove UV-induced (6-4) photoproducts with an excision pattern similar to that of bacteria. This is not surprising since homologs of the *uvrA*, *uvrB*, *uvrC* and *uvrD* protein encoding genes have been found in the genome of this archaeon. However, no homolog of this bacterial-type of NER pathway was found in other archaeal genomes with the exception of that of *Halobacterium* NRC-1.[41] The possibility of lateral gene transfer of bacterial-type NER pathways will be discussed below.

Molecular Mechanisms

Whereas for replication, many conserved proteins found in the Eukarya seem to be present in the Archaea, the situation is rather different for repair proteins. Using comparative genomic analysis, we found that almost all the known conserved DNA repair pathways are represented in the Archaea, but many conserved repair genes from those pathways are missing in some or in all of the archaeal genomes sequenced so far.[7] Biochemical studies of hyperthermophilic recombinant enzymes have greatly complemented the comparative genome analyses in elucidating DNA repair pathways in Archaea. In addition, the ability to purify and characterize thermostable variants of DNA modifying enzymes enables biochemical studies that would be difficult or impossible to carry out with the mesophilic homologs of these proteins. Below are examples where the structure function characterization of hyperthermophilic proteins enabled insights to be gained as to the role of homologs in much more complex eukaryotic systems. Most of those studies have been carried out with either *Sulfolobus*, an aerobic acidophile growing optimally at 80°C and pH 2 to 4, or *Pyrococcus*, an anaerobic heterotroph growing optimally at 100°C.[42,43]

Repair Recombination

The basis for the extraordinary DNA break repair capabilities of hyperthermophiles that implies exceptionally active reciprocal recombination systems[6] is being elucidated. The homologous recombination (HR) system, which is a key to their ds break repair capabilities, is emerging from studies of the RecA/Rad51 archaeal homolog, RadA,[44-50] the Mre11/rad50 complex in *P. furiosus*[51-54] and Holliday junction resolving enzymes in *Sulfolobus solfataricus*[55] and *P. furiosus*.[56,57] DNA ds break repair is a complex process that requires multiple enzymatic and structural activities to efficiently detect and process the broken DNA ends. In HR the end is first digested by a nuclease to form a 3' single stranded DNA tail. This tail is paired with a homologous DNA segment to allow strand invasion, homologous recombination and/or DNA repair synthesis.[2]

The role of the archaeal homolog, RadA, in DNA repair was first established by Sandler et al[44,58] and Woods and Dyall-Smith.[59] In vitro studies of RadA from *S. solfataricus, P. furiosus, P. islandicum* and *D. amylolyticus* further demonstrated its functional and structural resemblance to the RecA and Rad51 proteins.[44-50] These studies showed that the RadA protein is a DNA-dependent ATPase, forms nucleoprotein filaments on DNA, promotes formation of joint molecules and can catalyze DNA pairing and strand exchange. A second Rad protein has been found in the genomes sequence of *P. furiosus* and all the other euryarchaeal genome sequence so far.[7,46,47] It is called RadB and its function remains obscure. Rad B is closely related to the eukaryotic Rad51/DMC1 protein family, but while RadA is clustered with the eukaryotic enzymes, RadB branched separately.[7] Its interaction with one of the two subunits (DP1) of polII and RadA, suggested that it could be the functional homolog of the eukaryotic Rad55/Rad57 complex that promotes replication protein A-associated strand exchange activity by Rad51.[60] However, RadB also regulates the cleavage activity of the Hjc resolvase in *P. furiosis* suggesting that it could be involved in the resolution of Holliday junction during homologous recombination.[46,61] Recent studies with *E. coli*[61] suggest that homologous recombination systems are essential in the reestablishment of inactivated replication forks under normal growth conditions, performing an "essential housekeeping function". This housekeeping function might be of even greater importance in hyperthermophiles where the DNA is constantly subjected to damages due to the exposure of the cells at high temperature.[25]

Single-stranded DNA binding protein (SSB in Bacteria and RPA in Eukarya) plays essential roles in DNA replication, recombination and repair by binding and protecting ss DNA.[2] Archaeal single-stranded binding proteins have been characterized from both kingdoms, the Crenarchaeota and the Euryarchaeota, and while they differ from each other, they are all more closely related to the Eukaryal RFA than the bacterial SSB.[62-64] *P. furiosus* RPA and *S. solfataricus* SSB have been shown to to bind specifically ssDNA, and to greatly stimulate RadA-mediated strand-exchange reactions in vivo and in vitro.[62-64] In addition, *P. furiosus* RPA co-immunopreciptated with the recombination proteins RadA and Hjc (Holliday junction resolvase), clearly indicating its important role in DNA recombination.[62-64]

In Bacteria, the major pathway for ds break processing is carried out by the well characterized RecBCD complex.[2] In some bacterial systems, other pathways such as the recFOR pathway and the SbcC-SbcD complex are involved in ds break repair, although the role of the later is not well understood.[2] In Archaea, comparative genomic analyses identified homologs of the eukaryotic Mre11 and Rad50 proteins. In fact homologs of Mre11/SbcC and Rad50/SbcD are found in all domains of life, and are essential for genome integrity.[65]

Genetics studies in Eukarya indicate that the Mre11/Rad50 (MR) complex plays a key role in ds break repair. Rad50 contains a bipartite ATPase domain and an antiparallel coiled-coil domain. ATP promotes the association of two Rad50 ATPase domains, and it promotes DNA binding to the complex.[54] Mre11 has ssDNA exonuclease, 3' to 5' dsDNA exonuclease and hairpin opening activities.[66] Hopfner et al[52] resolved the crystal structure of the *P. furiosus* Mre11, and described the architecture of the Mre11/Rad50 complex. It is a heterotetrameric DNA processing head at the end of a double coiled-coil linker, containing two DNA-binding/processing active sites. The tight complex of Mre11 and the Rad50 ATPase domain suggest a mechanism for direct control of Mre11 exonuclease activity by conformational changes in the Rad50 ATPase domain.[52] However the role of the coiled-coil DNA domain of Rad50 remained to be elucidated. A study by De Jager et al[67] revealed a great flexibility in the Rad50 coiled-coil structure in a human Mre11/Rad50 complex. The crystal resolution of the *P. furiosus* Rad50 coiled-coil revealed that Rad50 functions as an ATP-modulated DNA cross-linker.[66] The crystal structure and electron micrographs suggest that this coiled-coil structure forms a zinc-mediated bridge between two DNA-binding heads, providing a link between two homologous DNA fragments, initiating and stabilizing displacement loops.[66]

The Mre11 activity does not seems suitable for 5' to 3' processive degradation of DNA ends before recombination, because it degrades DNA ends 3' to 5', opposite to the necessary direction. Therefore, the formation of the 3' DNA tail probably requires an additional nuclease. In Eukarya, the MR complex acts in association with a third protein, Nbs1 in human and Xrs2 in yeast.[2] The Nbs1 and Xrs2 are functional homologs, although in contrast to Nbs1, Xrs2 is not highly conserved.[68] The Mre11/Rad50/Nbs1 complex has been proposed as a candidate for the generation of 3' ssDNA tails, and Tauchi et al[68] showed that Nbs1 is essential for HR repair in higher vertebrate cells. A novel 5'-3' nuclease, NurA, has been characterized from the archaeon, *S. acidocaldarius.*[69] It has both a ss endonuclease activity and 5' to 3' exonuclease activity on ss and dsDNA. The authors suggest that NurA could be the third partner in the MR complex that processes ds breaks into 3' single stranded DNA tails in hyperthermophilic archaea.[69]

Holliday junction resolving enzymes (Hjc) specifically recognize four-way DNA junctions, resulting from the process of homologous recombination, cleaving them to generate recombinant DNA duplexes. Hjc have been isolated from most organisms from viruses to Bacteria and Eukarya, and have been recently characterized in two archaea *P. furiosus* and *S. solfataricus.*[70,71] This suggests that Holliday junctions are conserved intermediates for homologous recombination in the three domains of life. The archaeal resolvases do not have sequence similarity with any known proteins but are highly conserved within their domain.[55] They are both branch-dependent endonucleases that resolved synthetic Holliday junctions in presence of divalent metals.[70,71] The crystal structure of *P. furiosus* Hjc revealed a dimeric molecule with a positively charged surface containing highly conserved amino acid residues[72] Its fold is similar to that of the type II restriction endonucleases, including the conformation of the canonical catalytic residues. Mutational analysis and the *P. furiosus* Hjc, and further structural studies, suggest that the flexible N-terminal section of the molecule enhances the stability of the enzyme-junction DNA complex, and contributes to the correct positioning of the cleavage site of the DNA to the catalytic site of the enzyme.[72] The crystal structure of Hjc from *S. solfataricus*[73] is similar to that of *P. furiosus* Hjc with the exception of two regions, the N-terminal segment and Lys30/Lys31 loop, that were structurally disordered.[73] Using biophysical methods, Kvaratskhelia et al[74] report that multiple *S. solfataricus* Hjc dimers bind to each synthetic four-way junctions, producing significant distortion of the junction structure, resulting in higher order complexes and DNA cleavage inhibition. This auto-inhibition can the relieved by adding competitor duplex DNA or Sso7d, a ds DNA binding protein. The authors suggest that it may represent a mechanism by which the nuclease activity of Hjc is repressed in the absence of Holliday junctions.[74] The elucidation of the precise recognition and cleavage mechanisms must await the resolution of the structure of a Hjc-Holliday junction complex.

Error-Prone Repair

Despite efficient repair mechanisms, DNA lesions often persist, impeding the progression of DNA replication folks. Organisms have evolved specialized DNA polymerases that are able to progress through the DNA lesions. This pathway, termed translesion synthesis, error-prone repair or lesion bypass synthesis, is inherently mutagenic because of the miscoding nature of most damaged nucleotides.[75]

The Y-family of polymerases facilitates translesion replication for a variety of DNA lesions. They belong to the UmuC/dinB/Rev1/Rad30 superfamily of proteins, and share very little similarity with other polymerases family identified so far. UmuC proteins are only found in bacteria and in the archaeon *Halobacterium* NRC-1, whereas proteins from the Rad30 branch are only found in eukaryotes.[76,77] In contrast, the DinB family is found in the three domains of life. However, in the Archaea, it has only been found in the three *Sulfolobus* and the *Halobacterium* NRC-1 genomes. This raises questions concerning the type of polymerase or mechanism for translesion replication in the other archaea. The *S. solfataricus* P2 protein, called Dpo4 (DNA polymerase IV) was characterized in vivo and its crystal structure has been determined.[78] In addition to being able to replicate through abasic sites, which is a property of the DinB polymerases, Dpo4 can facilitate translesion bypass of UV lesions such as thymidine dimers and 6-4 pyrimidine-pyrimidome lesion.[78] In that respect it is more analogous to eukaryotic polymerase polη and could play the role of both DinB-like and Rad30-like polymerases in the Archaea. This is not surprising considering the reduction in the number of homologs found in Archaea when compared to Eukarya for replication and transcription. The 3-D structure of Dpo4 shows a widened active site able to accommodate 2 nucleotides and is a good model to understand the repair mechanism of lesions such as pyrimidine dimers. This is of significance since mutations in polymerase polη have been associated with human cancers and Xeroderma pigmentosa, a DNA repair defect that causes light hypersensitivity.[78]

Nucleotide Excision Repair

Nucleotide excision repair is a generic repair process that requires several successive steps: recognition of the lesion, cleavage of the strand containing it, removal of an oligonucleotide, and resynthesis with high fidelity at the site. Commonly, incisions bracket the site of the lesions closely. This process is very highly conserved in the Bacteria, as the UvrABCD pathway, and has been studied in yeast and human where at least eight proteins have been identified with activities corresponding to the bacterial system. The Xeroderma pigmentosa syndrome has been attributed to any of several mutations affecting this pathway. Several associated with NER in eukarya, the XPF/rad1, XPB/rad25, XPG/rad2 and XPD/rad3 human/yeast homologs, occur in NER functions in Archaea. One exception is *M. thermoautotrophicum*. Recent findings show that apurinic sites can be recognized by cell free extracts of this thermophilic methanogen,[40] indicating that the full suite of NER enzymes are present. In this case, however, NER is carried out by "bacterial" UvrABCD homologs that most likely are the result of interdomain lateral gene transfer. Interestingly, no damage recognition has been found in the archaeal genomes.

The recent finding that a type IB topoisomerase from the hyperthermophile *M. kandleri* has apurinic/apyrimidinic (AP) site-processing activities is very significant.[20] This enzyme not only makes incisions in undamaged DNA while acting as a topoisomerase, it is apparently functional in base excision DNA repair. It incises the phosphodiester backbone at the AP site, and consequently, at the AP endonuclease cleaved AP site, removes the 5' 2-deoxyribose 5-phosphate moiety so that a single-nucleotide gap with a 3'-hydroxyl and 5'-phosphate can be filled by DNA polymerase and ligase. Therefore, additional activities associated with type I topoisomerases may be part of the solution to the problem of the missing archaeal NER genes.

A novel structure specific endonuclease, Hef (helicase-associated endonuclease for fork-structured DNA) that has been characterized in *P. furiosus*.[79] The protein contains 2 domains. The N-terminal domain is similar to that of DEAH helicases whereas the C-terminal resembles eukaryotic XPF nucleases that are involved in NER. Although, no helicase activity was detected in recombinant Hef protein, Komori et al[79] showed that it has an endonuclease activity that cleaves at the 5'-side of nicked or branched DNA duplex. This protein has been found in six of the eight euryarchaeal genomes sequenced so far, and proteins consisting only of the C-terminal domain, similar to XPF proteins, have been found in the Crenarchaeal genomes. This data, all together suggest a possible dual role for Hef in the restoration of arrested replication folks and in nucleotide excision repair.[79]

Mismatch Repair

Mismatch repair is one of the most versatile and highly conserved DNA repair processes. It ensures genetic accuracy by preventing the accumulation of mutations that can be potentially deleterious to the organism. Experimental data from the hyperthermophilic acidophile *S. acidocaldarius*,[80,81] and evidence from the comparison of the three *Pyrococcus* genomes,[82] indicate relatively low mutation rates considering the DNA damaging and genotoxic nature of the habitats of these extremophiles. In addition, Bell and Grogan[83] obtained mutant strains of *S. acidocaldarius* that display extremely high rates of spontaneous mutation, i.e., a bacterial-like mutator phenotype. This data taken with the fact that the mutants did not show a large increase in sensitivity to DNA-damaging treatments, demonstrates that genetic accuracy-enforcement mechanisms can be inactivated by mutation in Archaea, and it raises the question of the biochemical nature of this mechanism.

Homologs of the conserved mismatch repair proteins, MutS and MutL are found in both the bacterial and eukaryal domains, including hyperthermophilic Bacteria. However the archaeal genomes sequenced to date do not reveal any evidence of mismatch repair systems, with two exceptions. *Halobacterium* NRC1 genome contains MutS/MutL protein encoding genes,[41] and that of *M. thermoautotrophicum* displays a *mutS* gene.[84] MutS-like encoding genes are also found in most archaeal genomes, including that *M. thermoautotrophicum* and in the *Pyrococcus spp.*,[82] but their deduced amino acid sequence is more related to that of eukaryotic MutSII proteins that are involved in chromosome segregation, than to bacterial MutS. In addition, MutL homolog were not identified in any genome apart from that of *Halobacterium* NRC-1, where it is likely the result of lateral gene transfer from bacteria.[41]

However, the absence of a homolog does not necessarily imply the absence of the activity. Aravind et al[65] observed that despite the limited number of conserved domains such as ATPases, DNA binding and protein-protein interaction domains, very few repair protein orthologs are conserved across the three domains of life. It is therefore possible that Archaea have novel versions of repair proteins, which might account for the inability to detect some of the conserved repair proteins when using sequence similarity searches on archaeal genomes. Those novel repair proteins might be found amongst the large number of conserved hypothetical genes that have been reported in archaeal genomes. As an example, Koonin's group[85] using a whole genome context analysis, identified a five-gene core in hyperthermophilic archaea with predicted functions that strongly suggest a role in DNA repair mechanisms, including a DNAse, a helicase and a novel DNA polymerase.[85]

The alternative to novel repair systems in the Archaea is that other repair processes, such as base excision repair (BER), could carry out some of the functions that are assumed to be carry out by MutS and MutL in bacteria.[83] Genetic and biochemical analysis of archaeal mutator phenotypes might allow the identification of the enzymes responsible for mismatch repair in those microorganisms.

Other Pathways

Direct Damage Reversal (DDR)

The most studied mechanism for DDR is photoreactivation (PHR). In this process, DNA photolyase catalyzes the repair of pyrimidine dimers in UV-damaged DNA, a reaction that requires visible light. Class I photolyases (*E. coli*, yeast) contain 1,5-dihydroFAD (FADH2) plus a pterin derivative (5,10-methenyltetrahydropteroylpolyglutamate). In class II photolyases (*Streptomyces griseus*, *Scenedesmus acutus*, *Anacystis nidulans*, *Myxococcus xanthus*, marsupials and

M. thermoautotrophicum) the pterin chromophore is replaced by an 8-hydroxy-5-deazaflavin derivative.[86]

The two classes of PHR enzymes exhibit a significant amino acid sequence similarities. Interestingly, *M. thermoautotrophicum* has a homolog of this "higher eukaryote" PHRII type photolyase although this strain is a methanogen and is limited to strictly anaerobic conditions. Yasui et al[87] reported a new class of photolyase with low similarity to the bacterial type photolyase, now called PHRII. PHRII also occurs in *Myxococcus xanthus*, goldfish and marsupials. Photolyase genes encoding PHRI homologs have also been identified in the genome sequences of *Halobacterium* NRC-1, (2 copies), and *Sulfolobus* spp. It is tempting to speculate that the LUCA contained copies of both PhrI and PhrII, and that PHR gene loss has been a frequent occurrence at all levels, involving both PHRI and PHRII.

Base Excision Repair (BER)

In this repair mechanism, an altered base is detected and removed by hydrolysis of the glycosidic bond between the base and the deoxyribose moiety. The DNA N'-glycosylases are key enzymes in this process, and a wide variety have been described. The families of N-glycosylases differ in their recognition of both the bases and their context as well as the specific type of damage or alteration. For example, the recognition of T:G mismatches in DNA is achieved by a specific mismatch glycosylase that excises thymine.[88] The product of N-glycosylase action, an AP site, is recognized and handled by several types of repair system including excision repair, recombination repair and long patch repair.

Homologous DNA glycosylase-like or mismatch N-glycosylases (MUG) have been characterized in *T. maritima*, the cryptic plasmid of *M. thermoautotrophicum*, *Pyrobaculum aerophylum* and *Aeropyrum pernix*.[89-92] These enzymes recognize U/G and T/G mismatches, or uracil in DNA, and are of particular importance in hyperthermophiles where the rate of cytosine and 5-methylcytosine deamination is greatly enhanced, with the potential of generating C to A transition mutations. Interestingly, Slesarev's group[20] recently reported AP endonuclease and DNA lyase activity in the Topo V (type IB) of *M. kandleri* contributing further to base excision repair in this hyperthermophile (see section on nucleotide excision repair).

The recent report of a 2.0 Å resolution crystal structure of the *A. fulgidus* thermophilic MIG (Mismatch Glycosylase) enzyme suggests that MIG distorts the target thymine nucleotide by twisting the thymine base approximately 90 degrees away from its normal anti position within DNA.[88] T:G mismatch repair can be initiated by a specific mismatch glycosylase (MIG) that is homologous to the helix-hairpin-helix (HhH) DNA repair enzymes. The authors proposed that functionally significant differences exist in DNA repair enzyme extrahelical nucleotide binding and catalysis that are characteristic of whether the target base is damaged or is a normal base within a mispair. These results explain why pure HhH DNA glycosylases and combined glycosylase/AP lyases cannot be interconverted by simply altering their functional group chemistry, and how broad-specificity DNA glycosylase enzymes may weaken the glycosydic linkage to allow a variety of damaged DNA bases to be excised.[88]

A variant of this adaptive recognition characteristic of glycosylases is seen in the uracil N-glycosylase from *A. fulgidus* (Afung) that recognizes uracil, which is produced at high levels by heat-induced cytosine deamination in hyperthermophiles.[93] The recognition and subsequent excision are adaptive in *A. fulgidus*, since it is stimulated at high growth and supra-optimal temperatures. Base excision repair of DNA alkylation damage is initiated by a methylpurine DNA glycosylase (MPG) function.[93] Such enzymes have previously been characterized from bacteria and eukarya, but not from archaea. Birkeland et al[94] identified activity for the release of methylated bases from DNA in cell-free extracts of *A. fulgidus*, an archaeon growing optimally at 83°C. An open reading frame homologous to the *alkA* gene of *E. coli* was overexpressed and identified as a gene encoding an MPG enzyme (M(r) = 34 251), hereafter designated *afalk*A. The purified AfalkA protein differs from *E. coli* AlkA by excising alkylated bases only, from DNA, in the following order of efficiency: 3-methyladenine (m(3)A) >> 3-methylguanine approximately 7-methyladenine >> 7-methylguanine. Although the rate of enzymatic release of m(3)A is highest in the temperature range of 65-75°C, it is only reduced by 50% at 45°C, a temperature that does not support growth of *A. fulgidus*. At temperatures above 75°C, nonenzymatic release of methylpurines predominates. The results suggest that the biological function of AfalkA is to excise m(3)A from DNA at sub-optimal and maybe even mesophilic temperatures. This hypothesis is further supported by the observation that the *afalk*A gene function suppresses the alkylation sensitivity of the *E. coli* tag *alk*A double mutant.[88]

Did Basic DNA Repair Mechanisms Evolve More Than Once

Comparative sequence analyses of proteins from the three domains of life prompted Eugene Koonin and coworkers to propose that DNA replication evolved twice independently, once in the bacterial lineage and the other in the archaeal/eukaryal lineage.[95] Furthermore, the universal conservation of some components of the DNA repair machinery suggested that the LUCA had DNA, but did not replicate in the same way modern cells do.[95] Leipe et al[95] proposed that LUCA was actually a RNA-centered organism with a possible DNA replication intermediate. Regardless of the information storage mode, the enzymatic basis for nucleic acid repair of strand breaks, and recombination repair, must have been present in order to preserve the fidelity of information transfer from generation to generation.

Bacterial and archaeal DNA repair systems share many common components, including glycosylases, several helicases and the B-type polymerases. It is possible to conjecture a primitive mode of DNA handling that combined DNA replication and DNA repair. If DNA replication in its modern form did evolve more than once,[95] then basic repair components such as the repair helicase in *M. kandleri*, which combines activities associated with base excision repair,[96] and the newly described branch-specific nuclease complex described by Komori et al[79] may have great significance. They may represent relics of such a combined system that required components of excision repair, base excision, and the resolution of recombinant DNA molecules during repair. Leipe et al[97] have suggested that the replicative helicase in bacteria is derived from *recA* paralogs, providing another instance of possible overlap between repair and replication, and suggesting a mechanism for building DNA repair enzymes by modular increments from preexisting components. If the mismatch repair systems, which appear to be active in the Archaea but cannot be detected by similarity searches, can be characterized and appear to be unlike the Bacterial/Eukaryal MutS/L systems, this would be prima facie evidence that at least one major system has multiple origins. The same argument can be used to support a separate origin of the photolyase systems in Archaea/Eukaryotes and in Bacteria, or the NER system in Bacteria and that of Eukarya. Consequently, it is likely that some of the more diverse systems for DNA repair such as NER, MMR, BER evolved independently in the bacterial and archaeal/eukaryal lineages. If those systems where already present in the LUCA, then they may have been lost and supplanted by lineage-specific variants, as has occurred with the PHR enzymes.

Aravind et al[65] observed that despite the limited number of conserved domains such as ATPases, DNA binding and protein-protein interaction domains, very few repair protein orthologs are conserved across the three domains of life. Factors such as the chemical environment of an organism, as well as the higher order of organization of the DNA within the organism most likely had a great influence on the composition of the DNA repair systems that were retained. Of all the known repair proteins, RecA/Rad51/RadA seems to be the only one present in every genome analyzed.[98] This is indeed surprising considering the key role of DNA repair processes in maintaining the integrity of the cell genetic material. Hjc resolvases are another group of enzymes critical in HR. The evolution of these enzymes may have resulted from lateral gene transfer, lineage-specific gene loss and nonorthologous gene displacement.[99] Sequence analysis of Hjc resolvase and other nucleases, suggests that the hjc resolvase function evolved independently from at least four distinct structural folds. In Bacteria, the main ancestral resolvase belongs to the RuvC superfamily (RNaseH fold), whereas archaeal Holliday junction resolvases (AHJR) are part of a newly defined class of endonucleases that evolved from a nuclease fold also found in DNA-specific restriction enzyme Mrr, RecB and *P. horikoshii*-type ATPase (PHAC) protein families.[99] Both of these nuclease folds are largely absent in eukaryotes where the identity of the proteins required for HR branch resolution remains unknown.

Lateral gene transfer (LGT) can also be a factor that influences the actual lay out of DNA repair proteins in the three domains of life. For example, homologs of the bacterial-type NER pathway, *uvrA/B/C/D*, were found in the genomes of *M. thermoautotrophicum* and *Halobacterium* NRC-1.[41,84] Those are the only archaea in which those genes have been found, strongly suggesting the possibility of LGT from bacteria. *Halobacterium* NRC-1 is also the only archaean with the mismatch repair proteins, MutS and MutL, whereas only MutS was found in the genome of *M. thermoautotrophicum*. This suggests LGT of this operon from the Bacteria. UmuC proteins are only found in bacteria and in the archaeon *Halobacterium* NRC-1, whereas proteins from the Rad30 branch are only found in eukaryotes.[77] In contrast, the DinB family is found in the three domains of life. However, in the Archaea, it has only been found in the three *Sulfolobus* and the *Halobacterium* NRC-1 genomes, again strongly suggesting LGT. *Halobacterium* NRC-1 seems to have a larger number of bacterial-acquired genes than any other archaea, might be the result of a life style in a mesophilic environment where many bacteria can also be found. However, our assessment of gene origin should be somewhat cautious, and consider that the LUCA might have encoded some of those repair genes that were then lost in early eukaryal evolution.

Conclusion

The DNA repair systems in thermophiles can be viewed as minimal systems that get the task of genome maintenance done rapidly and with high fidelity, as is required by the lifestyle of those organisms. The retention of DNA repair systems is determined by the types of DNA damage likely to be incurred by the lineage, and therefore the repair systems in hyperthermophiles appear to use many DNA modifying proteins in dual roles to carry out repair and replication, such as the TopoIB enzymes of *M. kandleri*[20] and the translesion bypass polymerases in *Sulfolobus* spp,[78] Archaea with lower optimal growth temperatures, such as *Halobacterium* spp and *M. thermoautotrophicum*, with more "headspace" in their genomes[100] have acquired mainly bacterial systems. This might resemble the early stages of the process whereby the early Eukarya built up the complexity and versatility of their DNA repair repertoires to the multiply redundant systems seen today in Eukarya such as yeast and human.

Acknowledgements

We acknowledge support from NASA, NSF, DOE, HFSP and the Knut and Alice Wallenberg Foundation.

References

1. Ward JF. DNA damage and repair. In: Nickoloff JA, Hoekstra MF, eds. Totowa, NJ: Hunana Press Inc., 1998:65-84
2. Friedberg EC, Walker GC, Siede W. DNA Repair and Mutagenesis. Washington, D.C: ASM Press, 1995.
3. Woese CR, Kandler O, Wheelis ML. Towards a natural system of organisms: Proposal for the domains Archaea, Bacteria, and Eucarya. Proc Nat Acad Sci USA 1990; 87:4576-4579.
4. Bohlke K, Pisani FM, Rossi M et al. Archaeal DNA replication: spotlight on a rapidly moving field. Extremophiles 2002; 6:1-14.
5. Stetter KO, Fiala G, Huber R et al. Hyperthermophilic microorganisms. FEMS Microbiol Rev 1990; 75:117-124.
6. DiRuggiero J, Santangelo N, Nackerdien Z et al. Repair of extensive ionizing-radiation DNA damage at 95°C in the hyperthermophilic archaeon Pyrococcus furiosus. J Bacteriol 1997; 179:4643-4645.
7. DiRuggiero J, Brown JR, Bogert AP et al. DNA repair systems in Archaea: Mementos from the last universal common ancestor? J Mol Evol 1999; 49:474-484.
8. Grogan DW. The question of DNA repair in hyperthermophilic archaea. Trends Microbiol 2000; 8:180-185.
9. Di Giulio M. The universal ancestor was a thermophile or a hyperthermophile. Gene 2001; 281:11-7.
10. Pace NR. Origin of life—facing up to the physical setting. Cell 1991; 65:531-3.
11. Barns SM, Delwiche CF, Palmer JD et al. Perspectives on archaeal diversity, thermophily and monophyly from environmental rRNA sequences. Proc Natl Acad Sci USA 1996; 93:9188-93.

12. Galtier N, Lobry JR. Relationships between genomic G+C content, RNA secondary structures, and optimal growth temperature in prokaryotes. J Mol Evol 1997; 44:632-636.
13. Santos H, da Costa MS. Compatible solutes of organisms that live in hot saline environments. Environ Microbiol 2002; 4:501-509.
14. Slesarev AI, Mezhevaya KV, Makarova KS et al. The complete genome of hyperthermophile Methanopyrus kandleri AV19 and monophyly of archaeal methanogens. Proc Natl Acad Sci USA 2002; 99:4644-4649.
15. Marguet E, Forterre P. Protection of DNA by salts against thermodegradation at temperatures typical for hyperthermophiles. Extremophiles 1998; 2:115-122.
16. Bouthier De LA Tour C, Portemer C, Kaltoum H et al. Reverse gyrase from the hyperthermophilic bacterium Thermotoga maritima: properties and gene structure. J Bacteriol 1998; 180:274-281.
17. Confalonieri F, Elie C, Nadal M et al. Reverse gyrase: A helicase-like domain and a type I topoisomerase in the same polypeptide. Proc Nat Acad Sci USA 1993; 90:4753-4757.
18. Borges KM, Bergerat A, Bogert AP et al. Characterization of the reverse gyrase from the hyperthermophilic archaeon Pyrococcus furiosus. J Bacteriol 1997; 179:1721-6.
19. Forterre P. A hot story from comparative genomics: Reverse gyrase is the only hyperthermophile-specific protein. Trends Genet 2002; 18:236-237.
20. Belova GI, Prasad R, Kozyavkin SA et al. A type IB topoisomerase with DNA repair activities. Proc Natl Acad Sci USA 2001; 98:6015-6020.
21. Sandman K, Bailey KA, Pereira SL et al. Archaeal histones and nucleosomes. Methods Enzymol 2001; 334:116-129.
22. Marc F, Sandman K, Lurz R et al. Archaeal histone tetramerization determines DNA affinity and the direction of DNA supercoiling. J Biol Chem 2002; 277:30879-30886.
23. Wardleworth BN, Russell RJ, Bell SD et al. Structure of Alba: an archaeal chromatin protein modulated by acetylation. EMBO J 2002; 21:4654-4662.
24. Bell SD, Botting CH, Wardleworth BN et al. The interaction of Alba, a conserved archaeal chromatin protein, with Sir2 and its regulation by acetylation. Science 2002; 296:148-151.
25. Peak MJ, Robb FT, Peak JG. Extreme resistance to thermally induced DNA backbone breaks in the hyperthermophilic Archaeon Pyrococcus furiosus. J Bacteriol 1995; 177:6316-6318.
26. Kawashima T, Amano N, Koike H et al. Archaeal adaptation to higher temperatures revealed by genomic sequence of Thermoplasma volcanium. Proc Natl Acad Sci USA 2000; 97:14257-62.
27. Ruepp A, Graml W, Santos-Martinez ML et al. The genome sequence of the thermoacidophilic scavenger Thermoplasma acidophilum. Nature 2000; 407:508-13.
28. Cann IK, Komori K, Toh H et al. A heterodimeric DNA polymerase: evidence that members of Euryarchaeota possess a distinct DNA polymerase. Proc Natl Acad Sci USA 1998; 95:14250-14255.
29. Gueguen Y, Rolland JL, Lecompte O et al. Characterization of two DNA polymerases from the hyperthermophilic euryarchaeon Pyrococcus abyssi. Eur J Biochem 2001; 268:5961-5969.
30. Matsunaga F, Forterre P, Ishino Y et al. In vivo interactions of archaeal Cdc6/Orc1 and minichromosome maintenance proteins with the replication origin. Proc Nat Acad Sci USA 2001; 20:11152-11157.
31. Kopylov VM, Bonch-Osmolovskaya EA, Svetlichnyi VA et al. g-irradiation resistance and UV-sensitivity of extremely thermophilic archebacteria and eubacteria. Mikrobiologiya 1993; 62:90-95.
32. Gerard E, Jolivet E, Prieur D et al. DNA protection mechanisms are not involved in the radioresistance of the hyperthermophilic archaea Pyrococcus abyssi and P. furiosus. Mol Genet Genomics 2001; 266:72-78.
33. Grogan DW. Exchange of genetic markers at extremely high temperatures in the archaeon Sulfolobus acidocaldarius. J Bacteriol 1996; 178:3207-3211.
34. Ghane F, Grogan D. Chromosomal marker exchange in the thermophilic archaeon Sulfolobus solfataricus. Microbiology 1998; 144:1649-1657.
35. Reilly MS, Grogan DW. Characterization of Intragenic Recombination in a hyperthermophilic archaeon via conjugational DNA exchange. J Bacteriol 2001; 183:2943-2946.
36. Grogan DW. Photoreactivation in an archaeon from geothermal environments. Microbiology 1997; 143:1071-1076.
37. Wood ER, Ghane F, Grogan DW. Genetic responses of the thermophilic archaeon Sulfolobus acidocaldarius to short-wavelength UV light. J Bacteriol 1997; 179:5693-5698.
38. Fitt PS, Sharma N, Castellanos G. A comparison of liquid-holding recovery and photoreactivtion in halophilic and nonhalophilic bacteria. Biochem Biophys Acta 1983; 739:73-78.
39. McCready S. The repair of ultraviolet light-induced DNA damage in the halophilic archaebacteria, Halobacterium cutirubrum, Halobacterium halobium and Haloferax volcanii. Mutation Res 1996; 364:25-32.

40. Ogrunc M, Becker DF, Ragsdale SW et al. Nucleotide excision repair in the third kingdom. J Bacteriol 1998; 180:5796-5798.
41. Ng WV, Kennedy SP, Mahairas GG et al. Genome sequence of Halobacterium species NRC-1. Proc Natl Acad Sci USA 2000; 97:12176-12181.
42. Fiala G, Stetter KO. Pyrococcus furiosus sp. nov. represents a novel genus of marine heterotrophic archaebacteria growing optimally at 100°C. Arch Microbiol 1986; 145:56-61.
43. Zillig W, Stetter KO, Wunderl S et al. The sulfolobus "caldariella" group: taxonomy of the basis of the structure of DNA-dependent RNA-polymerases. Arch Microbiol 1980; 125:259-269.
44. Sandler SJ, Satin LH, Samra HS et al. recA-like genes from three archaean species with putative protein products similar to Rad51 and Dmc1 proteins of the yeast Saccharomyces cerevisiae. Nucleic Acids Res 1996; 24:2125-2132.
45. Komori K, Miyata T, Daiyasu H et al. Domain analysis of an archaeal RadA protein for the strand exchange activity. J Biol Chem 2000.
46. Komori K, Miyata T, DiRuggiero J et al. Both RadA and RadB are involved in homologous recombination in Pyrococcus furiosus. J Biol Chem 2000; 275:33782-33790.
47. Reich CI, McNeil LK, Brace JL et al. Archaeal RecA homologues: different response to DNA-damaging agents in mesophilic and thermophilic Archaea. Extremophiles 2001; 5:265-275.
48. Spies M, Kil Y, Masui R et al. The RadA protein from a hyperthermophilic archaeon Pyrobaculum islandicum is a DNA-dependent ATPase that exhibits two disparate catalytic modes, with a transition temperature at 75 degrees C. Eur J Biochem 2000; 267:1125-37.
49. Seitz EM, Brockmann JP, Sandler SJ et al. RadA protein is an archaeal RecA protein homolog that catalyzes DNA strand exchange. Gen Develop 1998; 12:1248-1253.
50. Seitz EM, Kowalczykowski SC. The DNA binding and pairing preferences of the archaeal RadA protein demonstrate a universal characteristic of DNA strand exchange proteins. Mol Microbiol 2000; 37:555-60.
51. Hopfner KP, Craig L, Moncalian G et al. The Rad50 zinc-hook is a structure joining Mre11 complexes in DNA recombination and repair. Nature 2002; 418:562-6.
52. Hopfner KP, Karcher A, Craig L et al. Structural biochemistry and interaction architecture of the DNA double-strand break repair Mre11 nuclease and Rad50-ATPase. Cell 2001; 105:473-85.
53. Hopfner KP, Karcher A, Shin D et al. Mre11 and Rad50 from Pyrococcus furiosus: Cloning and biochemical characterization reveal an evolutionarily conserved multiprotein machine. J Bacteriol 2000; 182:6036-6041.
54. Hopfner KP, Karcher A, Shin DS et al. Structural biology of Rad50 ATPase: ATP-driven conformational control in DNA double-strand break repair and the ABC-ATPase superfamily. Cell 2000; 101:789-800.
55. Lilley DM, White MF. The junction-resolving enzymes. Nat Rev Mol Cell Biol 2001; 2:433-43.
56. Komori K, Sakae S, Daiyasu H et al. Mutational analysis of the Pyrococcus furiosus holliday junction resolvase Hjc revealed functionally important residues for dimer formation, junction DNA binding and cleavage activities. J. Biol Chem 2000.
57. Komori K, Sakae S, Fujikane R et al. Biochemical characterization of the hjc holliday junction resolvase of Pyrococcus furiosus. Nucleic Acids Res 2000; 28:4544-51.
58. Sandler SJ, Hugenholtz P, Schleper C et al. Diversity of radA genes from cultured and uncultured archaea: comparative analysis of putative RadA proteins and their use as a phylogenetic marker. J Bacteriol 1999; 181:907-15.
59. Woods WG, Dyall-Smith ML. Construction and analysis of a recombinant-deficient (radA) mutant of Haloferax volcanii. Mol Microbiol 1997; 23:791-797.
60. Hayashi I, Morikawa K, Ishino Y. Specific interaction between DNA polymerase II (PolD) and RadB, a Rad51/Dmc1 homolog, in Pyrococcus furiosus. Nucleic Acids Res 1999; 27:4695-4702.
61. Cox MM, Goodman MF, Kreuzer KN et al. The importance of repairing stalled replication forks. Nature 2000; 404:37-41.
62. Komori K, Ishino Y. Replication Protein A in Pyrococcus furiosus is involved in homologous DNA recombination. J Biol Chem 2001; 276:25654-25660.
63. Wadsworth RI, White MF. Identification and properties of the crenarchaeal single-stranded DNA binding protein from Sulfolobus solfataricus. Nucleic Acids Res 2001; 29:914-20.
64. Haseltine CA, Kowalczykoski SC. A distinctive single-strand DNA-binding protein from the Archaeon Sulfolobus solfataricus. Mol Microbiol 2002; 43:1505-15.
65. Aravind L, Walker DR, Koonin EV. Conserved domains in DNA repair proteins and evolution of repair systems. Nucleic Acids Res 1999; 27:1223-1242.
66. Hopfner KP, Putnam CD, Tainer JA. DNA double-strand break repair from head to tail. Curr Opin Struct Biol 2002; 12:115-22.

67. de jager M, van Noort J, van Gent DC et al. Human Tad50/Mre11 is a flexible complex that can tether DNA ends. Mol Cell 2001; 8:1129-1135.
68. Tauchi H, Koayashi J, Morishima KI et al. Nbs1 is essentil for DNA repair by homologous rcombination in higher vertebate cells. Nature 2002; 20:93-98.
69. Constantinesco F, Forterre P, Elie C. NurA, a novel 5'-3' nuclease gene linked to rad50 and mre11 homolgs of thermophilic Archaea. Embo J 2002; 3:537-542.
70. Komori K, Sakae S, Shinagawa H et al. A holliday junction resolvase from Pyrococcus furiosus: functional similarity to Escherichia coli RuvC provides evidence for conserved mechanism of homologous recombination in Bacteria, Eukarya, and Archaea. Proc Natl Acad Sci USA 1999; 96:8873-8.
71. Kvaratskhelia M, White MF. An archaeal holliday junction resolving enzyme from Sulfolobus solfataricus exhibits unique properties. J Mol Biol 2000; 295:193-202.
72. Nishino T, Komori K, Tsuchiya D et al. Crystal structure of the archaeal holliday junction resolvase Hjc and implications for DNA recognition. Structure 2001; 9:197-204.
73. Bond CS, Kvaratskhelia M, Richard D et al. Structure of Hjc, a holliday junction resolvase, from Sulfolobus solfataricus. Proc Natl Acad Sci USA 2001; 98:5509-14.
74. Kvaratskhelia M, Wardleworth BN, Bond CS et al. Holliday junction resolution is modulated by archaeal chromatin components in vitro. J Biol Chem 2002; 277:2992-2996.
75. Friedberg E. Out of the shadows and into the light: the emergence of DNA repair. Trends Biochem Sci 1995; 20:381.
76. Ling H, Boudsocq F, Woodgate R et al. Crystal structure of a Y-family DNA polymerase in action: a mechanism for error-prone and lesion-bypass replication. Cell 2001; 107:91-102.
77. Ohmori H, Friedberg EC, Fuchs RP et al. The Y-family of DNA polymerases. Mol Cell 2001; 8:7-8.
78. Boudsocq F, Iwai S, Hanaoka F et al. Sulfolobus solfataricus P2 DNA polymerase IV (Dpo4): An archaeal DinB-like DNA polymerase with lesion-bypass properties akin to eukaryotic poleta. Nucleic Acids Res 2001; 29:4607-16.
79. Komori K, Fujikane R, Shinagawa H et al. Novel endonuclease in Archaea cleaving DNA with various branched structure. Genes Genet Syst 2002; 77:227-41.
80. Jacobs KL, Grogan DW. Rates of spontaneous mutation in an archaeon from geothermal environments. J Bacteriol 1997; 179:3298-3303.
81. Grogan DW, Carver GT, Drake JW. Genetic fidelity under harsh conditions: Analysis of spontaneous mutation in the thermoacidophilic archaeon Sulfolobus acidocaldarius. Proc Natl Acad Sci USA 2001; 98:7928-7933.
82. Maeder DL, Weiss RB, Dunn DM et al. Divergence of the hyperthermophilic archaea Pyrococcus furiosus and P. horikoshii inferred from complete genomic sequences. Genetics 1999; 152:1299-1305.
83. Bell GD, Grogan DW. Loss of genetic accuracy in mutants of the thermoacidophile Sulfolobus acidocaldarius. Archaea 2002; 1:45-52.
84. Smith DR, Doucette-Stamm LA, Deloughery C et al. Complete genome sequence of Methanobacterium thermoautotrophicum DH: functional analysis and comparative genomics. J Bacteriol 1997; 179:7135-7155.
85. Makarova KS, Aravind L, Grishin NV et al. A DNA repair system specific for thermophilic Archaea and bacteria predicted by genomic context analysis. Nucleic Acids Res 2002; 30:482-96.
86. Jorns MS. DNA photorepair: chromophore composition and function in two classes of DNA photolyases. Biofactors 1990; 2:207-211.
87. Yasui A, Eker AP, Yasuhira S et al. A new class of DNA photolyases present in various organisms including aplacental mammals. EMBO J 1994; 13:6143-6151.
88. Mol CD, Arvai AS, Begley TJ et al. Structure and activity of a thermostable thymine-DNA glycosylase: evidence for base twisting to remove mismatched normal DNA bases. J Mol Biol 2002; 315:373-84.
89. Chung JH, Suh MJ, Park YI et al. Repair activities of 8-oxoguanine DNA glycosylase from Archaeoglobus fulgidus, a hyperthermophilic archaeon. Mutat Res 2001; 486:99-111.
90. Sartori AA, Fitz-Gibbon S, Yang H et al. A novel uracil-DNA glycosylase with broad substrate specificity and an unusual active site. EMBO J 2002; 21:3182-3191.
91. Fondufe-Mittendorf YN, Harer C, Kramer W et al. wo amino acids replacements change the substrate preference of DNA mismatch glycosylase Mig.MthI from T/G to A/G. Nucleic Acids Res 2002; 30:614-621.
92. Yang H, Fitz-Gibbon S, Marcotte EM et al. Characterization of a thermostable DNA Glycolase specific for U/G and T/G mismatches from the hyperthermophilic Archeon Pyrobaculum aerophilum. J Bacteriol 2000; 182:1272-1279.

93. Knaevelsrud I, Ruoff P, Anensen H et al. Excision of uracil from DNA by the hyperthermophilic Afung protein is dependent on the opposite base and stimulated by heat-induced transition to a more open structure. Mutat Res 2001; 487:173-90.
94. Birkeland NK, Anensen H, Knaevelsrud I et al. Methylpurine DNA glycosylase of the hyperthermophilic archaeon Archaeoglobus fulgidus. Biochem 2002; 41:12697-705.
95. Leipe DD, Aravind L, Koonin EV. Did DNA replication evolve twice independently? Nucleic Acids Res 1999; 27:3389-401.
96. Slesarev AI, Mezhevaya KV, Makarova KS et al. The complete genome of hyperthermophile Methanopyrus kandleri AV19 and monophyly of archaeal methanogens. Proc Nat Acad Sci USA 2002; 99:4644-4649.
97. Leipe DD, Aravind L, Grishin NV et al. The bacterial replicative helicase DnaB evolved from a RecA duplication. Genome Res 2000; 10:5-16.
98. Eisen JA, Hanawalt PC. A phylogenomic study of DNA repair genes, proteins, and processes. Mutat Res 1999; 435:171-213.
99. Aravind L, Makarova KS, Koonin EV. Survey and summary: Holliday junction resolvases and related nucleases: identification of new families, phyletic distribution and evolutionary trajectories. Nucleic Acids Res 2000; 28:3417-32.
100. Lawrence JG, Ochman H. Reconciling the many faces of lateral gene transfer. Trends Microbiol 2002; 10:1-4.

Chapter 12

Extant Variations in the Genetic Code

Manuel A.S. Santos and Mick F. Tuite

Introduction

The discovery in the 1960s of an identical genetic code in *Escherichia coli*, viruses and mammalian cells suggested that all living organisms use the same genetic code. The existence of a universal genetic code prompted Crick[1] to propose the "Frozen Accident Theory" which states that the genetic code does not evolve. This theory was based on the assumption that in the last common ancestor, life-forms had reached a level of complexity that would not tolerate alterations in the identity of their codons. That is, once proteins had acquired a certain level of functionality, any alteration in codon identity would introduce structural and functional disruption with a high probability that this would be lethal or highly detrimental.

The discovery in the late 1970s that, in vertebrate mitochondria, the Ile-AUA and UGA-stop codons were decoded as Met and Trp respectively, questioned both the universality and the apparent frozen state of the genetic code, the central tenants of Crick's hypothesis.[2] This discovery prompted the formulation of two new hypotheses to explain the origin of the 'deviant' genetic code of mitochondrial genomes; one stating that mitochondrial codes represent primitive forms of the genetic code which existed before fixation of the universal code,[3] and the other stating that mitochondrial codon reassignments were allowed to evolve from the universal code because mitochondrial genomes encode a much smaller number of proteins and are thus better able to tolerate changes in codon identity.[4] However, the discoveries over the last 20 years, of genetic code changes in the mitochondrial genomes of a wide range of organisms (Table 1) and in both prokaryotic and eukaryotic genomes (Table 2); (Fig. 1) indicate that several different variant genetic codes most likely evolved from the "frozen" universal genetic code. If this were not so, one would have to consider that different mitochondria evolved from different primitive prokaryotes, an assumption which is not corroborated by molecular phylogeny, experimental data or theoretical models. In this article, we discuss the current thinking on how and why certain genetic code alterations evolved and will bring together neutral and nonneutral theories in an attempt to explain the contribution of different mechanisms to the evolution of alternative genetic codes.

The spectrum of described and validated genetic code alterations (Table 3) suggest that certain codon families are more prone to undergo codon reassignments than others. In mitochondrial systems, genetic code alterations have been described in a variety of metazoans, fungi, red algae, green plants, alveolates, stramenopiles, haptophytes and euglenozoans (for a recent review see ref. 5 and Figs. 1 and 5). In eukaryotic nuclear systems, while stop codons have been reassigned in green algae, ciliates and diplomonads, the only known sense-to-sense codon reassignment is the Leu-CUG codon found in a large proportion of the fungal species of the genus *Candida*. Interestingly, of the three stop codons —UAA, UGA and UAG— bacteria apparently only reassign the UGA-stop to Trp, however *Mycoplasma* spp. do not use the Arg-CGG codon and *Micrococcus* spp. do not use the Arg-AGA and the Ile-AUA codons (Table 2).

The Genetic Code and the Origin of Life, edited by Lluís Ribas de Pouplana.

Table 1. Codon reassignment in mitochondrial genomes

	Standard Code	Code Alteration 1		Code Alteration 2
1	UGA Stop →	Trp	→	-
2	AUA Ile →	Met	→	-
3	AGR Arg →	Ser	→	-
4	AUA Ile →	Met	→	Ile
5	AAA Lys →	Asn	→	-
6	AGR Arg →	Ser	→	Gly
7	UAA Stop →	Tyr	→	-
8	CUN Leu →	Thr	→	-
9	CGN Arg →	Unassigned	→	-
10	AGR Arg →	Unassigned	→	Stop
11	AGA Arg →	Unassigned	→	Gly
12	AGR Arg →	Ser	→	Unassigned
13	AGA Arg →	Unassigned	→	Ser
14	UAG Stop →	Leu	→	-
15	UAG Stop →	Ala	→	-
16	UCA Ser →	Stop	→	-

A total of 14 codons are known to have been reassigned in mitochondrial genomes of Metazoa, fungi, red algae, green plants, Alveolates, Stramenopiles, Haptophytes and Euglenozoans (reviewed by Knight[5]). The stop and AGR codons are the most commonly reassigned (code alteration 1) while several codons have been reassigned to a second different meaning (code alteration 2).

Table 2. Genetic code alterations in eukaryotic and prokaryotic genomes

Codon	Standard Code		Code Alteration	Organisms
UAR	Stop	→	Gln	**Ciliates:** *Zosterograptus* sp., *Naxella* sp., Spirotrichs, Oligohymenophorans; *Condylostoma magnum*; **Diplomonads; Gree algae:** *Acetabularia* spp. *Batophora cesthedi*
UGA	Stop	→	Cys	**Ciliates:** *Euplotes* spp.
CUG	Leu	→	Ser	**Fungi:** *Candida* spp.
AGA	Arg	→	Unassigned	**Bacteria:** *Micrococcus* spp.
AUA	Ile	→	Unassigned	**Bacteria:** *Micrococcus* spp.
UGA	Stop	→	Trp	**Bacteria:** *Bacillus subtilis*; Ciliates: *Coipoda* spp.; *Heteotrichs.*
CGG	Arg	→	Unassigned	**Bacteria:** *Mycoplasma* spp.
UGA	Stop	→	Unassigned	**Ciliates:** *Myctotherus ovalis, Pseudomicrothorax dubius*

Nine out of the 64 codons are known to have been reassigned independently in a number of organisms (reviewed by Knight[5]).

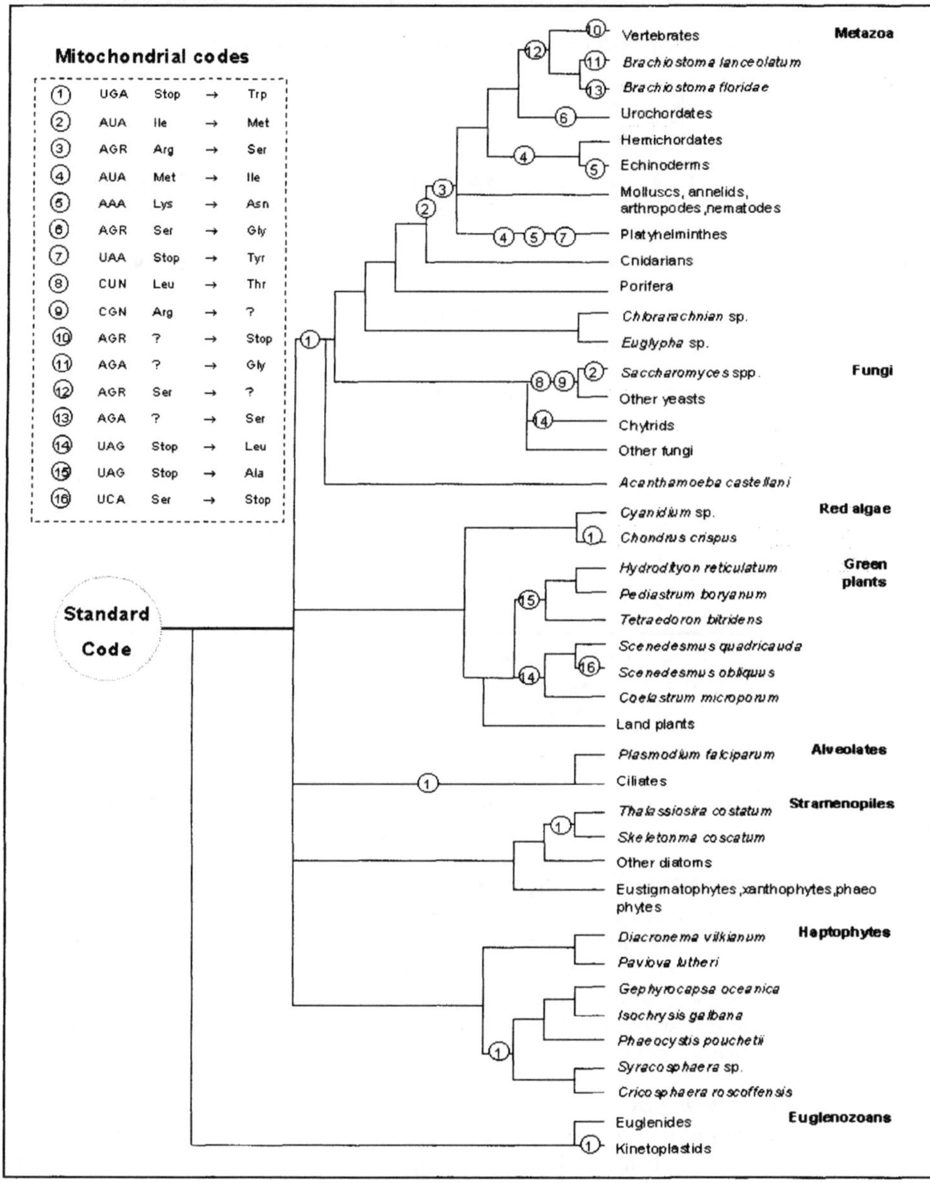

Figure 1. A summary of known extant variations in the genetic code of mitochondria. Note: This figure was adapted from Fig. 2 with courtesy of Nature Reviews in Genetics.[5]

The two Arg codons AGA/AGG, the three termination codons and the Ile codon AUA are the most commonly reassigned codons and, apart from these, other noteworthy examples are the rather dramatic mitochondrial reassignment of the entire CUN codon family from Leu to Thr in the yeast *Saccharomyces cerevisiae* and the unassignment of the Arg-CGN codon family in the mitochondrial genome of the yeast species *Kluyveromyces thermotolerans*. It is important to note that codons that are reassigned in mitochondrial genomes may also be reassigned in

Table 3. A summary of known extant variations in the genetic code

UUU	Phe	UCU	Ser	UAU	Tyr	UGU	Cys
UUC	Phe	UCC	Ser	UAC	Tyr	UGC	Cys
UUA	Leu	**UCA**	**Ser Stop**	**UAA**	**Stop Tyr Gln**	**UGA**	**Ter Trp Cys**
UUG	Leu	UCG	Ser	**UAG**	**Stop Leu Ala Gln**	UGG	Trp
CUU	**Leu Thr**	CCU	Pro	CAU	His	**CGU**	**Arg UN**
CUC	**Leu Thr**	CCC	Pro	CAC	His	**CGC**	**Arg UN**
CUA	**Leu Thr**	CCA	Pro	CAA	Gln	**CGA**	**Arg UN**
CUG	**Leu Thr Ser**	CCG	Pro	CAG	Gln	**CGG**	**Arg UN**
AUU	Ile	ACU	Thr	AAU	Asn	AGU	Ser
AUC	Ile	ACC	Thr	AAC	Asn	AGC	Ser
AUA	**Ile MetUN**	ACA	Thr	**AAA**	**Lys Asn**	**AGA**	**Arg UN Ser Gly Ter**
AUG	Met	ACG	Thr	AAG	Lys	**AGG**	**Arg UN Ser Gly Stop**
GUU	Val	GCU	Ala	GAU	Asp	GGU	Gly
GUC	Val	GCC	Ala	GAC	Asp	GGC	Gly
GUA	Val	GCA	Ala	GAA	Glu	GGA	Gly
GUG	Val	GCG	Ala	GAG	Glu	GGG	Gly

Codon reassignments (bold) occur in several codon families with many codons being reassigned more than once. Codons that are reassigned in mitochondria are also reassigned in bacteria and in the nuclear genomes of eukaryotes suggesting that similar evolutionary forces drive both types of reassignment.

nuclear genomes, suggesting that similar evolutionary forces drive evolution of codon reassignments in both systems.

Mechanisms of Codon Reassignment

The diversity of codon reassignments described above suggests that such reassignments might have evolved through distinct molecular mechanisms and that they may be more widespread in nature than expected. Nevertheless, the spectrum of established extant (i.e., still existing) genetic code alterations has prompted the formulation of two main theories in an attempt to explain how organisms may be able to survive change in their genetic code assignments; the "Codon Capture Theory" and the "Ambiguous Intermediate Theory".

The "Codon Capture Theory"

This theory advocates that genetic code changes are neutral and arise from evolutionary fluctuations in the GC/AT balance of an evolving genome. The theory proposes that, as the balance shifts from a predominance of GC to AT pairs, or vice versa, certain codons —rich in GC or AT— will disappear from the genome. This hypothesis is supported by the findings that in *Micrococcus luteus*, whose genome is 74% G + C, two AU rich codons with As in the 3' position, namely AGA and AUA, are not used (i.e., are unassigned) while in *Mycoplasma capricolum*, whose genome is only 25% G + C, the GC-rich Arg codon CGG is unassigned.[6-9] A central tenant of this theory is that the initial step of codon disappearance is critical for codon reassignment to occur in that it allows for the tRNA(s) that decode the disappeared codon(s) to also be lost from the genome. In such a scenario, if the eliminated codon reappears through mutational drift, then the translational machinery would be unable to decode it forcing the ribosome to stall whenever the reintroduced codon appeared at the ribosomal A-site. Thus, the only way the ribosome could continue translation would be for the reintroduced codon to be decoded by a near cognate tRNA i.e., a tRNA whose anticodon is closely related in sequence (usually 2 out of 3) to the now disappeared tRNA anticodon. This later step is also critical in codon reassignment in that a near cognate tRNA would capture, albeit at low efficiency, the "new" codon reintroduced by mutational drift. Since the miscoding tRNA would belong to a noncognate amino acid family, the codon would therefore be reassigned to a different amino acid.

The attractiveness of the 'Codon Capture Theory' lies in the premise that codon reassignment would not lead to wholesale changes in the amino acid sequence of the encoded proteins during mRNA translation and, consequently, there would be no negative effect arising from such an event i.e., it would be neutral. Despite this, it is intriguing to note that some codons apparently remain unassigned in certain bacterial and mitochondrial genomes, while some reassigned codons do not follow the GC/AT genome balance rule. More importantly, recent genome sequencing projects show that, at least in eukaryotic nuclear and bacterial systems, G + C content is not evenly distributed along the length of the genome making it difficult for a codon to disappear completely from a genome. Also important is that in mitochondria, which have genomes with a very high AT content, some codon reassignments violate the "Codon Capture Theory". For example, all of the following A-ending codons - Arg-AGA Ile-AUA, stop-UAA/UGA, Lys-AAA, Ser-UCA and Leu-CUA - have undergone reassignment indicating that evolutionary fluctuations in G + C content alone is not the underlying mechanism for codon reassignment in mitochondrial genomes.[5,10-13]

The "Ambiguous Intermediate Theory"

This theory postulates that codon reassignment is driven by selection through a mechanism that requires ambiguity in the decoding of a codon.[10] It suggests that tRNA structural change, which can introduce such decoding ambiguity, is the key element in the mechanism of codon reassignment. That is, a codon would first be decoded ambiguously by two tRNAs, a cognate and a mutant near-cognate tRNA and then, in a selection driven process, the mutant tRNA would capture the codon being reassigned. The theory does not require codon disappearance from the genome and suggests that ambiguous mutant tRNAs, i.e., ambiguous with respect to codon recognition, introduce some sort of selective advantage that allows for its selection.

Like the "Codon Capture Theory", the "Ambiguous Intermediate Theory" does not satisfactorily explain all known examples of codon reassignment or unassignment, however, of the 15 different described and validated codon reassignments nine can be explained by this theory[5,10] suggesting that tRNA structural change leading to decoding ambiguity, might play an important role in the evolution of codon reassignment. The theory is supported by the reassignment of stop codons via codon misreading mediated by wild-type tRNAs in various organisms and by the double identity of a novel $tRNA^{Ser}$ (which can be charged with either Ser or Leu); in a variety of *Candida* spp. in vivo.[10,14-15]

A Unifying Model

The "Codon Capture" and "Ambiguous Intermediate" theories are not mutually exclusive. The impact of evolutionary shifts in overall GC or AT content of a genome on codon usage[12,16-18] is an important evolutionary force driving codon reassignment in that reduction of codon usage decreases the toxic effects of codon reassignment by reducing the mutational load associated with ambiguous translation of the codon being reassigned.[11] Therefore, reduced usage of a particular codon might provide a first step in a chain of molecular events that lead to the eventual reassignment of that codon. If so, ambiguous decoding, as postulated by the "Ambiguous Intermediate Theory", could represent a late step in reassignment pathways while biased genome replication events, resulting from mutation in DNA polymerases and/or DNA repair systems and that lead to incorporation As and Gs in the nascent strands in response to Cs and Ts in template strands respectively, might represent the critical initial events that trigger evolution of alternative genetic codes.

That almost all rarely used codons maintain their standard identity in all scrutinised genomes suggests that rarely used codons are not specifically prone to reassignment. This leads to the hypothesis that the G + C content of a genome is an important force in lowering usage of specific codon(s) and of codon unassignment, but plays a much less significant role in subsequent reassignment events. The latter events have to be mediated by tRNA mutation and genetic code ambiguity as postulated by the "Ambiguous Intermediate Theory". Furthermore, recent studies on codon reassignment in mitochondrial systems indicate that the prevalence of a codon's third base better predicts codon usage than does GC content. This apparent failure of GC content to predict codon usage is due to the fact that A and T, and C and G are uncorrelated in mitochondrial genomes. In addition, low codon frequencies are related to reassignment, but are not necessary, nor sufficient, for reassignment. This indicates that other evolutionary forces may be at play in mitochondrial genetic systems with ambiguous decoding being the most likely candidate to trigger codon reassigment.[13]

Conceptually, the "Codon Capture" and the "Ambiguous Intermediate" theories represent opposite faces of the same coin. Thus, while the "Codon Capture Theory" relies on biased DNA replication to alter the frequency of usage of the codons third base, which in extreme cases drives codons to extinction, the "Ambiguous Intermediate Theory" relies on point mutations in the translational machinery, in particular in tRNAs (for sense codon reassignment) and release factors (for the reassignment of nonsense codons), thereby introducing ambiguous decoding that subsequently triggers codon reassignment in a selection driven process.

The Selective Forces Driving Evolution of Alternative Genetic Codes

The key questions that must be answered if we are to fully understand the evolution of alternative genetic codes is: "Why and how do they evolve"? That the evolution of alternative genetic codes of mitochondria and *Mycoplasma* spp. is, in both cases, apparently linked to an evolutionary increase in the AT content of the genome coupled with genome size reduction, suggests that, at least in these two cases, alternative codes may arise as a consequence of the evolutionary forces shaping genome structure.[19-21] That the endosymbiotic lifestyle of *Mycoplasma* spp. puts their genomes under very specific evolutionary constraints that can lead to gene disintegration and ultimately genome size reduction[22] and that mitochondrial genomes evolved from much larger genomes related to the α-proteobacteria[23] and retained only a very small part of the ancestor genome, shows that genome reduction has important consequences for the evolution of alternative genetic codes. This is because it puts the overall translational machinery under high mutational pressure in order to reduce the size and number of translational factors. That animal mitochondria do not have the tRNA species to decode the Ile-AUA and the Arg-AGA/G codons (resulting in their being decoded as Met and Ser respectively), and that mitochondrial tRNAs from various organisms have atypical tRNAs with short D and T-loops, supports this hypothesis.[24]

The high frequency of human diseases caused by mutations in mitochondrial tRNAs also suggests that high mutational rates can have an important impact on the overall tRNA population.[25] However, such a high mutation rate does not appear to be the primary reason for the frequent appearance of genetic code changes in mitochondrial systems.[13] Nevertheless, it is worth noting that genome reduction is achieved through accumulation of deletions and mutations leading to gene loss and genome degradation.[22]

In order to fully understand the evolution of alternative genetic codes we also need to know what competitive advantage they bring —if any— in order to allow for their selection. Based on the assumption that most genetic code alterations evolve through tRNA or release factor structural change,[5] one possibility is that certain mutations in these translational factors might alter their decoding properties allowing for expression of hidden open reading frames (ORFs) through alternative decoding.[13] This hypothesis is corroborated by the finding that alterations in the modified nucleosides of tRNAs, in particular those nucleosides present in the anticodon-arm of a tRNA, are very important in the fine tuning of tRNA decoding.[26,27] For example, alterations in the pattern of modified nucleosides at position 37, 3′ to the anticodon

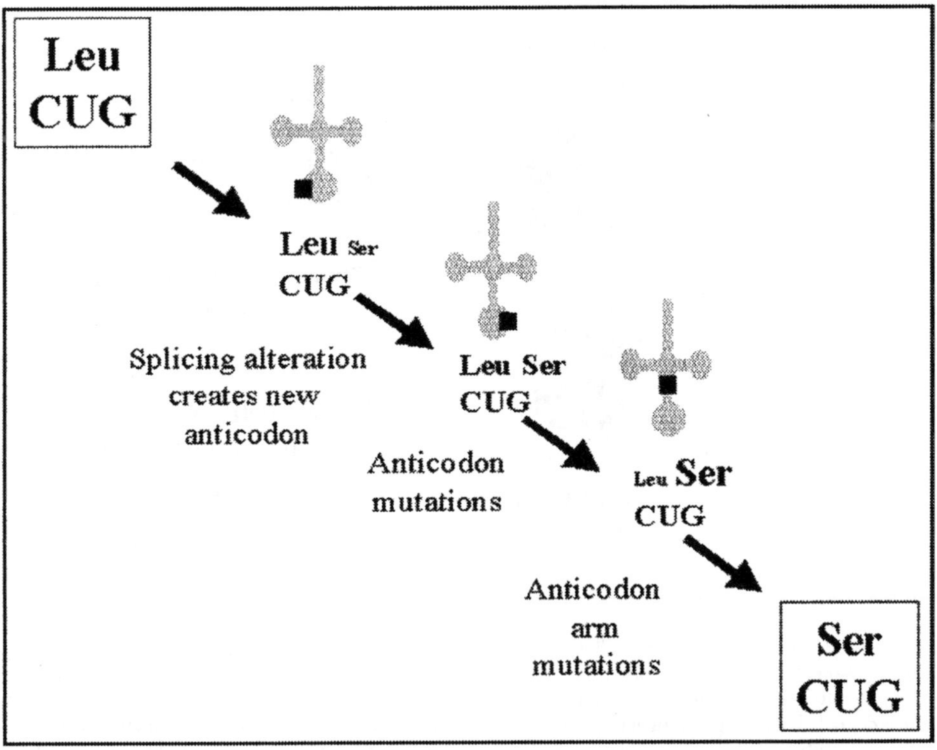

Figure 2A. Evolutionary pathway of CUG reassignment in *Candida spp.* The Ser-tRNA$_{CAG}$ that decodes CUG in *Candida* spp. most likely evolved from a tRNA^{Ser} through an altered splicing event which introduced a leucine 5′-CAG-3′anticodon in a Ser-tRNA$_{IGA}$.[57] Sequence alignments of the anticodon-arm of Ser-tRNA$_{CAG}$ from various *Candida* species and structural probing of the tRNA in solution[39] show that this class of tRNA^{Ser} has atypical anticodon-arms which are under strong mutational pressure. The diagram illustrates a nonneutral step-wise reassignment pathway involving a sequence of tRNA mutations.[30] Reprinted by permission from Nature Reviews Genetics 2001; 2:49-58; Macmillan Magazines Ltd.

sequence, promote frameshifting, while single mutations outside the anticodon-arm promote both sense codon and stop codon misreading.[28,29] Therefore, expression of hidden ORFs and the suppression of deleterious nonsense and missense mutations might provide new functionalities of critical importance for adaptation and survival under specific physiological conditions. Whether or not this provides a strong enough evolutionary advantage to trigger codon reassignment remains an open question.

Attempts have been made to reconstruct the pathway that leads to CUG codon reassignment (Leu to Ser) found in *Candida* spp., in the closely related yeast *S. cerevisiae*. These studies have shown that engineered CUG ambiguity (i.e., being decoded both as Leu and Ser in the same cell) induces a novel cellular stress response. This response creates a preadaptation condition that protects cells from lethal environmental challenges such as high doses of cadmium, arsenite and cycloheximide.[30] These translationally ambiguous cells are also more tolerant to heat, oxidants and salt suggesting that this stress response triggered by translational ambiguity has pleiotropic effects that allow for adaptation to new ecological niches (Fig. 2). That the stress response in *S. cerevisiae* is also induced by antibiotic-induced mistranslation,[31] supports these findings.

Interestingly, engineered CUG codon ambiguity in *S. cerevisiae* leads to an induction of the expression of genes encoding molecular chaperones suggesting that stress tolerance brought

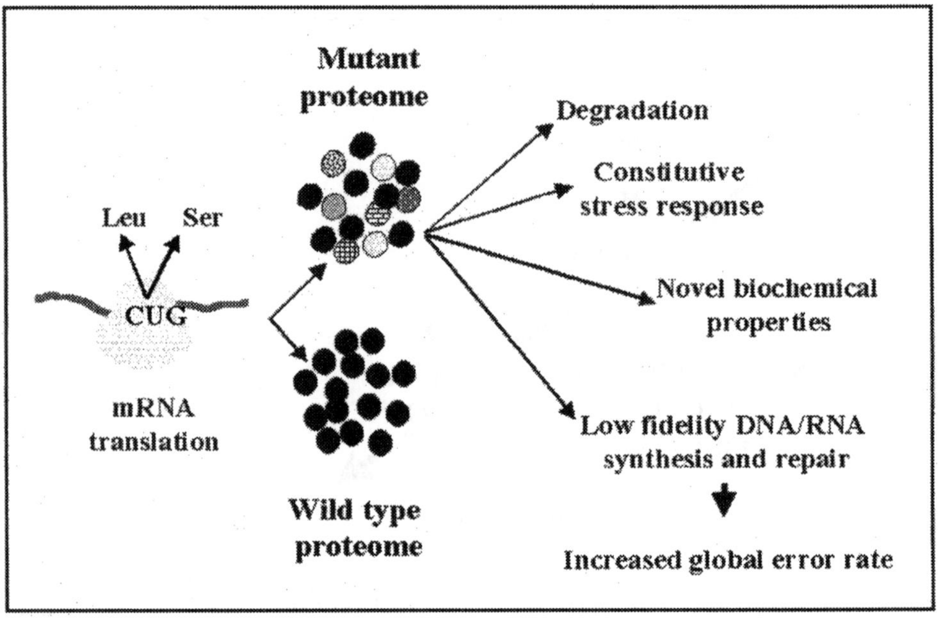

Figure 2B. Evolutionary pathway of CUG reassignment in *Candida spp.* The cellular consequences of CUG reassignment. Most of the *Candida* proteome is not affected due to the low usage of the CUG codon in *Candida* mRNAs. The mutant proteome synthesized by ambiguous CUG decoding does not fold properly and is degraded or aggregates triggering the cellular stress response. Novel biochemical properties might also arise from ambiguous decoding albeit at very low level. However, if DNA replication and repair systems are mutated, it is highly likely that subpopulations of cells arise due to induction of an hypermutable phenotype. This might significantly increase adaptation potential to new ecological niches. Reprinted by permission from Nature Reviews Genetics 2001; 2:49-58; Macmillan Magazines Ltd.

about by CUG ambiguity relies on the same cellular machinery as that involved in protection of cells against heat-shock and other environmental stresses.[30] That is, genetic code alterations might be strictly linked to adaptation to new environments. However, the deleterious effects of codon reassignment suggest that it might only be tolerated when alternative cellular mechanisms fail to provide the capacity for overcoming the environmental challenge. This interplay between the competitive edge brought about by the stress response and the environment may be critical in determining whether the ambiguous codon (or codons) becomes reassigned or if it maintains the standard identity. This deduction is corroborated by the finding that the CUG-decoding ser-$tRNA_{CAG}$ in *Candida* spp. appeared before the species level divergence of *S. cerevisiae* and *C. albicans*. This implies that the ancestor of these two yeast species had already undergone a period of genetic instability, which resulted in CUG reassignment in *Candida* spp. whereas in *Saccharomyces* spp there has been a reversion back to the standard genetic code.[32]

Another important aspect of codon reassignment that must be considered is its occurrence in asexual populations that have lost the potential to create genetic diversity through meiotic recombination e.g., in *C. albicans*. If one considers that several different protein isoforms can result from the translation of a single mRNA as a consequence of ambiguous decoding, then it is tempting to speculate that such an event would lead to an expansion of the proteome and that some of the additional proteins so-synthesised by ambiguous decoding might have novel functionalities (Fig. 2B). Ambiguous decoding also has the potential to significantly increase global decoding error rates by hitting the cellular information maintenance and expression machineries, i.e., proteins involved in DNA synthesis and repair, the ribosome and mRNA processing factors. Sub-populations of hypermutable cells might therefore result from ambiguous decoding as has been reported in the case of a strain of *E.coli* expressing a mutant $tRNA^{Gly}$ which decodes the Asp codons GAU and GAC as Gly.[33] That hypermutable bacterial strains such as *E. coli* 057:H7 have a significantly higher environmental adaptation potential than nonhypermutable strains[34,35] suggests that genetic code ambiguity can in fact create genetic diversity, both directly and indirectly, driving more rapid evolution of novel phenotypes.

While the translational misreading of sense codons can have pleiotropic effects on protein function, stop codon readthrough is of more limited scope leading simply to a C-terminal extension. However, despite stop codon readthrough being deleterious at high level,[36] a number of eukaryotic, bacterial and viral proteins are naturally and abundantly expressed through programmed stop codon readthrough events mediated by naturally-occurring (i.e., nonmutant) tRNAs (reviewed by Farabaugh[37]). Thus, under certain conditions, either to correct a nonsense mutation in an essential gene or to express novel ORFs, stop codon readthrough can be advantageous. As discussed above, the issue is whether or not the selective advantage introduced is strong enough to trigger codon reassignment.

Structural Alterations in the Translation Machinery Are Required for Codon Reassignment

Genetic code changes are mediated by structural alterations in one or more of the components of the translational machinery, in particular tRNAs or termination release factors (RFs). Such alterations are introduced in these molecules most likely through a number of specific mutations, but no general rule seems to provide a unifying mechanism for evolution of codon reassignment. The examples described below however provide us with important insights into the type of structural changes required for codon reassignment.

CUG Reassignment in Candida Spp

In *Candida* spp. the Ser-$tRNA_{CAG}$ which decodes the Leu codon CUG as Ser, has undergone a novel mutation in the anticodon loop, 5′ to the anticodon triplet, changing the conserved uridine at position 33 (U_{33}) to guanosine (G_{33},[38] and Fig. 3). This novel mutation is likely to have been critical for CUG codon reassignment by lowering the decoding efficiency of the tRNA and simultaneously preventing efficient recognition of the Ser-$tRNA_{CAG}$ by the Leu-tRNA synthetase, i.e., it also acts as a leucylation identity antideterminant.[15] These proposed roles for G_{33} are in fact critical for the evolution of CUG reassignment in that lowering CUG decoding efficiency would have minimized the deleterious effect of ambiguous CUG decoding during the early stages of CUG reassignment.[30] As the reassignment progressed, leucylation efficiency of the Ser-$tRNA_{CAG}$ should have decreased such that the reassignment could proceed to completion (see Fig. 2A). That G_{33} decreases leucylation by altering the structure of the anticodon-arm of the Ser-$tRNA_{CAG}$[39] shows that this unusual feature of the tRNA may also have played an important role in the late stages of CUG reassignment.[14,15] The Leu-tRNA synthetase is known to contact the anticodon-arm of Leu-tRNAs and makes specific contacts with both the middle codon nucleotide A_{35} and m^1G_{37} with the methyl group of the latter being a major identity determinant for the recognition by the Leu-tRNA synthetase.[15] Interestingly, in one *Candida* spp., namely *Candida cylindracea*, the CUG has been fully reassigned to Ser because m^1G_{37} has been replaced by A_{37} (Fig. 3). It is therefore likely that, in most other *Candida* species, CUG ambiguity occurs due to the presence of m^1G_{37} although this has only been formally demonstrated for a couple of species.[15] The reason why the CUG codon has become fully reassigned in *C. cylindracea* while the other *Candida* species still maintain a double identity (i.e., Ser and Leu) for this codon, is unclear. However, low level decoding of the CUG codon as Leu in those *Candida* species which have a m^1G_{37} - containing Ser-$tRNA_{CAG}$ suggests that CUG ambiguity is functionally relevant.

In addition to the features described above, the Ser-$tRNA_{CAG}$ from various *Candida* species also have divergent anticodon arms (Fig. 3). That tRNAs that decode the same codons in

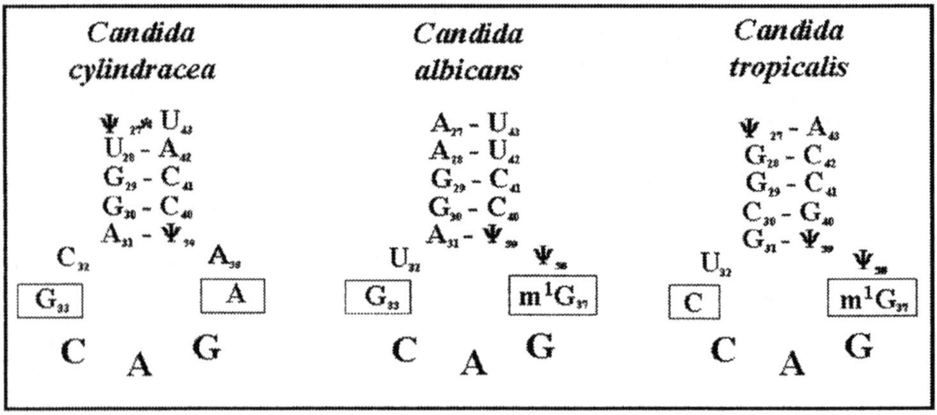

Figure 3. Anticodon-arms of three different Ser-$tRNA_{CAG}$ which mediate CUG reassignment in *Candida* species. The boxed nucleotides represent positions which have been shown experimentally to play important roles in the evolution of CUG reassignment. m^1G_{37} and A_{35} are identity determinants for the Leu-tRNA synthetase and allow for mischarging of the m^1G_{37}-containing tRNAs with leucine in vivo. Position 33 is a conserved uridine in all cytoplasmic elongator tRNAs and plays a critical role in the U-turn of the anticodon-loop. G_{33} lowers the decoding efficiency of the ser-$tRNA_{CAG}$ which apparently was very important in the early stages of CUG reassignment in that it minimised the toxic effects of serine CUG decoding.[14]

different organisms have identical or very similar anticodon-arms, and that decoding efficiency and accuracy are apparently modulated by the anticodon-arm - the so called extended-anticodon[40] - highlights further the novelty of these tRNAs. It remains to be established whether the mutations in the anticodon-arm reflect coevolution of each tRNA: aminoacyl-tRNA synthetase pair or if different *Candida* species have different levels of CUG ambiguity as dictated by the particular anticodon-stem. In *C. zeylanoides* for example, the proportion of Leu incorporated at the CUG codon in relation to Ser incorporation is approximately 1:20 and this proportion increases if a pyrimidine is introduced at position 33.[15] This observation suggests that the *C. tropicalis* ser-$tRNA_{CAG}$, which has C_{33} instead of G_{33} (Fig. 3), might decode the CUG codon with much higher level of ambiguity, a prediction that remains to be experimentally confirmed.

Structural Changes Required for Mitochondrial Codon Reassignment

The reassignment of the CUN codon family from Leu to Thr in yeast mitochondria is a further example of a genetic code change mediated by novel tRNAs and aminoacyl-tRNA synthetases. In *S. cerevisiae* mitochondria, the Thr-$tRNA_{UAG}$, which decodes the CUN codon family, has a 6bp anticodon stem instead of the canonical 5bp stem, and a 6 nucleotide anticodon-loop instead of the canonical 7. In another yeast species *Torulopsis glabrata*, the anticodon-loop has the canonical 7 nucleotides, however the anticodon stem has 6 base-pairs like the *S. cerevisiae* Thr-$tRNA_{UAG}$ (Fig. 4). Furthermore, the sixth base pair of the anticodon-stem is a non Watson-Crick U-U base pair suggesting that the tertiary structure of the anticodon-arm may be flexible. This contrasts with the cytoplasmic Thr-$tRNA_{UGU}$, which decodes the standard Thr codon family ACN and has a canonical 5 bp anticodon-stem and a 7 nucleotide anticodon-loop. In this tRNA the fifth bp of the anticodon-stem is also a U-U suggesting that these tRNAs have unusually structured anticodon-stems (Fig. 4).

An important feature of CUN reassignment in yeast mitochondria is the existence of a Thr-tRNA synthetase specific for the mitochondrial Thr-$tRNA_{UAG}$. This novel aminoacyl-tRNA synthetase does not recognize the standard cytoplasmic Thr-$tRNA_{UGU}$ which is recognized by a different Thr-tRNA synthetase.[41] Taken together, these features are consistent with CUN reassignment having evolved through a series of steps, which required both a new $tRNA^{Thr}$ isoacceptor and a new aminoacyl-tRNA synthetase. This is —at present— the only well docu-

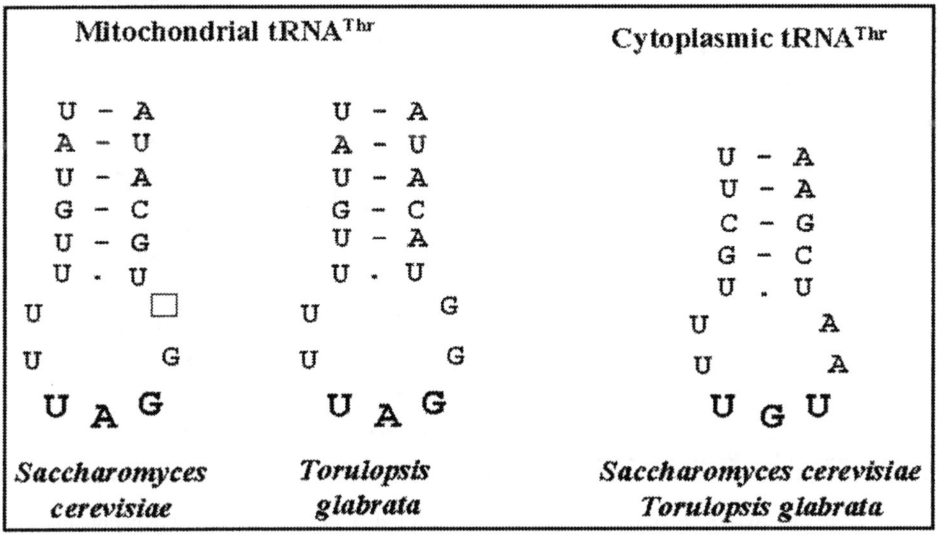

Figure 4. Structural alteration of the tRNAs involved in CUN reassignment in yeast mitochondria. The anticodon-arm structures of the mitochondrial $tRNA^{Thr}$ involved in CUN reassignment from Leu to Thr in *Saccharomyces cerevisiae* and *Torulopsis glabrata* and the cytoplasmic $tRNA^{Thr}$ which decodes the standard Thr ACA codon in both *S. cerevisiae* and *Toruplopsis glabrata*. The *S. cerevisiae* mitochondrial $tRNA^{Thr}$ has a deletion at position 38, 3′ of the anticodon (indicated by the open box) and a noncanonical 6 base-pair anticodon-stem, while the *T. glabrata* homologue only has the noncanonical 6 base-pair anticodon stem. The cytoplasmic $tRNA^{Thr}$ also has the U-U nonWatson-Crick base pair at the bottom of the anticodon-stem that is characteristic of this family of isoacceptor tRNAs. Yeast mitochondria also encode a unique Thr-tRNA synthetase, which only charges these particular tRNAs whereas the cytoplasmic $tRNA^{Thr}$ is charged by a different Thr-tRNA synthetase.

mented case where an aminoacyl-tRNA synthetase is important for codon reassignment and it highlights the importance of these enzymes in the evolutionary pathways of codon reassignment.

Stop Codon Reassignment

Stop codons have been reassigned in both eukayotic nuclear and mitochondrial genomes aswell as in bacterial genomes (reviewed by Osawa[9] and Knight et al[5]). Stop codons are apparently reassigned by naturally-occurring suppressor tRNAs that can translate one or other stop codon at low efficiency in addition to their cognate codon(s). This suggests that the ability to translate stop codons may play a crucial role in the early stages of nonsense-to-sense codon reassignment. These nonsense suppressor tRNAs arise in many cases in standard tRNAs by one of three mechanisms: through mutations outside the anticodon-arm, via tRNA editing events or alteration in the pattern of modified nucleosides thereby expanding the decoding capabilities of the tRNA.[29,42,43] For example, $tRNAs^{Gln}$ able to read the UAA and UAG termination codons have A_{24} in the D-arm and a G-C or G-U base pair at the anticodon helix position 27-43, both of which are known to stimulate first and third codon position wobble[29] (Fig. 7). A different mechanism exists in *Leishmania tarentolae* mitochondria where an apparent tRNA editing event in a cytoplasmic imported $Trp\text{-}tRNA_{CCA}$ has converted it into a $Trp\text{-}tRNA_{UCA}$, which is now able to decode the UGA stop as Trp[42] (Fig. 6).

In addition to tRNA change mediating stop codon reassignments, another component of the translational machinery that has played an important role in the evolution of alternative genetic codes are the RFs whose stop codon recognition specificities are generally altered in cases where stop codons have become reassigned[44-46] (see Table 1). Recent studies aimed at elucidating the molecular mechanism of codon recognition by the eukaryotic RF, eRF1,[47]

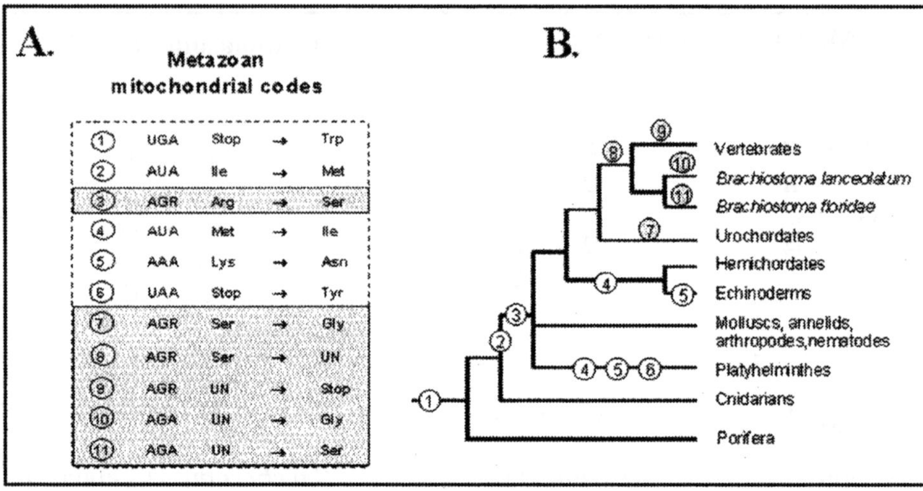

Figure 5. Multiple AGR reassignments in Metazoan mitochondria. Reassignment of AGR codons in metazoan mitochondrial codes represent six of the eleven known examples of codon reassignment in Metazoan mitochondrial codes. Unassigment of AGR codons is apparently required prior to the reassignment of these codons to Stop. UN = unassigned. Adapted from Knight et al, 2001.

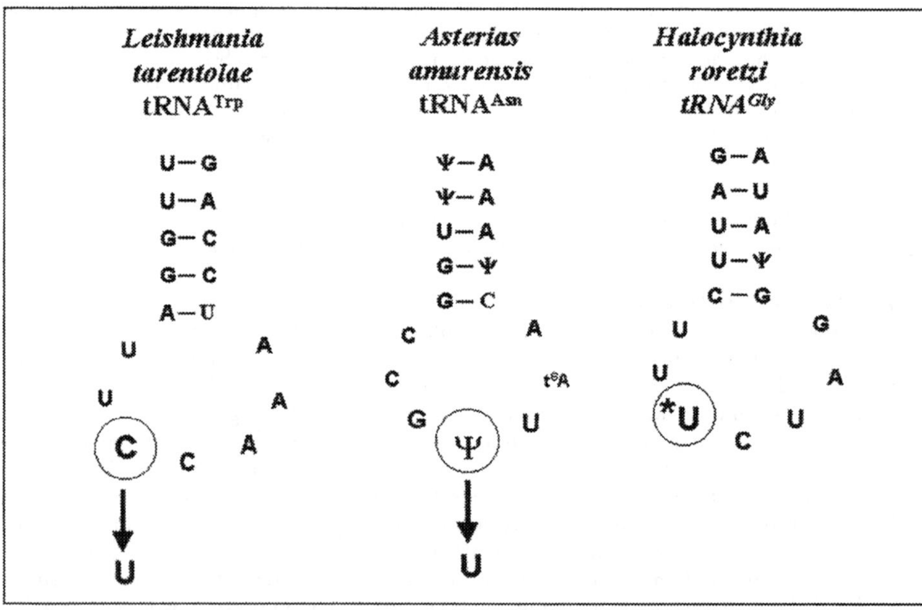

Figure 6. tRNA structural alterations in the evolution of codon reassignment. The secondary structure of the anticodon-arms of the *Leishmania tarentolae* $tRNA^{Trp}$, the *Asterias amurensis* $tRNA^{Asn}$ and the *Halocynthia roretzi* $tRNA^{Gly}$. In *L. tarentolae*, C to U editing converted the CCA anticodon of the cytoplasmic $Trp\text{-}tRNA_{CCA}$, which decodes the TGG codon, into a UCA anticodon which is able to decode the UGA stop codon in the mitochondria. In *A. amurensis*, a ψ in the middle position of the anticodon expanded the decoding capacity of the $Asn\text{-}tRNA_{GUU}$ from two (AAC/AAU) to three codons allowing it to also decode the lysine AAA codon. The *H. roretzi* tRNA has a novel modification in the anticodon first base (designated U*) which allows it to decode the Arg codons AGA and AGG as Gly.[42,58-59]

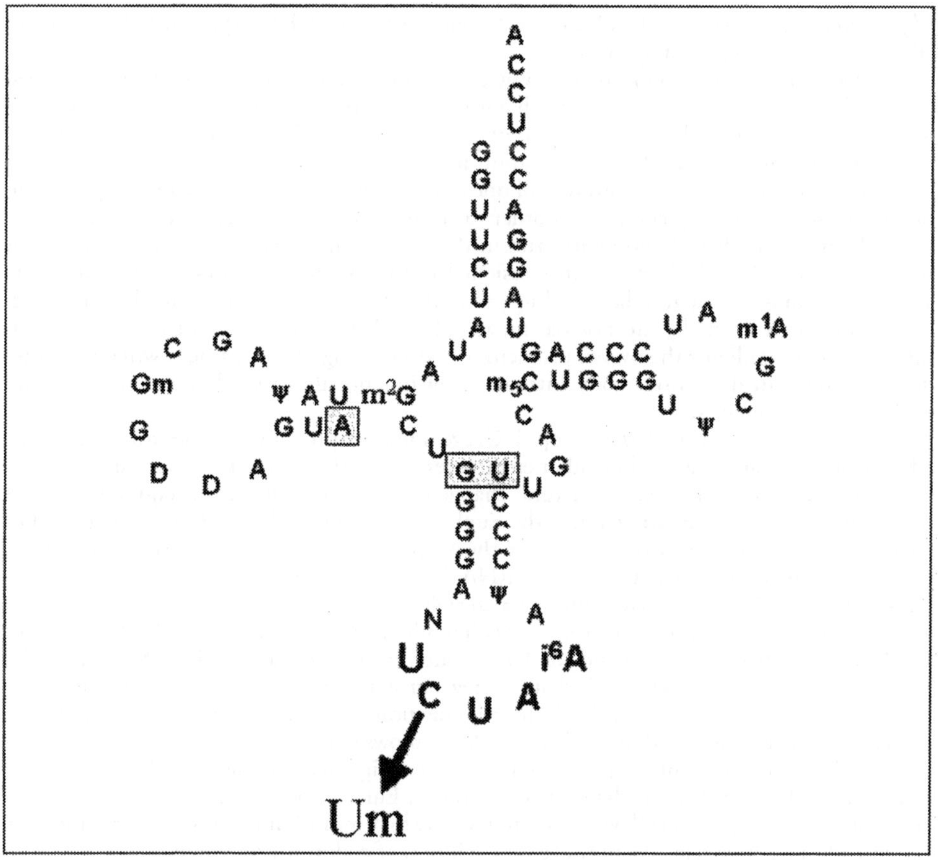

Figure 7. tRNA structural alterations involved in stop codon capture. In ciliates, naturally-occurring tRNA^{Gln} are able to decode the UAG and UAA stop codons due to the presence of A24 in the D-stem and a G-U or G-C base pair at the top of the anticodon-stem (indicated by the shaded boxes). These enhance first and third position wobble thereby allowing UAG/UAA recognition. Low level recognition of these codons by naturally-occurring tRNAs is the first step in stop codon reassignment in that it allows for stop codon capture by the suppressor tRNAs whose codons mutate to increase decoding efficiency.

together with molecular phylogenetic analysis of eRF1 from organisms that reassign stop codons,[44,48] show that the mechanism of stop codon recognition is rather complex. eRF1:codon recognition involves the stop codon binding to three cavities on the surface of eRF1[47,48] and an alteration of the specificity of the eRF1: codon interaction might require substitution of several key amino acids.

The recently solved crystal structure of the human eRF1 shows that it mimics a tRNA molecule and that the three clearly identifiable structural domains, 1, 2 and 3, mimic the anticodon loop domain, the aminoacyl-acceptor stem and the T-stem, respectively.[49] In particular, domain 1 encompassing the amino terminus of the molecule, contains the conserved amino acid sequence Asn-Ile-Lys-Ser (NIKS) which has been implicated in stop codon recognition. In addition, residues I32 and L123 of eRF1 are also thought to be directly involved in codon recognition.[45,47,50] Interestingly, in *Tetrahymena* spp., where UAA and UAG are decoded as Gln, I32 and L123 are replaced by V32 and F123, respectively, while in *Euplotes* and

Blepharisma spp., where the UGA stop is decoded as Cys and Trp respectively, I32 is maintained but L123 is replaced by I123.[50]

It is likely that stop codon recognition by eRF1 requires more than the residues mentioned above,[45] however the spectrum of eRF1 changes in those organisms in which stop codons have been reassigned suggest that reassignment of a stop codon requires a rather complex series of amino acid changes that lead to an alteration in the structure of the translation termination machinery such that it loses recognition of one or two stop codons while maintaining recognition of the remaining stop codon(s) in order to ensure proper translation termination.

In the mitochondria of vertebrates and in the green plant *Scenedesmus obliquus*, both the Arg-AGR and the Ser-UCA codons are reassigned as stop codons, indicating that certain amino acid substitutions can expand the decoding properties of release factors allowing them to recognize codons that normally are not recognized. That these eRF1 mutations introduce new functionalities highlights the overall problem of codon reassignment, namely, while most random mutations are deleterious, some introduce new functionalities which represent significant evolutionary advantages.

The observation that in *Euplotes* spp. UGA reassignment to cysteine does not involve any additional tRNA and that, unlike other eukaryotes, it encodes two eRF1 genes, suggests that, at least in this species, the eRF1 structural change might be the only factor required for UGA reassignment.[46,51,52] In in vitro studies, the Euplotes eRF1 clearly does not recognise the UGA codon but does recognise the other two.[46] In the other cases, the existence of naturally-occurring nonsense supressor tRNAs, such as the $tRNA^{Gln}$ in *Tetrahymena* spp., suggests stop codon reassignment might have evolved through a coevolutionary mechanism in which both tRNAs and eRF1 have played important roles. This hypothesis is supported by the finding that the UGA codon in *Bacillus subtilis* and the UAG codon in the Archaebacterium *Methanosarcina barkeri* are ambiguously translated, that is, they can signal translation termination or can be decoded as Trp or Lys respectively.[53,54] The physiological significance of Trp-UGA decoding in *B.subtilis* is not yet established, nevertheless it clearly shows that stop codon reassignment evolves through decoding ambiguity as postulated by the "Ambiguous Intermediate Theory" (see section The "Ambiguous Intermediate Theory" above). Elucidation of the cellular role of UGA ambiguity in this organism will be important if we are to determine how stop codon readthrough can provide a positive selective pressure for evolution of codon reassignment.

The Role of Aminoacyl-tRNA Synthetases in Codon Reassignment

The standard 20 aminoacyl-tRNA synthetases are divided into two distinct classes (class I and II) and each class is further subdivided into three subclasses a, b and c. There is a clear correspondence of Ia and IIa, Ib and IIb, and Ic with IIc in terms of the chemical properties of the amino acids they charge (Table 4). For example, hydrophobic amino acids such as Val, Ile and Leu (Ia) are placed across from Ala and Pro (IIa) and residues that are sterically similar in shape are matched, e.g., Val and Thr.[55] Remarkably, sense-to-sense codon reassignments follow this subclass division. For example, AUA from Ile (Ia) to Met (Ia); CUN from Leu (Ia) to Thr or Ser (IIa), AGR from Arg (Ia) to Ser or Gly (IIa) and AAA from Lys (Ib) to Asn (IIb). Considering the low number of known and fully validated codon reassignments, it is difficult to make generalisations, however one could speculate that class Ia enzymes might mischarge tRNAs normally charged by a IIa enzyme more easily than IIb or IIc. Hence the reassignments might also reflect structural and evolutionary constraints imposed by the aminoacyl-tRNA synthetase on tRNA charging.

Future Prospects and Implications for Functional Genomics

There are two important practical implications of extant variations in the genetic code; in expression of authentic, biologically active recombinant proteins in the favoured expression hosts which have been endowed with the universal genetic code (e.g., *E. coli*, *S. cerevisiae*), and in the in silico decoding of genome sequence data.[56] The recombinant expression problem can be readily circumvented by replacement of the offending reassigned codon(s) in the gene or

Table 4. Subclass division of aminoacyl tRNA synthetases

Class I Subclass		Class II Subclass		Code Changes
Ia	ArgRS	**IIa**	SerRS	**Arg → Ser**
	LeuRS		ThrRS	**Leu → Thr**
	ValRS		AlaRS	**Ile → Met**
	IleRS		GlyRS	**Ser → Gly**
	CysRS		ProRS	
	MetRS		HisRS	
Ib	GluRS	**IIb**	AspRS	
	GlnRS		LysRS	
	LysRS		AsnRS	*Lys →Asn*
Ic	TyrRS	**IIc**	PheRS	
	TrpRS			

Members of each class of aminoacyl-tRNA synthetases can be further subdivided into three subclasses; I, II and III. Genetic code alterations only occur within each subclass, that is, within subclasses Ia and IIa or Ib and IIb as indicated by the bold and italicized amino acids (adapted from Ribas de Pouplana and Schimmel[55])

cDNA, with codon(s) which will be appropriately decoded to give the authentic polypeptide sequence. This can of course only be done if the identity of all 64 codons has been unambiguously confirmed for the species from which the gene or cDNA was originally isolated. Comparative DNA and protein sequence analysis would seem to be the only option to achieve this. As discussed above, one additional problem could relate to the apparent ambiguous decoding of the CUG codon as either Leu or Ser in certain *Candida* species. Is this the only example of codon ambiguity in nature?

Confirmation of codon assignments is vital if we are to be able to fully exploit the flood of genome sequencing data, in particular annotation of ORFs. Although there has been little effort to date to systematically confirm codon assignments, apart from the widely exploited "model" organisms (see www.ncbi.nlm.nih.gov/entrez/query.fcgi? db= Taxonomy), given the increasing number of examples of exceptions to the universal code, this must be made a requirement before release of any genome sequence if it is to be of any use.

Note Added in Proof

While preparing this manuscript, the editor (Lluís Ribas de Pouplana) called our attention to a recent observation made by Ryckelynck[60] and colleges consistent with one of our predictions (see section "The Role of Aminoacyl-tRNA Synthetases in Codon Reassignment"). These authors report that yeast aspartyl-tRNA synthetase (subclass IIb) mischarges $tRNA^{Glu}$ (normally charged by an aaRS of the Ib subclass) significantly more than other noncognate tRNAs.

Acknowledgments

Research in the Santos laboratory is supported by the Portuguese Foundation for Science and Technology. Grants: Praxis/P/BIA/11139/98 and Praxis/C/SAU/14123/98. Research in the Tuite laboratory is supported by the Biotechnology and Biological Sciences Research Council (BBSRC), and by The Wellcome Trust. We thank Bernard Davenport for critical reading of the manuscript, and Dr. R.D. Knight (Princeton) for communicating information prior to publication.

References

1. Crick FH. The origin of the genetic code. J Mol Biol 1968; 38:367-379.
2. Barrel BG, Bankier AT, Drouin J. A different genetic code in human mitochondria. Nature 1979; 282:189-194.
3. Hasegawa M, Miyata T. On the antisymmetry of the amino acid code table. Origins of Life 1980; 10:265-270.
4. Jukes TH. Amino acid codes in mitochondria as possible clues to primitive codes. J Mol Evol 1981; 18:15-17.
5. Knight RD, Freeland SJ, Landweber LF. Rewiring the keyboard: Evolvability of the genetic code. Nature Rev Genet 2001; 2:49-58.
6. Osawa S, Jukes TH, Watanabe K et al. Recent evidence for evolution of the genetic code. Microbiol Rev 1992; 5:229-264.
7. Oba T, Andachi Y, Muto A et al. CGG: An unassigned or nonsense codon in Mycoplasma capricolum. Proc Natl Acad Sci USA 1991; 88:921-925.
8. Muto A, Andachi Y, Yuzawa H et al. The organization and evolution of transfer RNA genes of Mycoplasma capricolum. Nucleic Acids Res 1990; 18:5037-5043.
9. Osawa S. Evolution of the Genetic Code. New York: Oxford University Press, 1995.
10. Schultz DW, Yarus M. On the malleability in the genetic code. J Mol Evol 1996; 42:597-601.
11. Yarus M, Schultz DW. Further comments on codon reassignment. J Mol Evol 1997; 45:3-6.
12. Knight RD, Freeland SJ, Landweber LF. A simple model based on mutation and selection explains trends in codon and amino acid usage and GC composition within and across genomes. Genome Biol 2001; 2:1-13.
13. Knight RD, Landweber LF, Yarus M. How mitochondria redefine the code. J Mol Evol 2001; 53:299-313.
14. Santos MAS, Perreau VM, Tuite MF. Transfer RNA structural change is a key element in the reassignment of the CUG codon in Candida albicans. EMBO J 1996; 15:5060-5068.
15. Suzuki T, Ueda T, Watanabe K. Polysemous codon: A codon with multiple amino acid assignment caused by dual specificity of tRNA identity. EMBO J 1997; 16:1122-1134.
16. Sharp PM, Matassi G. Codon usage and genome evolution. Curr Opin Genet Dev 1994; 4:851-860.
17. Muto A, Osawa S. The guanosine and cytosine content of genomic DNA and bacterial evolution. Proc Natl Acad Sci USA 1987; 84:166-169.
18. Sueoka N. Directional mutation pressure and neutral molecular evolution. Proc Natl Acad Sci USA 1988; 85:2653-2657.
19. Anderson AGE, Kurland CG. An extreme codon preference strategy: codon reassignment. Mol Biol Evol 1991; 8:530-544.
20. Anderson AGE, Kurland CG. Genomic evolution drives the evolution of the translation system. Biochem Cell Biol 1995; 73:775-787.
21. Anderson AGE, Kurland CG. Reductive evolution of resident genomes. Trends Microbiol 1998; 6:263-268.
22. Silva FJ, Amparo L, Moya A. Genome size reduction through multiple events of gene disintegration in Buchnera APS. Trends Genet 2001; 17:615-618.
23. Gray MW, Burger G, Lang BF. The origin and early evolution of mitochondria. Genome Biol 2001; 2:1-5.
24. Watanabe K, Osawa S. tRNA sequences and variations in the genetic code. In: Soll D, RajBhandary U, eds. tRNA: Structure, Biosynthesis and Function. Washington: ASM Press, 1995:225-250.
25. Sternberg D, Chatzoglou E, Laforet P et al. Mitochondrial DNA transfer RNA gene sequence variations in patients with mitochondrial disorders. Brain 2001; 124:984-994.
26. Urbonavicius J, Qian Q, Durand JMB et al. Improvement of reading frame maintenance is a common function for several tRNA modifications. EMBO J 2001; 20:4863-4873.
27. Björk GR, Jacobsson K, Nilsson K et al. A primordial tRNA modification required for the evolution of life. EMBO J 2001; 20:231-239.
28. Hagervall TG, Tuohy TMF, Atkins JF et al. Deficiency of 1-methylguanosine in tRNA from Salmonella typhimurium induces frameshifting by quadruplet translocation. J Mol Biol 1993; 232:756-765.
29. Schultz DW, Yarus M. tRNA structure and ribosomal function. I. tRNA nucleotide 27-43 mutations enhance first position wobble. J Mol Biol 1994; 235:1381-1394.
30. Santos MAS, Cheesman C, Costa V et al. Selective advantages created by codon ambiguity allowed for the evolution of an alternative genetic code in Candida spp. Mol Microbiol 1999; 31:937-947.
31. Grant CM, Firoozan M, Tuite MF. Mistranslation induces the heat-shock response in the yeast Saccharomyces cerevisiae. Mol Microbiol 1989; 3:215-220.

32. Sugita T, Nakase T. Nonuniversal usage of the leucine CUG codon and the molecular phylogeny of the genus Candida. Syst Appl Microbiol 1999; 22:79-86.
33. Murphy HS, Humayun Z. Escherichia coli cells expressing a mutant glyV (glycine tRNA) gene have a UVM-constitutive phenotype: Implications for mechanisms underlying the mutA or mutC mutator effect. J Bacteriol 1997; 179:7507-7514.
34. Torkelson J, Harris RS, Lombardo MJ et al. Genome-wide hypermutation in a subpopulation of stationary-phase cells underlies recombination-dependent adaptive mutation. EMBO J 1997; 16:3303-3311.
35. Rosche WA, Foster PL. The role of transient hypermutators in adaptive mutation in Escherichia coli. Proc Natl Acad Sci USA 1999; 96:6862-6867.
36. Stansfield I, Kushnirov VV, Jones KM et al. A conditional-lethal translation termination defect in a sup45 mutant of the yeast Saccharomyces cerevisiae. Eur J Biochem 1997; 245:557-563.
37. Farabaugh PJ. Programmed Alternative Reading of the Genetic Code. Heidelberg: Molecular Biology Intelligence Unit, Springer Verlag, 1997.
38. Santos MAS, Keith G, Tuite MF. Nonstandard translational events in Candida albicans mediated by an unusual $tRNA^{Ser}$ with a 5'-CAG-3' (leucine) anticodon. EMBO J 1993; 12:607-616.
39. Perreau VM, Keith G, Holmes MW et al. The Candida albicans CUG-decoding seryl-tRNA has an atypical anticodon stem-loop structure. J Mol Biol 1999; 293:1039-1053.
40. Yarus M. Translational efficiency of transfer RNA's: Uses of an extended anticodon. Science 1982; 218:646-652.
41. Pape LK, Tzagoloff A. Cloning and characterization of the gene for the yeast cytoplasmic threonyl-tRNA synthetase. Nucleic Acids Res 1985; 13:6171-6183.
42. Alfonzo JP, Blanc V, Estevez AM et al. C to U editing of the anticodon of imported mitochondrial $tRNA^{Trp}$ allows decoding of the UGA stop codon in Leishmania tarentolae. EMBO J 1999; 18:7056-7062.
43. Zerfass K, Beier H. Pseudouridine in the anticodon G ψ A of plant cytoplasmic tRNA(Tyr) is required for UAG and UAA suppression in the TMV-specific context. Nucleic Acids Res 1992; 20:5911-5918.
44. Inagaki Y, Doolitle WF. Evolution of the eukaryotic translation termination system: origins of release factors. Mol Biol Evol 2000; 17:882-889.
45. Lozupone CA, Knight RD, Landweber LF. The molecular basis of nuclear code change in ciliates. Curr Biol 2001; 11:65-74.
46. Kervestin S, Frolova L, Kisselev L et al. Stop codon recognition in ciliates: Euplotes release factor does not respond to reassigned UGA codon. EMBO Rep 2001; 2:680-684.
47. Bertram G, Bell HA, Ritchie DW et al. Terminating eukaryote translation: domain 1 of release factor eRF1 functions in stop codon recognition. RNA 2000; 6:1236-1247.
48. Inagaki Y, Blouin C, Doolittle WF et al. Convergence and constraint in eukaryotic release factor 1 (eRF1) domain1: the evolution of stop codon specificity. Nucleic Acids Res 2002; 30:532-544.
49. Song H, Mugnier P, Das AK et al. The crystal structure of human eukaryotic release factor eRF1-Mechanism of stop codon recognition and peptidyl-tRNA hydrolysis. Cell 2000; 100:311-321.
50. Lehman N. Molecular evolution: Please release me, genetic code. Curr Biol 2001; 11:R63-R66.
51. Grimm M, Brunen-Nieweler C, Junker V et al. The hypotrichous ciliate Euplotes octocarinatus has only one type of $tRNA_{Cys}$ with GCA anticodon encoded on a single macronuclear DNA molecule. Nucleic Acids Res 1998; 26:4557-4565.
52. Liang A, Brunen-Nieweler C, Muramatsu T et al. The ciliate Euplotes octocarinatus expresses two polypeptide release factors of the eRF1 type. Gene 2001; 262:161-168.
53. Matsugi J, Murao K, Ishikura H. Effect of Bacillus subtilis tRNA(Trp) on readthrough rate at an opal UGA codon. J Biochem (Tokyo) 1998; 123:853-858.
54. Paul L, Ferguson DJ, Krzycki JA. The trimethylamine methyltransferase gene and multiple dimethylamine methyltransferase genes of Methanosarcina barkeri contain in-frame and read-through amber codons. J Bacteriol 2000; 182:2520-2529.
55. Ribas de Pouplana L, Schimmel P. Aminoacyl-tRNA synthetases: Potential markers of genetic code development. Trends Biochem Sci 2001; 26:591-596.
56. O'Sullivan JM, Davenport JB, Tuite MF. Codon reassignment and the evolving genetic code: Problems and pit-falls in post-genome analysis. Trends in Genetics 2001; 17:20-22.
57. Yokogawa T, Suzuki T, Ueda T et al. Serine tRNA complementary to the nonuniversal serine codon CUG in Candida cylindracea: Evolutionary implications. Proc Natl Acad Sci USA 1992; 89:7408-7411.
58. Tomita K, Ueda T, Watanabe K. The presence of pseudouridine in the anticodon alters the genetic code: a possible mechanism for assignment of the AAA lysine codon as asparagine in echinoderm mitochondria. Nucleic Acids Res 1999; 27:1683-1689.

59. Kondow A, Suzuki T, Yokobori S et al. An extra tRNAGly (U*CU) found in ascidian mitochondria responsible for decoding nonuniversal codons AGG/AGA as glycine. Nucleic Acids Res 1999; 27:2554-2559.
60. Ryckelynck M, Giegé, R, Frugier, M. Yeast $tRNA^{Asp}$ charging accuracy is threatened by the N-terminal extension of Aspartyl-tRNA synthetase. J Biol Chem 2003; 278(11):9683-90.

Chapter 13

Adaptive Evolution of the Genetic Code

Rob D. Knight, Stephen J. Freeland and Laura F. Landweber

All known genetic codes use 4 bases and 20 amino acids, but many other bases and amino acids have been synthesized and/or found in organisms. The coding relationships between particular trinucleotides and amino acids can and have evolved, as shown by variants in both mitochondrial and nuclear lineages. Here we review the evidence that various aspects of the genetic code, including its composition, its degeneracy and the assignments of particular codons to particular amino acids are in some sense optimal, chosen over alternatives by natural selection. We also examine several specific proposals about how the code evolved prior to its fixation in the last common ancestor of extant life. Although the pattern of codon assignments appears nearly optimal, other claims for adaptive features are more speculative and many interesting questions remain unresolved.

Framing the Questions

The most fundamental divide between theories of genetic code evolution hinges on whether the code is fixed or evolvable. Intuitions about the mutability of the code separate models that assume a direct reason for every codon assignment from the beginning (stereochemical and mathematical theories) from those that assume that the code can and has changed during the course of early evolution (adaptive and coevolutionary theories). Here we focus on the latter, all of which fundamentally assume that the code has changed **for some reason**, whether to increase catalytic diversity by adding amino acids or to minimize the phenotypic impact of genetic errors made during replication and translation. If we accept that the code is only one of a vast number of possibilities then it becomes legitimate to ask why we have this code rather than its alternatives: in other words, what is it good for?

Not every feature of an organism is an adaptation. Some features are determined by the laws of physics;[1] others arise as side-effects of other adaptive choices.[2] In order to demonstrate that a trait *t* is 'an adaptation for' a property *p*, it is necessary to show

A. that variation in *t* actually does cause variation in *p*, and
B. that the fitness advantage attributable to heritable variation in *p* led to the fixation of *t* in a population in which it was originally polymorphic.[3] Thus, the claim that *t* is an adaptation for *p* is strong, but often testable.

The first point is uncontroversial: the genetic code and its pattern of codon/amino acid assignments defines the distribution and characteristics of all mutation, which in turn is the source of all heritable biological variation. The idea that the components and organization of the genetic code are adaptive has therefore been challenged primarily on the second point, that the code actually was selected over alternatives.

Is the Choice of Coding Components Optimal?

The modern code links L-α-amino carboxylic acids to codons made of nucleic acids based on D-ribose and purines and pyrimidines with a particular hydrogen-bonding pattern, but we need not take this for granted. Could the set of amino acids and bases, or even the peptide and

The Genetic Code and the Origin of Life, edited by Lluís Ribas de Pouplana.

nucleic acid backbones, have been selected from a range of possibilities as the most stable, least energetically costly, or most catalytically active solutions? Although several interesting possibilities have been suggested, they remain necessarily speculative. No naturally occurring variation is known in molecular coding components beyond base and amino acid modification, thus we have no hint as to the appropriate class of possibilities to evaluate. There are hundreds of different amino acids produced in cells,[4] perhaps a couple of dozen plausible nucleic acid backbones, and perhaps half a dozen plausible alternatives to peptide backbones, as well as the possibility of codes with more than 20 amino acids and more than 4 bases, the contrast class of possible codes that this section considers is prohibitively large ($> 10^{100}$).

Selection of Nucleic Acid Constituents

The idea that RNA preceded DNA is well-established in the literature as the RNA World hypothesis,[5] and the difficulty of ribonucleotide reduction suggests that DNA arose only after proteins were already being used as sophisticated catalysts.[6] However, D-ribose is not the only possible backbone that can support complementary base pairing. Analogs of RNA using sugars with fewer[7-9] and more (see ref. 10 for detailed review) carbons can be synthesized; ribose backbones do not even allow more stable base pairs than do other pentoses, which might suggest that the pairing strength has been 'optimized' rather than maximized,[10] or, alternatively, that pairing strength was not of primary selective importance. No sugar is stable under mainstream predictions of prebiotic conditions,[11] and it is possible that the first backbones were based on alternative chemistries. PNA, peptide nucleic acid,[12] has the nucleotide bases bonded to a peptide backbone; it forms stable base pairs (even in heteroduplex with DNA or RNA),[13] and can be plausibly synthesized prebiotically.[14] It is possible that an early system based on PNA was displaced by RNA,[15] perhaps because the charged backbone reduces aggregation and because the 2′-OH group allows a wider range of catalytic activity.[9,16] However, such suggestions remain purely speculative.

Given RNA as the information storage macromolecule, the four standard bases are not the only possibilities. A variety of alternatives with different hydrogen bond donor and acceptor patterns have been synthesized that support new complementary base pairs;[17,18] some of these can even be incorporated by standard polymerases.[17,19,20] However, many possible base pairs are unstable because of increased tautomerism.[17] Whilst adenine is easily produced by HCN polymerization,[21] many nonstandard purines are produced under similar conditions,[22] perhaps indicating that adenine was preferable for some reason. None of the standard bases are stable under prebiotic conditions, however, suggesting that either life arose rapidly or alternative bases were originally used.[23]

Selection of Amino Acid Constituents

The choice of amino acids in the code may also be adaptive: many functional groups, such as halides, carbonyls, phosphates, and sulfonates, are not represented in the standard amino acids, although amino acids containing these groups can be synthesized (and some, such as citrulline, are even common in cells). There is only partial overlap between the amino acids used in the standard code and those available by prebiotic synthesis, suggesting that some were invented later as metabolism grew more complex,[24] and perhaps that some primordial amino acids were eliminated from the code because they were, overall, less useful in proteins.[25] The most extensive investigation into the set of coding and noncoding amino acids is that of Weber and Miller,[26] who rule out several prebiotically plausible amino acids on structural grounds (for instance, ornithine, which is an analog of lysine but one methylene group shorter, is unstable to cyclization by lactam formation). However, they are unable to account for the absence of several amino acids common in prebiotic synthesis, such as norleucine, norvaline, pipecolic acid, and alpha-aminobutyric acid; similarly, complex amino acids such as Trp, Arg, and His are 'justified' by assuming that their functional groups are necessary for catalytic activity, and that these are the simplest amino acids that contain those functional groups.

An alternative perspective[27,28] suggests that amino acids were chosen from metabolic pathways as those that were more useful for protein synthesis than as intermediates. Specifically, some amino acids were excluded because they were incompatible with accurate translation (e.g., norleucine is misincorporated for Met in modern cells). However, this reasoning seems backwards: it is more likely that some amino acids are misincorporated in modern cells **because** cells have not historically been exposed to high levels of them, especially since aminoacyl-tRNA synthetases can discriminate between amino acids as similar as leucine and isoleucine with near-perfect accuracy. By analogy, the evolutionary history of oxygen metabolism illustrates that even extreme toxicity is a relative and often transitory phenomenon over an evolutionary timescale! Furthermore, many amino acids that are found in the code, such as Asp, Arg, Glu, Asn, and Gln, are central in extant metabolism, so the argument that some amino acids are 'more useful' as intermediates is unconvincing.

A more radical question is why we have amino acids at all: there is no reason to believe that catalytic side-chains could only be carried by a peptide backbone. One possible alternative is thioesters resulting from the condensation of thiols with carboxylic acids, which may have been common in thermal vents.[29] Other possibilities include beta-amino acids, hydroxy acids, amides produced by diamino and dicarboxylic acid monomers, and esters. However, the relative stability of the amide backbone, along with the conformational rigidity enforced by the alpha-amino linkage, may account for the choice of alpha-amino acids.[26]

Selection of Chemical Alphabet Size

A subtly different question is whether the number of bases and/or amino acids is adaptive. Various authors have suggested a primordial code based on a system of fewer than 4 nucleotides. One of the earliest speculations was for an all-purine code in which Inosine paired with Adenine.[30] However, the most common line of speculation has been for a binary primordial code that utilized a single purine-pyrimidine pair.[31] A subtle twist on this idea is that regardless of the bases present, primordial translation machinery may have only distinguished purine from pyrimidine.[32] To date, the only attempt to rigorously model the size of the nucleic acid alphabet comes from Szathmary.[33,34] Assuming an RNA-world origin, he calculates that 4 bases are better than 2 or 6 as a tradeoff between catalytic sophistication and replicative fidelity, based on estimates of the likelihood of arbitrary catalysis (measuring functional diversity) and the actual pairing energies of the standard bases versus nonstandard alternatives such as those synthesized by Piccirilli et al.[17] Such an explanation implicitly predates genetic coding, but the size of the nucleotide alphabet is clearly heavily interrelated with the size of the amino acid alphabet. For example, a triplet codon binary code could only unambiguously code 16 amino acids, but the addition of another base pair (or an increased ability to distinguish within the purines and pyrimidines) would necessarily increase coding potential to up to 64 amino acids. The only attempt to explore the optimality of the current amino acid alphabet size comes from reference 35. This model, which does much to foreshadow Szathmary's subsequent investigations of the nucleotide alphabet, assumed that addition of a new amino acid offers a catalytic advantage but only at the cost of disrupting existing protein synthesis, and found that a 20-amino acid code could plausibly represent an optimum. However, Wong himself emphasized the preliminary nature of his model, and to date no-one has extended his findings to a more rigorous study.

More broadly, although it seems likely that the genetic code evolved from an earlier form with fewer codons and amino acids,[31] there is no agreement in the literature about how this occurred. For instance, the first codons have been proposed to be RRY,[31] RNY,[36,37] GNN,[38-40] all-(A,U),[41] all-(G,C),[42,43] all-purine,[44] NYN vs. NRN,[32,45] GCU alone,[46] etc. The fundamental difficulty is that the genetic code consists of a small, highly connected set of elements, and it is easy to see patterns that are only weakly supported by evidence. Recently, Trifonov has compiled from the literature 40 conjectures about the pathway of code evolution, averaged the rank orders together, and called the result a 'consensus temporal order' of the amino acids.[47]

Trifonov is encouraged by the generally low pair wise correlations between estimates (only about 20 of the 720 pair wise correlations are above 0.5), suggesting that this lack of relatedness implies that the measures are truly independent. However, the consensus rank order is 88% correlated with the rank abundance of amino acids in proteins (which itself is almost completely determined by the number of codons,[48] and 80% correlated with molecular weight (the two measures are themselves 73% correlated). This suggests that different authors intuitively agree that larger amino acids, which tend to have fewer codons, were probably late entries into the code, but disagree on every other point!

Summary

In summary, there is clear reason to believe that some amino acids (such as ornithine), bases (such as bromouracil, which has such a high rate of tautomerism that it is a powerful mutagen), and backbones (such as PNA, which is uncharged and therefore tends to aggregate, and which has no backbone hydroxyls to participate in catalysis) were not used for genetic coding. However, the evidence is too sketchy to conclude that the actual components of the coding system replaced inferior alternatives as an adaptive or optimal choice. There is a stronger case that the size of the nucleotide alphabet represents an evolutionary optimum **if all 4 nucleotides evolved prior to protein coding**. Equivalent analysis for the size of the amino acid alphabet is less well developed.

An Optimal Pattern of Code Degeneracy?

Assuming 4 bases, triplet codons and 20 amino acids as a framework for the code, it is clear that at least some amino acids must be assigned more than one codon. In fact, this is true for all amino acids except Met and Trp. There are two senses in which this pattern of degeneracy within the code could be adaptive in terms of mutation/mistranslation. The first sense merely asks whether the symbols are arranged to minimize (or maximize) the chance of reaching a different symbol (amino acid meaning) by substituting a single nucleotide; this line of investigation prompted the earliest claims for an adaptive arrangement of codon assignments (section *An Optimal Pattern of Codon Assignments*) but with hindsight appears to be weak evidence given biochemical considerations. Specifically, Crick's Wobble Hypothesis,[49] which suggests that 3rd-position blocks are split between purine-ending and pyrimidine-ending codons (and not, for instance, between A+C-ending and G+U-ending codons) because G recognizes both C and U at the 3rd position (and, similarly, U pairs with both A and G), is borne out by detailed studies of pairing in modern tRNAs. Moreover, wobble pairing is remarkably successful in predicting which codon doublets will be split: those where the first two positions are G or C always form a family box, while those where the first two positions are A or U are always split between two or more amino acids.[50-52] No naturally occuring variant codes deviate from this rule,[53] which may reflect the different bond strengths of AU and CG base pairs. Thus the overall pattern of degeneracy may reflect chemical constraints, although the small number of known variants makes such conclusions highly uncertain: it remains possible that the conformation of the 'wobble base' in the tRNA, which allows this ambiguous misreading, may be a derived state and hence itself an adaptation.[54]

The second sense in which redundancy might be adaptive asserts that there are metabolic or functional reasons that proteins have a particular composition, and that code degeneracy is optimized to reflect this composition (for instance assigning more codons to heavily used amino acids). The appropriate contrast class for this section is all codes that assign at least one codon to each of the 20 amino acids and Stop (not preserving the pattern or distribution of degeneracy found within the standard code). To a first approximation, assuming each codon is assigned to an amino acid independently, this is (21^{64}-20^{64}) or about 10^{84} possible codes.[55] It is obvious that the standard code does not look like one randomly drawn from this class, since different codons for the same amino acid nearly always start with the same first- and second-position base: this not be expected by chance.[38] Still, it is unclear whether this pattern, which has the effect of reducing missense substitutions, was selected for the purpose rather than being a combination of chance and tRNA wobble.

Coding Theory Explanations for Redundancy

Purely mathematical explanations for degeneracy, which test whether the genetic code matches one of several formally optimal codes, have been far less successful. For example, the optimal way of encoding a probability distribution of states (in terms of minimizing message length) is in a Huffman code, in which the most likely outcomes are assigned shorter symbols. Although the use of tRNA adaptors might in principle allow this (if anticodons of different lengths were used), in fact all actual codons are the same length, except for stop codons which are read by protein release factors and are now known to have a strong 4th-base context effect.[56] Similarly, the genetic code might have been a Baudot code, in which adjacent symbols are generated by sliding a reading frame across a cycle of bits,[57] or a Gray code, an example of a minimum change code in which binary encodings of objects are arranged in a ring such that changes in progressively less significant digits of the encoding lead to substitution of increasingly similar objects.[58] However, the evidence that the standard genetic code is optimal in any of these formal senses is unconvincing: it is far from clear that amino acid properties can meaningfully be measured as a ring rather than on a linear scale, and the claim that 'subcodes' for particular amino acids are optimal[57,59-62] reduces to nothing more than the claim that codons for a single amino acid are, in general, adjacent. Though some of the earliest claims for an optimal arrangement of codon assignments made exactly this point in much simpler language,[63,64] this aspect of the standard code can, as ever, be adequately explained by the biochemical limitations of wobble pairing.

It has been suggested that the combination of triplet codons and 20-21 symbols is a 'hardware optimum' minimizing the number of components required for translation,[65] but this appears to be numerology based on the fact that e^3 is close to 20. Moreover, the genetic code does **not** maximize the entropy of codon assignments (required for optimization in this sense), since different amino acids are assigned grossly inequitable numbers of codons. Additionally, the number of tRNAs actually used varies from species to species, but should be a universal, low number were the code really optimized to minimize the number of components required for translation: while this effect is extreme in mitochondria, there is no evidence that it has influenced code evolution even there.[66] The frequency with which particular codons and amino acids are used varies widely between species, although much of the variance can be accounted for by changes in base composition. In particular, over 80% of the variance in frequency in even as large and chemically active amino acid as arginine is explained by genome GC content.[67] Thus, although the number of codons assigned to each amino acid does correlate with overall abundance in proteins,[68] it is most likely that this is because neutral mutation leads proteins to reflect the code rather than the reverse.[48,69] Similarly, although smaller and less complex amino acids tend to be assigned more codons,[70,71] the range of amino acid usage in different species makes it unlikely that this codon assignment is an adaptation to minimize the metabolic cost of making proteins.

There have been several group-theoretic explanations for the code's structure based on symmetry breaking,[72-74] but these share the flawed assumption that variant codes diverged while the code was still partially ambiguous (rather than from an already complete standard code). More fundamentally, such theories offer no reason as to why symmetry at this level would be biologically relevant; it is possible to invent an algebra that recaptures **any** dichotomous classification, and so these techniques can only describe, rather than explain, the code structure. Effectively then, such explanations use algebra to express the observation that the distribution of amino acid assignments within the standard code is decidedly non-random: an observation that dates back to the early 1960s.

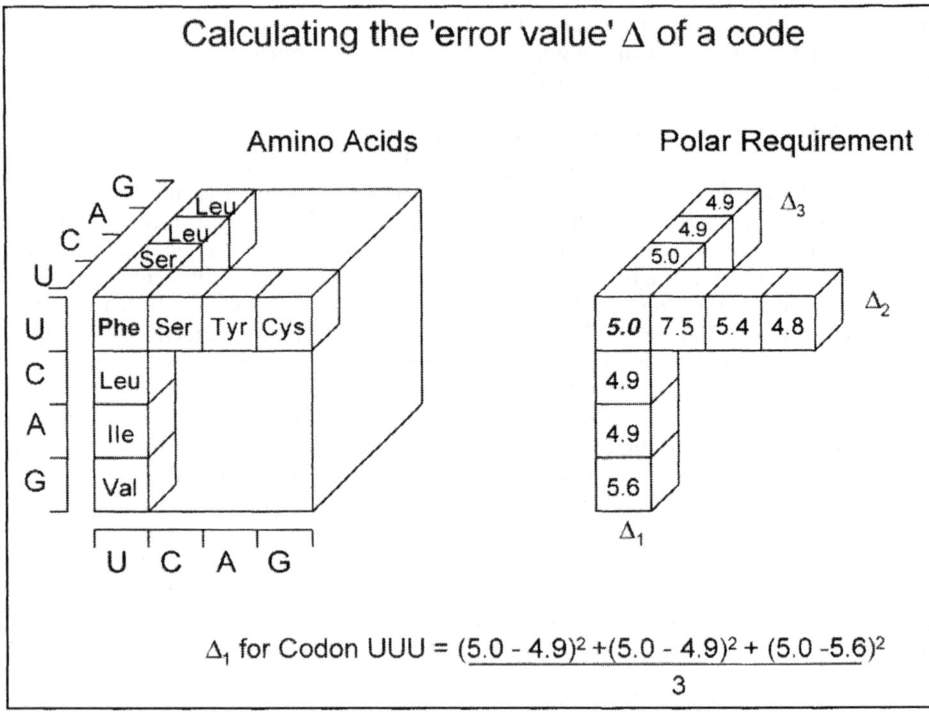

Figure 1. Calculating the error value of a code. Assign a value (in this case, Polar Requirement) to each amino acid, or to each pair of amino acids, to generate a distance matrix showing the magnitude of effect of each type of substitution. Then average the effects of all the substitutions over the whole code, optionally weighting for different types of mutation, or using a different modular power (e.g., linear instead of squared deviations). In this example, the distance for the UUU to CUU Phe to Leu transition is $|5.0 - 4.9| = 0.1$, which contributes to the first-position transition error. Distance matrices can be constructed from a linear measure, such as polar requirement, or from substitution matrices calculated from observed substitutions in proteins.

Biological Explanations for Redundancy

Finally, a pattern of degeneracy that minimized nonsynonymous substitutions might not be ideal for evolution, which relies on mutations as the raw material for adaptation. An analysis of the likelihood that single-base mutations substitute amino acids of different polarity, and of the average shortest path length for the interconversion of pairs of amino acids, suggested that the code represents "tradeoff between robustness and flexibility",[75] although, as always, it is difficult to tell a tradeoff between two dubious adaptations from what would be expected by chance.[2] Similarly, codon assignments may be adapted as a tradeoff between misreading frequency and speed of translation,[76] but in the absence of comparable data from random codes (and a compelling reason why the tradeoff would produce the observed values) it is impossible to assess the claim (Fig. 1).

In summary, there is no strong evidence to date that the number of codons assigned to each amino acid is adaptive, or that the pattern of degeneracy makes the code formally optimal in terms of its error-resistance properties: in fact, the code has enough redundancy to be used as an error-detecting code,[58] but this capacity is not used. All variant codes change the pattern of redundancy, so if a particular pattern **were** adaptive it would be possible for it to be fixed by natural selection. However, the most convincing explanations for the code's redundancy to date are based on chemistry: Crick's wobble hypothesis explains why 3rd positions are split between purines and pyrimidines, while Lagerkvist's observations on degeneracy and doublet GC content explain which codon blocks are split. Although these proximal (mechanical) explanations may represent the modern mechanisms by which adaptation was effected, the adaptive explanations thus far proposed become far less compelling on closer scrutiny.

An Optimal Pattern of Codon Assignments?

Over and above the non-random degeneracy of the code, researchers quickly noted that physiochemically similar amino acids are assigned to codons that share 2 out of 3 bases in common. In particular, the identity of the second base within a codon is well-correlated with the hydrophobicity of the amino acid to which it is assigned.[45,63,64,77-82]

These patterns prompted others to launch what has subsequently become the most extensively explored claim for an adaptive genetic code: that the pattern of codon assignments has been shaped by natural selection to reduce errors in replication and/or translation. Of particular note, the Translation Error model[83,84] suggested that the primary selective pressure was a reduction in deleterious phenotypic impact of translation errors. This was a significantly more sophisticated adaptive hypothesis than the Lethal Mutation model[63,64] in that Translation Error considers the physiochemical similarity of different amino acids, whereas Lethal Mutation was limited to a consideration of synonymous versus non synonymous substitutions.

Mutation and mistranslation differ in one other subtle but important detail: for both, transition errors between the two purines or the two pyrimidines are more frequent than transversions from purine to pyrimidine or vice versa, but translation alone introduces the concept of reading frame-dependent error. Woese noted that, although the 3rd and 1st position bases were optimized (in the sense that single base errors tended to substitute amino acids of similar hydrophobicity), 2nd position changes tended to be non-conservative, indicating a reading frame-dependent effect;[85] this was consistent with the relative frequency of streptomycin-induced misreadings in model polypeptides.[86] However, some caution is warranted. Although a general **mutational** transition bias may be derived from biochemical principles of nucleotide tautomerization,[87-89] recorded patterns of mistranslation may be due to the peculiarities of codon/anticodon pairings in modern tRNA and mRNA within the context of the modern ribosome: this might not accurately reflect the situation when coding was established,[54] and the degree of code evolution that occurred between the emergence of 'modern' coding machinery and the fixation of the standard code is unknown.

In a paper that foreshadowed much of the current debate, Alff-Steinberger explicitly tested the average effect of point substitutions at different positions using Monte Carlo simulations, comparing the actual code to 200 alternative codes made by shuffling the amino acid properties among the 20 blocks of synonyms found in the canonical code, and comparing the average magnitude of errors induced by single-base misreadings at each position. Testing a wide range of amino acid physiochemical properties, he found that the standard code outperformed the majority of alternatives for all of them: the 3rd-position base was most highly optimized relative to random codes (unsurprisingly, as he included the silent substitutions that dominate here), followed by the 1st position base, and there was no evidence for optimization in the 2nd-position base, consistent with the relative effects of translation error.[90] This clear evidence that the standard code minimized errors better than random codes was ignored for over 20 years before it was replicated in part (and apparently unknowingly) by Haig and Hurst.[91] This might in part be due to an apparent error in Alff-Steinberger's calculations: 2 of us have independently attempted to corroborate his work, but have been unable to replicate any of his quantitative results. However, it seems more likely that the subsequent delay in developments to the adaptive exploration of the code reflects the deeper context of research into the nature of the genetic code.

Early Objections to Adaptive Patterns of Codon Assignment

Prior to these early theories of adaptive codon assignments, there existed a surprisingly widespread belief that the standard code could have taken no other form, (e.g., ref. 92, but see also ref. 93 for a review of early theories). Such views originated from assumptions of "direct templating" between nucleic acids and amino acids during translation such that the code was a deterministic outcome of stereochemical interactions. They failed as the molecular machinery of the genetic code was unraveled: Crick's putative 'adaptor' molecules[94] were identified as transfer RNA molecules, which physically dislocate amino acids from their corresponding codons. Although some research sought to adapt deterministic stereochemical theories to postulate fits between amino acids and their cognate tRNAs,[85,95-97] the very existence of tRNA adaptors paved the way for an evolved and evolvable genetic code.[98] Indeed, Crick essentially predicted the "codon capture" mechanism of codon reassignment (ref. 99: see below) and recognizing possible adaptive value in code reorganization a generation before these ideas achieved widespread interest. However, further investigations of the amino acid assignments of the code revealed its apparent universality in organisms as diverse as *E. coli*, humans and yeast, prompting Crick's "frozen accident" (non-adaptive) explanation for genetic code structure:

'This theory states that the code is universal because at the present time any change would be lethal, or at least very strongly selected against. This is because in all organisms (with the possible exception of certain viruses) the code determines (by reading the mRNA) the amino acid sequences of so many highly evolved protein molecules that any change to these would be highly disadvantageous unless accompanied by many simultaneous mutations to correct the "mistakes" produced by altering the code.

This accounts for the fact that the code does not change. To account for it being the same in all organisms one must assume that all life evolved from a single organism (more strictly, from a single closely interbreeding population). In its extreme form, the theory implies that the allocation of codons to amino acids at this point was entirely a matter of "chance."[31]

In other words, once a genetic code achieved sufficient scope and accuracy for a cell to rely on its protein products then regardless of its selective value relative to other possibilities, the deleterious impact of variation would preclude further change. It was this simple rationale and the persistence of apparent universal coding that probably account for the generation-length delay in developing Alff-Steinberger's results.

Naturally Occurring Variants of the Standard Genetic Code

However, we now know that the standard code is not universal. The first naturally occurring variant genetic code was identified in 1979 within vertebrate mitochondria.[100] The variation is slight, and initial explanations sought to downplay the significance of the find. Specifically, it was explained as either a genetic fossil of an ancestral code from which the 'universal' code subsequently emerged,[101-103] or that vertebrate mitochondria encoded so few protein products that subtle code changes might be uniquely tolerable in this context.[103,104] Both explanations sought to sidestep the "freezing" argument by implying an unusually low reliance on coded protein products. However, numerous further discoveries of non-standard codes, all secondarily derived variations on the 'universal' or 'canonical' code unambiguously indicate a different interpretation: the code can and has changed within nuclear and organelle lineages. Although the standard genetic code is indeed common to most organisms, the fact that this is not universally true begs an explanation of:

A. why it was not one of the known variant codes that gave rise to all extant life instead of the 'canonical' code, and
B. what the possible range of genetic codes might look like.

All known variant codes share some properties. First, 4-codon 'family' blocks with the same initial doublet (e.g., CGN) can be either split or unsplit. In the canonical code, most split blocks are 2/2 between the pyrimidines and the purines (e.g., GAY Asp and GAR Glu); however, the AUN block is split 3/1 between Ile and Met. In variant codes, the main form of change seems to be variation back and forth between this type of 2/2 and 3/1 split, which can be explained, at least in terms of mechanism, by variation in chemical modification at the 'wobble' base at the first position of the tRNA anticodon.[53,105,106]

A qualitatively different type of variation is block reassignment. For example, yeast mitochondria assign the CUN block, normally Leu, to Thr;[107] the AGR codon block has been reassigned from Arg to Ser, Gly and Stop in the metazoa (reviewed in ref. 108). This type of reassignment is caused by duplication and mutation of tRNAs: typically the amino acid/acceptor interaction remains unchanged, but the anticodon gains new specificity. Although most block reassignments are compatible with known mechanisms of coding ambiguity,[109,110] where a tRNA expands its range to read additional codons that are also read by another tRNA, the identity of some changes (such as the CUN block reassignment) cannot be reconciled with this mechanism: no known anticodon could show codon recognition overlap between the two blocks. Consequently it might seem prudent to assume only that **all** the codons for a particular amino acid cannot be reassigned or made ambiguous simultaneously, on the assumption that this effective removal of an amino acid from the code would be prohibitively deleterious. However, even this minimal assumption may have been violated during the code's early evolution (section "An Optimal Pattern of Code Degeneracy"), and the question of whether today's sophisticated, proteinaceous world prohibits such loss is unclear; it is perhaps noteworthy that new code variants are still emerging every year, and that codon reassignment once seemed intuitively impossible.

Pathways to Codon Reassignment

The most widely believed pathway by which codon reassignment could occur, 'Codon Capture', proposes that changes in directional mutation causes certain codons to temporarily disappear from the genome; such codons might change their assignments (for instance, by mutation of tRNAs) with selective neutrality and be fixed once the direction of mutation reverts by selection pressure for translation of the newly abundant codons (even if the meaning is altered).[99,111-113] Although this model does not adequately describe all codon reassignments,[66] successive rounds of AT- and GC-pressure could potentially exchange the meaning of any two arbitrary codons.[54] Thus the modern translation system of tRNAs, aminoacyl-tRNA synthetases (aaRS), release factors and ribosomes can in principle support adaptive optimization of codon assignments.

More generally, it is likely that the standard genetic code evolved from a simpler ancestral form encoding fewer amino acids (section "Is the Choice of Coding Components Optimal?"); the extent to which this code expansion process continues in modern organisms remains unclear. Many amino acids, such as hydroxyproline and phosphoserine, are not incorporated during translation but are produced later by enzymes acting on the nascent polypeptides (sometimes reversibly). Selenocysteine, which is incorporated during translation[114] in certain species by a special tRNA[115] at UGA stop codons in certain sequence contexts (a hairpin recognition element upstream, and a 4th base context that is recognized only weakly by release factors),[56,116,117] may be an example of a recently added amino acid. The selenocysteine tRNA is originally recognized by seryl-tRNA synthetase and charged with Ser, after which a specific enzyme, selenocysteine synthase, recognizes the charged tRNA and converts the amino acid to selenocysteine.[118-120] This phenomenon has parallels with Asn and Gln, which, in some species, are mischarged by aspartyl- and glutamyl-tRNA synthetase respectively, and converted to the amide on the tRNA[121] (see refs. 122, 123 for reviews relating this fact to code evolution; for more recent references on the function and phylogenetic distribution of these enzymes see refs. 124, 125). Thus it is possible that tRNA-dependent modification is an early stage in cotranslational incorporation of amino acids, and that codon assignments are influenced by the order in which amino acids were added.[38,39,122,126-131] At a more theoretical level, selection for strains of bacteria that can use nonstandard amino acids has in some cases led to stable incorporation of non-canonical amino acids.[27,28,132]

Current Evidence for an Adaptive Pattern of Codon Assignments

Given that codons can be reassigned, and indeed have been in a wide variety of genomes, claims for an adaptive code can move from discussions of general plausibility to questions of extent and strength of evidence: to what extent is code structure optimized and how accurately can we characterize what was probably a complex and multidimensional optimization process? More explicitly, assuming the 4 bases, 64 codons, 20 amino acids and the (imperfectly understood) pattern of redundancy found within the standard code, we can define a set of possible codes that comprises 20! or 2.4×10^{18} members. Using this set, we can ask to what extent does the standard code reduce the average change in amino acid properties that results from single base substitutions relative to other possible codes?

In this context, Haig and Hurst[91] resurrected Alff-Steinberger's methodology to reach some rather different results. Alff-Steinberger's original study reported that the code also minimized errors in molecular weight, polar requirement, number of dissociating groups, pK of carboxyl group, isoelectric point, and alpha-helix-forming ability, with the 3rd position base most highly optimized and the 2nd position base not optimized at all.[90] Haig and Hurst reported that the code was highly optimized for polar requirement (1 in 10 000), somewhat less optimized for hydropathy (1 in 1000), and not optimized for molecular volume and isoelectric point.[91,133] Interestingly, polar requirement[85] is a measure of the hydrophobicity of free amino acids, while hydropathy[134] is a measure of the hydrophobicity of the side-chains alone. Unfortunately any significance associated with this difference is obscured by the facts that

i. no sensitivity data is available for either measure, rendering detailed comparisons difficult
ii. worse, many hydropathy values were arbitrarily adjusted relative to the experimental data (see ref. 134, pp. 109-110)!

Freeland and Hurst then extended this basic model to incorporate biases known to influence mutation and translation.[135] Measuring amino acid similarity in terms of polar requirement, they found that the perceived optimality of the code increases 1 order of magnitude when a mild transition bias is incorporated into the calculations, and a further order of magnitude when a base-position effect is incorporated: taking Friedman and Weinstein's observations of mistranslation for poly-U[136] and extrapolating them quantitatively to the code as whole, they found that the genetic code outperformed 999,999 out of a sample of 1 million random alternatives.

Further analysis revealed that these results are qualitatively robust to methodological variation (e.g., to variations in the power to which the difference between pairs of amino acids is raised when calculating quantitative differences between pairs of amino acids) is varied, or the PAM74-100 matrix,[137] a measure of the actual frequency of amino acid substitutions in distantly related proteins, is used.[138,139] In fact, about 100-fold fewer better codes are found with PAM74-100 than with polar requirement,[139] perhaps suggesting that the code is better adapted to the functional properties of amino acids within proteins than to the properties of the free amino acids. This point is critically important for determining when the code was most recently optimized.

Amongst the latest extensions of this analysis is the interesting observation is that the amino acids are arranged so as to give a smooth fitness landscape: in other words, the first- and second-position bases have a roughly consistent, additive effect on several amino acid properties, including polarity.[140] This has the effect of allowing 'fine-tuning' of amino acid properties by single-base mutational events. Interestingly, relative codon and amino acid usage in different species can be largely explained by a mutation-selection balance on individual bases, and the rate of change under directional mutation of each base at each position is highly correlated with the average effect of changing that base.[67] It is possible that this result holds because of the structure of the code, rather than in spite of it.

A different approach is to test which amino acid properties correlate with specific types of substitution matrix. Such matrices may be derived from the genetic code itself (which is arranged such that some amino acids are easier to interconvert than others) or formed from direct observations of the pattern of substitutions separating pairs of homologous proteins (which, at short mutational distances, largely reflect the code's structure). Studies based on these methods, although less clear-cut than those that test directly what fraction of codes minimize changes in particular properties under point substitution better than does the actual code, consistently show that the main property implicated in substitutability is hydrophobicity or measures tightly correlated with it.[137,141-147] Depending on the study, the most tightly correlated hydrophobicity scale may be one that is measured directly as a chemical property of the amino acids or side-chains,[85,134,142,148-150] or indirectly as relative abundance in the interiors/exteriors of proteins.[151-154] However, other potentially important factors (such as the size of the side-chains) do not turn out to be nearly as tightly correlated with the code structure. The converse of this approach has also been tried, with similar results: in this case, random amino acid indices were generated, the ones that best matched the genetic code substitution matrix were selected, and these scales were correlated with measures of polarity.[155] Similar results have been obtained by multivariate[156] and neural network[157,158] analyses that partition codons into classes based on multidimensional analysis of amino acid properties, and find that the second-position base defines classes linked to various hydrophobicity measures (or principal components thereof). Thus it should not be controversial that the code minimizes changes in hydrophobicity better than does almost every random permutation of amino acid properties among codon blocks.

Current Objections to Adaptive Patterns of Codon Assignment

Despite this overwhelming evidence, a small but vocal group of researchers still doubt that the code has been optimized to minimize the effects of genetic errors. This arises from two fundamental misunderstandings.

The first misunderstanding is to assume that, if the code has been optimized by natural selection, that it must be the best of all possible codes at minimizing the distance function. Consequently, better codes found by powerful computer search algorithms (or calculated from first principles), but which do not resemble the standard code, have been presented as prima facie evidence that the code is not optimal, and therefore cannot have been optimized.[159-165]

These analyses fail to take into account two important facts. First, the average effect of amino acid changes in proteins is unlikely to be perfectly recaptured by a single linear scale of physical properties, and so a code that minimizes a single one of these properties will not necessarily look anything like the actual code.[139] Second, although search algorithms can sample billions of different codes, evolution is unlikely to have had similar opportunity given the extreme cost of changing an already functional code, and so we might either expect the code to be trapped at a local, rather than global, optimum, or that the code adapts asymptotically and has not yet reached perfection. Thus, the fact that the code is not the best of all possible codes on a particular hydrophobicity scale does not mean that it has not evolved to minimize changes in hydrophobicity under point misreading, any more than the fact that the vertebrate retina is wired backwards means that the eye is not adapted for vision.

The second misunderstanding is to assume that the important property for measuring the extent of code optimization is not the fraction of codes to which the actual code is superior, but rather the distance still separating the actual code from the Panglossian ideal.[129,159,161-164] Figure 2 shows the difference between these two models, using an actual distribution of random codes derived from the PAM74-100 matrix.[139] The distribution is roughly Gaussian: better codes are rarer (and confer lesser relative advantage) as the code becomes more highly optimized. No-one has studied the accessibility of better codes as optimization proceeds, but it is likely that better codes get exponentially rarer at the tails of the distributions and so it would be necessary to cross huge fitness valleys to find a code that is better than a near-optimal one. However, the fact that such a small fraction of codes are better at minimizing errors than the actual code strongly suggests that selection has played a role in determining its structure.[166]

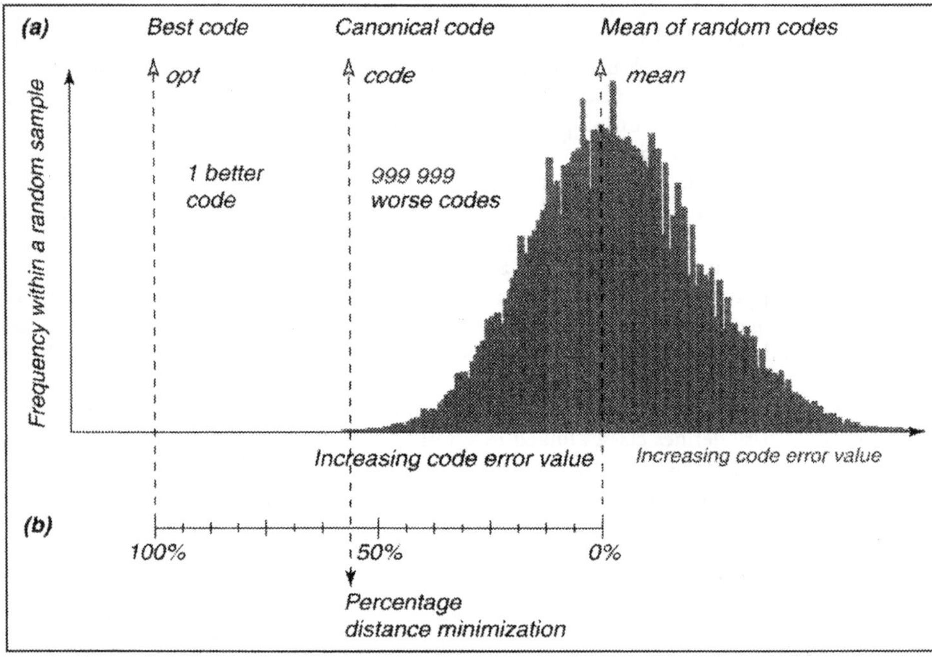

Figure 2. A comparison of methods. A) The statistical (sampling) approach provides a direct estimate of the probability that a code as good as that used by Nature would evolve by chance; B) the engineering approach measures code optimality on a linear scale. The figure illustrates the very different values that resulted from using the two methods for the same set of random codes.

A Metabolic Counter-Argument

The most serious challenge to the idea that codon assignments have been optimized for error minimization, at least in terms of published criticism, is the idea that the standard code still contains a strong historical signature of the pathway by which it expanded: surely, the argument goes, if codon assignments had been shuffled to reduce error, these nonadaptive patterns would have been erased?[35,126,127,129,159,161,162,167,168] Thus 'coevolution' between codon assignments and the amino acid complement has been viewed as a major alternative to adaptive models: coevolution accepts that the code did change, and that codon assignments were rearranged, but only to insert new amino acids as metabolism invented them and **not** to minimize the effects of mutation/mistranslation.

In fact, there is no necessary contradiction between history and adaptation. Adaptive traits typically reveal traces of their past: the bat's wing is no less adapted for flying, nor the human hand for grasping, because they are immediately recognizable as pentadactyl limbs. More fundamentally, however, historical patterns in the code can only cause difficulties for the adaptive hypothesis if they actually exist. Indeed, it is possible that the identity and codon assignments of new amino acid additions to the code were steered by selection for error minimization.[155]

More broadly, however, the picture of an expanding genetic code is not as clear as is sometimes portrayed. The idea that metabolically related amino acids are clustered together in the code is not new,[80] but the form of the idea that has received the most attention (cited more than 120 times as of early 2001) is that of Wong.[122] According to this model, new amino acids were formed by tRNA-dependent modification, much as Gln and Asn are formed from Glu and Asp in some bacteria, and thus took over a subset of codons from their metabolic precursors. Specifically, Wong identified 8 precursor-product pairs: Ser → Trp Ser →Cys, Val →Leu, Thr → Ile, Gln → His, Phe → Tyr, Glu → Gln, and Asp → Asn. Using the hypergeometric distribution, he estimated the probability that, for each of these pairs of amino acids, the number of codons of the product that were connected to at least one codon of the precursor would be as great as those actually observed. He then combined these probabilities using Fisher's method, to get an overall probability of 0.0002 of observing at least as much overlap between products and precursors by chance.[122]

Unfortunately, there are good reasons to disbelieve this result. The first is that even randomly generated codes contain many 'product-precursor pairs' by chance, especially if known pathways from *E. coli* are used instead of Wong's original pairs.[169,170] Although the 'codon correlation score' used in this study to assess the position of the actual code relative to random codes has been criticized,[167,168] the fact remains that it is difficult to show that patterns involving small groups of codons are real. There are more fundamental problems with Wong's analysis, however. First, some of the alleged product-precursor pairs involve running metabolic pathways backwards: the conversion from Thr to Met would have to proceed via the common intermediate homoserine, but getting to this state from Thr would require reversal of two steps normally coupled to ATP hydrolysis. Second, Wong assumes that all codons can change independently, yet no base at the wobble position of the anticodon in tRNA can distinguish NNU from NNC. Thus there are really only 48 codon blocks that can change independently, not 64, which means that adjacent NNY block represent only one event and not two. When these problems are corrected, the probability of observing as many adjacent product-precursor pairs by chance alone rises to 16.8%. When the test is limited to codon assignments found in the standard code (Wong makes 2 key assumptions about intermediate codon block assignments that are no longer seen today), the probability rises to 62%.[171] Thus this particular pathway of code evolution has no statistical support.

A more convincing biosynthetic pattern in the code is that amino acids with the same 1st-position base tend to be metabolically related.[38] Specifically, amino acids with A at the first position are derived from Asp; those with C at the first position are derived from Glu; those with U at the first position are derived from intermediates in glycolysis; and those with G at the first position are both at the heads of metabolic pathways and are plausibly prebiotic.[39,130] If amino acid assignments were somehow constrained such that their metabolic pathways determined their 1st-position base, then could the appearance of adaptation within codon assignments be explained on the grounds that biosynthetically related amino acids are likely to have similar properties?

These questions can be directly addressed by comparing the actual code to random codes generated by partitioning amino acids to classes related by 1st-position base in the actual code, and randomizing only within each of these classes. This reduces the number of possible codes from 20! to $5!^4$ or about 2×10^9 possible codes, a factor of about 10.[9] These sets are so different in size that they are effectively independent: even accepting the most extreme claims for optimality in the absence of restriction, every single one of the biosynthetically constrained set could be better than the actual code! However, this is not the case. Although the distributions of constrained and unconstrained possible codes are significantly different, the differences in estimated optimality are rather small: the standard code still appears better than nearly every constrained code, for both polar requirement and PAM74-100 and over a wide range of transition/transversion biases.[139,172] Consequently, even if history really did constrain the first-position base absolutely, error-minimizing optimization would still be required to explain the values of the second- and third-position bases assigned to the codons for each amino acid.

In summary, the code probably evolved from a simpler form, although the exact pathway is still unknown. However, even if the code contains traces of its evolutionary history (such as the association between first-position base and metabolic pathway), they cannot explain why it seems to minimize errors to the extent that it does.

Concluding Remarks

The existence of variant genetic codes proves that codon assignments are not fixed, but rather can and do change. It therefore seems reasonable to assume that the code has changed between the time tRNAs were invented and the last common ancestor of extant life. In light of the fact that almost no codes minimize genetic errors (in terms of the difference in polarity between the intended and accidentally inserted amino acid) better than does the standard code, it further seems reasonable that at least some of this change has been adaptive.

Although there is still debate about the extent of this adaptation, most current disagreement stems from inappropriate distance estimates: it is no use comparing the actual code to the best of all possible codes if there is no pathway by which the optimum could be reached, and it is unclear that any single measure of polarity will recapture the actual effect of substitutions in proteins sufficiently accurately that we would expect the code to be adapted to it alone. It is necessary to take into account the frequency distribution of possible codes, rather than to assume that optimization will be linear throughout the whole range. In these terms, the code appears nearly optimal with respect to changes in hydrophobicity, but not other parameters, over wide range of parameter space. Although it is attractive to assume that this is the result of natural selection for error minimization, alternative explanations cannot yet be ruled out: for example, stereochemical principles could assign similar amino acids to similar codons.[155] However, the view that the code is not highly ordered is no longer tenable.

Simple chemical rules based on GC-content explain the pattern of degeneracy in codon blocks, although it is unclear whether these are the (restrictive) causes of the degeneracy or merely proximal mechanisms by which adaptive rules based on translation speed and/or accuracy are enforced. Different species use markedly different amino acid compositions, which are highly correlated with their overall genome nucleotide composition, and codons are used at highly unequal frequencies, so it seems unlikely that the number of codons assigned to each amino acid gives a unique optimum amino acid composition for proteins.

Whether the choice of components of the genetic code was optimized by natural selection is less clear. There are good chemical reasons why certain bases and amino acids were not used, but the presence and absence of others is still a mystery. It seems likely that the standard code grew (at least in terms of the addition of novel, biosynthetically derived amino acids) from a simpler primordial form, but variation in claims for the detail of this process, and statistical re-evaluation of one of the best accepted versions contribute to a murky picture. Future evidence for the pathway of code expansion may do much to clarify the manner in which the standard code achieved its impressive error minimizing properties (e.g., through shuffling of codon assignments or simply through the selectively constrained addition of 'buffering' amino acids).

Finally, it is often overlooked that the molecular components of the genetic code may have changed significantly during the course of genetic code evolution. In particular, sophisticated proteins involved in the translation system, such as the aminoacyl-tRNA synthetases, cannot have preceded protein synthesis itself. Although the recent artificial selection of ribozymes that catalyze these reactions[173-176] support the idea that the protein synthetases may have usurped the role from earlier catalysts made of RNA[177,178] and a generalized form of this view might be expected if the RNA world existed as a simple and straightforward forerunner to our world of coded proteins. The extent to which these ill-characterized predecessors may have presented different selective pressures that shaped code evolution remains a mystery.

References

1. Thompson D. On Growth and Form. Cambridge: Cambridge University Press, 1917.
2. Gould SJ, Lewontin R. The Spandrels of San Marco and the Panglossian Paradigm: A critique of the adaptationist programme. Proceedings of the Royal Society of London 1979; 205:581-598.
3. Neander K. The teleological notion of "function". Australasian Journal of Philosophy 1991; 69.4:454-468.
4. Voet D, Voet JG. Biochemistry. 2 ed. New York: John Wiley & Sons, 1995.
5. Gilbert W. The RNA world. Nature 1986; 319:618.
6. Freeland SJ, Knight RD, Landweber LF. Do proteins predate DNA? Science 1999; 286(5440):690-2.
7. Weber AL. The triose model: Glyceraldehyde as a source of energy and monomers for prebiotic condensation reactions. Orig Life Evol Biosph 1987; 17(2):107-19.
8. Weber AL. Thermal synthesis and hydrolysis of polyglyceric acid. Orig Life Evol Biosph 1989; 19:7-19.
9. Joyce GF. Schwartz AW, Miller SL et al. The case for an ancestral genetic system involving simple analogues of the nucleotides. Proc Natl Acad Sci USA 1987; 84:4398-4402.
10. Eschenmoser A. Chemical etiology of nucleic acid structure. Science 1999; 284(5423):2118-24.
11. Larralde R, Robertson MP, Miller SL. Rates of decomposition of ribose and other sugars: Implications for chemical evolution. Proc Natl Acad Sci USA 1995; 92:8158-8160.
12. Nielsen PE, Egholm M, Berg RH et al. Sequence-selective recognition of DNA by strand displacement with a thymine-substituted polyamide. Science 1991; 254:1497-1500.
13. Hanvey JC, Peffer NJ, Bisi JE et al. Antisense and antigene properties of peptide nucleic acids. Science 1992; 258(5087):1481-5.
14. Nelson KE, Levy M, Miller SL. Peptide nucleic acids rather than RNA may have been the first genetic molecule. Proc Natl Acad Sci USA 2000; 97(8):3868-71.
15. Nielsen PE. Peptide nucleic acid (PNA): A model structure for the primordial genetic material? Orig Life Evol Biosph 1993; 23(5-6):323-7.
16. Joyce GF, Orgel LE. Prospects for understanding the origin of the RNA world. In: Gesteland RF, Atkins JF, eds. New York: Cold Spring Harbor Laboratory Press, 1993:1-25.
17. Piccirilli JA, Krauch T, Moroney SE et al. Enzymatic incorporation of a new base pair into DNA and RNA extends the genetic alphabet. Nature 1990; 343:33-37.
18. Wu Y, Ogawa AK, Berger M et al. Efforts toward expansion of the genetic alphabet: Optimization of interbase hydrophobic interactions. J Am Chem Soc 2000; 122:7621-7632.
19. Lutz MJ, Horlacher J, Benner SA. Recognition of a non-standard base pair by thermostable DNA polymerases. Bioorg Med Chem Lett 1998; 8(10):1149-52.
20. Ogawa AK, Wu YQ, McMinn DL et al. Efforts toward the expansion of the genetic alphabet: Information storage and replication with unnatural hydrophobic base pairs. J Am Chem Soc 2000; 122:3274-3287.
21. Oró J, Kimball P. Synthesis of purines under possible primitive earth conditions. I. Adenine from hydrogen cyanide. Arch Biochem Biophys 1961; 94:217-227.
22. Levy M, Miller SL. The prebiotic synthesis of modified purines and their potential role in the RNA world. J Mol Evol 1999; 48(6):631-7.
23. Levy M, Miller SL. The stability of the RNA bases: Implications for the origin of life. Proc Natl Acad Sci USA 1998; 95(14):7933-8.
24. Wong JT-F, Bronskill PM. Inadequacy of prebiotic synthesis as origin of proteinaceous amino acids. J Mol Evol 1979; 13:115-125.
25. Jukes TH. Arginine as an evolutionary intruder into protein synthesis. Biochem Biophys Res Comm 1973; 53(3):709-714.
26. Weber AL, Miller SL. Reasons for the occurrence of the twenty coded protein amino acids. J Mol Evol 1981; 17:273-284.
27. Budisa N, Minks C, Alefelder S et al. Toward the experimental codon reassignment in vivo: Protein building with an expanded amino acid repertoire. FASEB J 1999; 13(1):41-51.
28. Budisa N, Minks C, Medrano FJ et al. Residue-specific bioincorporation of non-natural, biologically active amino acids into proteins as possible drug carriers: Structure and stability of the per-thiaproline mutant of annexin V. Proc Natl Acad Sci 1998; 95:455-459.
29. de Duve C, Vital Dust. New York: Basic Books, 1995.
30. Orgel LE. Evolution of the genetic apparatus. J Mol Biol 1968; 38:381-393.
31. Crick FHC. The origin of the genetic code. J Mol Biol 1968; 38:367-379.
32. Fitch WM, Upper K. The Phylogeny of tRNA Sequences Provides Evidence for Ambiguity Reduction in the Origin of the Genetic Code. Cold Spring Harbor Symp Quant Biol 1987; 52:759-767.
33. Szathmary E. Four letters in the genetic alphabet: A frozen evolutionary optimum? Proc R Soc Lond B Biol Sci 1991; 245(1313):91-9.

34. Szathmary E. What is the optimum size for the genetic alphabet? Proc Natl Acad Sci USA 1992; 89(7):2614-8.
35. Wong JT-F. The evolution of a universal genetic code. Proc Natl Acad Sci USA 1976; 73(7):2336-2340.
36. Eigen M, Schuster P. The hypercycle: A Principle of Natural Self-Organization. New York: Springer, 1979.
37. Shepherd JC. Periodic correlations in DNA sequences and evidence suggesting their evolutionary origin in a comma-less genetic code. J Mol Evol 1981; 17(2):94-102.
38. Dillon LS. The Origins of the Genetic Code. Bot Rev 1973; 39:301-345.
39. Taylor FJR, Coates D. The code within the codons. Bio Systems 1989; 22:177-187.
40. Davis BK. Evolution of the genetic code. Prog Biophys Mol Biol 1999; 72(2):157-243.
41. Jiménez-Sánchez A. On the origin and evolution of the genetic code. J Mol Evol 1995; 41:712-716.
42. Hartman H. Speculations on the origin of the genetic code. J Mol Evol 1995; 40:541-544.
43. Lehman N, Jukes TH. Genetic code development by stop codon takeover. J Theor Biol 1988; 135(2):203-14.
44. Baumann U, Oró J. Three stages in the evolution of the genetic code. Bio Systems 1993; 29:133-141.
45. Fitch WM. The relation between frequencies of amino acids and ordered trinucleotides. J Mol Biol 1966; 16:1-8.
46. Trifonov E, Bettecken T. Sequence fossils, triplet expansion, and reconstruction of earliest codons. Gene 1997; 205(1-2):1-6.
47. Trifonov EN. Consensus temporal order of amino acids and evolution of the triplet code. Gene 2000; 261(1):139-151.
48. King JL, Jukes TH. Non-Darwinian evolution. Science 1969; 164(881):788-98.
49. Crick FH. Codon-anticodon pairing: the wobble hypothesis. J Mol Biol 1966; 19(2):548-555.
50. Lagerkvist U. "Two out of three": An alternative method for codon reading. Proc Natl Acad Sci USA 1978; 75(4):1759-1762.
51. Lagerkvist U. Codon misreading: a restriction operative in the evolution of the genetic code. American Scientist 1980; 68:192-198.
52. Lagerkvist U. Unorthodox codon reading and the evolution of the genetic code. Cell 1981; 23:305-306.
53. Knight RD, Freeland SJ, Landweber LF. Rewiring the keyboard: evolvability of the genetic code. Nat Rev Genet 2001; 2:49-58.
54. Szathmáry E. Codon swapping as a possible evolutionary mechanism. J Mol Evol 1991; 32:178-182.
55. Jungck JR. The genetic code as a periodic table. J Mol Evol 1978; 11:211-224.
56. Tate WP, Poole ES, Dalphin ME et al. The translational stop signal: Codon with a context, or extended factor recognition element? Biochimie 1996; 78:945-952.
57. Cullman G, Labouygues J. Noise immunity of the genetic code. Bio Systems 1983; 16:9-29.
58. Swanson R. A unifying concept for the amino acid code. Bull Math Biol 1984; 46:187-203.
59. Cullman G, Labouygues J. The logic of the genetic cdoe. Math Model 1987; 8:643-646.
60. Figureau A. Information theory and the genetic code. Orig Life 1987; 17:439-449.
61. Figureau A. Optimization and the genetic code. Orig Life Evol Biosph 1989; 19:57-67.
62. Figureau A, Pouzet M. Genetic code and optimal resistance to the effect of mutations. Orig Life Evol Biosph 1984; 14:579-588.
63. Sonneborn TM. Degeneracy of the genetic code: extent, nature, and genetic implications. In: Bryson V, Vogel HJ, eds. Evolving Genes and Proteins. New York: Academic Press, 1965:377-297.
64. Zuckerkandl E, Pauling L. Evolutionary divergence and convergence in proteins. In: Bryson V, Vogel HJ, eds. Evolving Genes and Proteins. New York: Academic Press, 1965.
65. Soto MA, Toha CJ. A hardware interpretation of the evolution of the genetic code. BioSystems 1985; 18:209-215.
66. Knight RD, Landweber LF, Yarus M. How Mitochondria Redefine the Code. J Mol Evol 2001; 53(4-5):299-313.
67. Knight RD, Freeland SJ, Landweber LF. A simple model based on mutation and selection explains trends in codon and amino-acid usage and GC composition within and across genomes. GenomeBiology 2001; 2(4):http://www.genomebiology.com/2001/2/4/research/0010/.
68. Mackay AL. Optimization of the genetic code. Nature 1967; 216(111):159-60.
69. Ota T, Kimura M. Amino acid composition of proteins as a product of molecular evolution. Science 1971; 174(5):150-3.
70. Dufton MJ. The significance of redundancy in the genetic code. J Theor Biol 1983; 102(4):521-6.
71. Dufton MJ. Genetic code synonym quotas and amino acid complexity: cutting the cost of proteins? J Theor Biol 1997; 187(2):165-73.

72. Antillon A, Ortega-Blake I. A group theory analysis of the ambiguities in the genetic code: on the existence of a generalized genetic code. J Theor Biol 1985; 112(4):757-69.
73. Bashford JD, Tsohantjis I, Jarvis PD. A supersymmetric model for the evolution of the genetic code. Proc Natl Acad Sci USA 1998; 95(3):987-92.
74. Hornos JE, Hornos YM. Algebraic model for the evolution of the genetic code. Physical Review Letters 1993; 71(26):4401-4404.
75. Maeshiro T, Kimura M. The role of robustness and changeability on the origin and evolution of genetic codes. Proc Natl Acad Sci USA, 1998; 95(9):5088-93.
76. Klump HH. The physical basis of the genetic code: the choice between speed and precision. Arch Biochem Biophys 1993; 301(2):207-9.
77. Speyer JF, Lengyel CB, Wahba AJ et al. Synthetic polynucleotides and the amino acid code. Cold Spring Harbor Symp Quant Biol, 1963; 28:559-567.
78. Volkenstein MV. Coding of polar and non-polar amino acids. Nature 1965; 207:294-295.
79. Woese CR. Order in the genetic code. Proc Natl Acad Sci USA 1965; 54:71-75.
80. Pelc SR. Correlation between coding triplets and amino acids. Nature 1965; 207:597-599.
81. Epstein CJ. Role of the amino-acid 'code' and of selection for conformation in the evolution of proteins. Nature 1966; 210:25-28.
82. Goldberg AL, Wittes RE. Genetic code: aspects of organization. Science 1966; 153:420-424.
83. Woese CR. On the evolution of the genetic code. Proc Natl Acad Sci USA 1965; 54:1546-1552.
84. Woese CR. The Genetic Code: The Molecular Basis for Genetic Expression. New York: Harper & Row, 1967.
85. Woese CR, Dugre DH, Dugre SA et al. On the fundamental nature and evolution of the genetic code. Cold Spring Harbor Symp Quant Biol 1966; 31:723-736.
86. Davies J, Gilbert W, Gorini L. Streptomycin, suppression, and the code. Proc Natl Acad Sci USA 1964; 51:883-890.
87. Topal MD, Fresco JR. Base pairing and fidelity in codon-anticodon interaction. Nature 1976; 263(5575):289-93.
88. Topal MD, Fresco JR. Complementary base pairing and the origin of substitution mutations. Nature 1976; 263(5575):285-9.
89. Suen W, Spiro TG, Sowers LC et al. Identification by UV resonance Raman spectroscopy of an imino tautomer of 5-hydroxy-2'-deoxycytidine, a powerful base analog transition mutagen with a much higher unfavored tautomer frequency than that of the natural residue 2'-deoxycytidine. Proc Natl Acad Sci USA 1999; 96(8):4500-5.
90. Alff-Steinberger C. The genetic code and error transmission. Proc Natl Acad Sci USA 1969; 64:584-591.
91. Haig D, Hurst LD. A quantitative measure of error minimization in the genetic code. J Mol Evol 1991; 33:412-417.
92. Gamow G, Possible mathematical relation between deoxyribonucleic acid and protein. Kgl Dansk Videnskab Selskab Biol Medd 1954; 22:1-13.
93. Haurowitz F. Chemistry and Biology of Proteins. New York: Academic Press, 1950.
94. Crick FHC. The structure of nucleic acids and their role in protein synthesis. Biochem Soc Symp 1957; 14:25-26.
95. Dunnill P. Triplet Nucleotide—amino acid pairing: A stereochemical basis for the division between protein and nonprotein amino acids. Nature 1966; 210:1267-1268.
96. Pelc SR, Welton MGE. Stereochemical relationship between coding triplets and amino-acids. Nature 1966; 209:868-872.
97. Woese CR, Dugre DH, Saxinger WC et al. The molecular basis for the genetic code. Proc Natl Acad Sci USA 1966; 55:966-974.
98. Crick FHC. The recent excitement in the coding problem. Progress in nucleic acids 1963; 1:163-217.
99. Osawa S, Jukes TH. Codon reassignment (codon capture) in evolution. J Mol Evol 1989; 28:271-278.
100. Barrell BG, Bankier AT, Drouin J. A different genetic code in human mitochondria. Nature 1979; 282(5735):189-94.
101. Jukes TH. Amino acid codes in mitochondria as possible clues to primitive codes. J Mol Evol 1981; 18(1):15-7.
102. Grivell LA. Molecular evolution. Deciphering divergent codes. Nature 1986; 324(6093):109-10.
103. Lewin B, Genes V. Oxford: Oxford University Press, 1994.
104. Hasegawa M, Miyata T. On the antisymmetry of the amino acid code table. Orig Life 1980; 10(3):265-70.
105. Muramatsu T, Nishikawa K, Nemoto F et al. Codon and amino-acid specificities of a transfer RNA are both converted by a single post-transcriptional modification. Nature 1988; 336(6195):179-81.

106. Senger B, Auxilien S, Englisch et al. The modified wobble base inosine in yeast tRNAIle is a positive determinant for aminoacylation by isoleucyl-tRNA synthetase. Biochemistry 1997; 36(27):8269-75.
107. Bonitz SG, Berlani R, Coruzzi G et al. Codon recognition rules in yeast mitochondria. Proc Natl Acad Sci USA 1980; 77(6):3167-70.
108. Osawa S, Ohama T, Jukes TH et al. Evolution of the mitochondrial genetic code. I. Origin of AGR serine and stop codons in metazoan mitochondria. J Mol Evol 1989; 29(3):202-7.
109. Schultz DW, Yarus M. Transfer RNA mutation and the malleability of the genetic code. J Mol Biol 1994; 235:1377-1380.
110. Schultz DW, Yarus M. On malleability in the genetic code. J Mol Evol 1996 42:597-601.
111. Jukes TH, Osawa S, Muto A et al. Evolution of Anticodons: Variations in the Genetic Code. Cold Spring Harbor Symposia on Quantitative Biology, 1987; 52:769-776.
112. Osawa S, Jukes TH. Evolution of the genetic code as affected by anticodon content. Trends Genet 1988; 4(7):191-198.
113. Osawa S, Jukes TH, Watanabe K et al. Recent evidence for evolution of the genetic code. Microbiol Rev 1992; 56(1):229-64.
114. Zinoni F, Birkmann A, Leinfelder W et al. Cotranslational insertion of selenocysteine into a formate dehydrogenase from Escherichia coli directed by a UGA codon. Proc Natl Acad Sci USA 1987; 84:3156-3160.
115. Leinfelder W, Zehelein E, Mandrand-Berthelot MA et al. Gene for a novel tRNA species that accepts L-serine and cotranslationally inserts selenocysteine. Nature 1988; 331(6158):723-5.
116. Zinoni F, Heider J, Bock A. Features of the formate dehydrogenase mRNA necessary for decoding of the UGA codon as selenocysteine. Proc Natl Acad Sci USA 1990; 87(12):4660-4.
117. Tate WP, Mansell JB, Mannering SA et al. UGA: A dual signal for 'stop' and for recoding in protein synthesis. Biochemistry (Mosc) 1999; 64(12):1342-53.
118. Commans S, Bock A. Selenocysteine inserting tRNAs: an overview. FEMS Microbiol Rev 1999; 23(3):335-51.
119. Lenhard B, Orellana O, Ibba M et al. tRNA recognition and evolution of determinants in seryl-tRNA synthesis. Nucleic Acids Res 1999; 27(3):721-9.
120. Forchhammer K, Boesmiller K, Bock A. The function of selenocysteine synthase and SELB in the synthesis and incorporation of selenocysteine. Biochimie 1991; 73(12):1481-6.
121. Schön A, Kannangara CG, Gough S et al. Protein biosynthesis in organelles requires misaminoacylation of tRNA. Nature 1988; 331:187-190.
122. Wong JT-F. A co-evolution theory of the genetic code. Proc Natl Acad Sci USA 1975; 72(5):1909-1912.
123. Di Giulio M. Origin of glutaminyl-tRNA synthetase: an example of palimpsest? J Mol Evol 1993; 37:5-10.
124. Becker HD Kern D. Thermus thermophilus: a link in evolution of the tRNA-dependent amino acid amidation pathways. Proc Natl Acad Sci USA 1998; 95(22):12832-7.
125. Tumbula DL, Becker HD, Chang WZ et al. Domain-specific recruitment of amide amino acids for protein synthesis. Nature 2000; 407(6800):106-10.
126. Di Giulio M. Some aspects of the organization and evolution of the genetic code. J Mol Evol 1989; 29:191-201.
127. Di Giulio M. On the relationships between the genetic code coevolution hypothesis and the physicochemical hypothesis. Z Naturforsch 1991; 46c:305-312.
128. Di Giulio M. On the origin of the genetic code. J Theor Biol 1997; 187:573-581.
129. Di Giulio M. The historical factor: the biosynthetic relationships between amino acids and their physiochemical properties in the origin of the genetic code. J Mol Evol 1998; 46:615-621.
130. Miseta A. The role of protein associated amino acid precursor molecules in the organization of genetic codons. Physiol Chem Phys Med NMR 1989; 21:237-242.
131. Wong JT-F. Coevolution of genetic code and amino acid biosynthesis. TIBS 1981; 6:33-36.
132. Wong JT-F. Membership mutation of the genetic code: loss of fitness by tryptophan. Proc Natl Acad Sci USA 1983; 80:6303-6306.
133. Haig D, Hurst LD. A quantitative measure of error minimization in the genetic code. J Mol Evol 1999; 49(5):708.
134. Kyte J, Doolittle RF. A simple method for displaying the hydropathic character of a protein. J Mol Biol 1982; 157(1):105-32.
135. Freeland SJ, Hurst LD. The genetic code is one in a million. J Mol Evol 1998; 47(3):238-248.
136. Friedman SM, Weinstein IB. Lack of fidelity in the translation of ribopolynucleotides. Proc Natl Acad Sci USA 1964; 52(988-996).

137. Benner SA, Cohen MA, Gonnet GH. Amino acid substitution during functionally constrained divergent evolution of protein sequences. Protein Eng 1994; 7(11):1323-32.
138. Ardell DH, On error minimization in a sequential origin of the standard genetic code. J Mol Evol 1998; 47(1):1-13.
139. Freeland SJ, Knight RD, Landweber LF et al. Early fixation of an optimal genetic code. Mol Biol Evol 2000; 17(4):511-518.
140. Aita T, Urata S, Husimi Y. From amino acid landscape to protein landscape: analysis of genetic codes in terms of fitness landscape. J Mol Evol 2000 50(4):313-23.
141. Koshi JM, Goldstein RA. Mutation matrices and physical-chemical properties: Correlations and implications. Proteins 1997; 27(3):336-44.
142. Wolfenden R, Andersson L, Cullis PM et al. Affinities of amino acid side chains for solvent water. Biochemistry 1981; 20(4):849-55.
143. Szathmáry E, Zintzaras E. A statistical test of hypotheses on the organization and origin of the genetic code. J Mol Evol 1992; 35:185-189.
144. Tomii K, Kanehisa M. Analysis of amino acid indices and mutation matrices for sequence comparison and structure prediction of proteins. Protein Eng 1996; 9(1):27-36.
145. Xia X, Li WH. What amino acid properties affect protein evolution? J Mol Evol 1998; 47(5):557-64.
146. Joshi NV, Korde VV, Sitaramam V. Logic of the genetic code: Conservation of long-range interactions among amino acids as a prime factor. J Genet 1993; 72:47-58.
147. Sitaramam V. Genetic code preferentially conserves long-range interactions among the amino acids. FEBS Lett 1989; 247(1):46-50.
148. Fauchere J, Pliska V. Hydrophobic parameters pi of amino acid side chains from the partitioning of N-acetyl-amino-acid amides. Eur J Med Chem 1983; 18(4):369-375.
149. Radzicka A, Young GB, Wolfenden R. Lack of water transport by amino acid side chains or peptides entering a nonpolar environment. Biochemistry 1993; 32(27):6807-9.
150. Rose GD, Wolfenden R. Hydrogen bonding, hydrophobicity, packing, and protein folding. Annu Rev Biophys Biomol Struct 1993; 22:381-415.
151. Robson B, Suzuki E. Conformational properties of amino acid residues in globular proteins. J Mol Biol 1976; 107(3):327-56.
152. Levitt M. Conformational preferences of amino acids in globular proteins. Biochemistry 1978; 17(20):4277-85.
153. Wertz DH, Scheraga HA. Influence of water on protein structure. An analysis of the preferences of amino acid residues for the inside or outside and for specific conformations in a protein molecule. Macromolecules 1978 11(1):9-15.
154. Nakashima H, Nishikawa K, Ooi T. Distinct character in hydrophobicity of amino acid compositions of mitochondrial proteins. Proteins 1990; 8(2):173-8.
155. Knight RD, Freeland SJ, Landweber LF. Selection, history and chemistry: the three faces of the genetic code. Trends Biochem Sci 1999; 24(6):241-7.
156. Sjöström M, Wold S. A multivariate study of the relationship between the genetic code and the physical-chemical properties of amino acids. J Mol Evol 1985; 22:272-277.
157. Jiménez-Montaño MA. On the syntactic structure and redundancy distribution of the genetic code. Bio Systems 1994; 32:11-23.
158. Tolstrup N, Toftgard J, Engelbrecht J et al. Neural network model of the genetic code is strongly correlated to the GES scale of amino acid transfer free energies. J Mol Biol 1994; 243(5):816-20.
159. Wong JT. Role of minimization of chemical distances between amino acids in the evolution of the genetic code. Proc Natl Acad Sci USA 1980; 77(2):1083-1086.
160. Goldman N. Further results on error minimization in the genetic code. J Mol Evol 1993; 37(6):662-4.
161. Di Giulio M. The extension reached by the minimization of the polarity distances during the evolution of the genetic code. J Mol Evol 1989; 29:288-293.
162. Di Giulio M. Genetic code origin and the strength of natural selection. J Theor Biol 2000; 205(4):659-61.
163. Di Giulio M. The origin of the genetic code. Trends Biochem Sci 2000; 25(2):44.
164. Di Giulio M, Capobianco MR, Medugno M. On the optimization of the physicochemical distances between amino acids in the evolution of the genetic code. J Theor Biol 1994; 168:43-51.
165. Judson OP, Haydon D. The genetic code: What is it good for? J Mol Evol 1999; 49:539-550.
166. Freeland SJ, Knight RD, Landweber LF. Measuring adaptation within the genetic code. Trends Biochem Sci 2000; 25(2):44-5.
167. Di Giulio M. The coevolution theory of the origin of the genetic code. J Mol Evol 1999; 48(3):253-5.
168. Di Giulio M, Medugno M. The robust statistical bases of the coevolution theory of genetic code origin. J Mol Evol 2000; 50(3):258-63.

169. Amirnovin R. An analysis of the metabolic theory of the origin of the genetic code. J Mol Evol 1997; 44:473-476.
170. Amirnovin R, Miller SL. Response. J Mol Evol 1999; 48:253-255.
171. Ronneberg TA, Landweber LF, Freeland SJ. Testing a biosynthetic theory of the genetic code: Fact or artifact? Proc Natl Acad Sci USA 2000; 97(25):13690-5.
172. Freeland SJ, Hurst LD. Load minimization of the code: history does not explain the pattern. Proc Roy Soc Lond B 1998, 265:1-9.
173. Illangasekare M, Sanchez G, Nickles T et al. Aminoacyl-RNA Synthesis Catalyzed by an RNA. Science 1995; 267:643-647.
174. Illangasekare M, Yarus M. Specific, rapid synthesis of Phe-RNA by RNA. Proc Natl Acad Sci USA 1999; 96(10):5470-5.
175. Illangasekare M, Yarus M. A tiny RNA that catalyzes both aminoacyl-RNA and peptidyl-RNA synthesis. Rna 1999; 5(11):1482-9.
176. Lee N, Bessho Y, Wei K et al. Ribozyme-catalyzed tRNA aminoacylation. Nat Struct Biol. 2000; 7(1):28-33.
177. Nagel GM, Doolittle RF. Phylogenetic Analysis of the Aminoacyl-tRNA Synthetases. J Mol Evol 1995; 40:487-498.
178. Wetzel R. Evolution of the aminoacyl-trna synthetases and the origin of the genetic code. J Mol Evol 1995; 40:545-550.

Chapter 14

Expanding the Genetic Code in Vitro and in Vivo

Thomas J. Magliery and David R. Liu

Introduction

Insight into biological function at almost every level, from catalysis to signal transduction to structure, requires a detailed understanding of proteins, biopolymers of remarkable di versity assembled from only twenty amino acid building blocks. Site-directed mutagenesis —the process by which an amino acid in a protein is swapped for one of the other 19 natural proteinogenic amino acids— has emerged as one of the most useful and powerful tools at the biochemist's disposal.[1,2] Not only does site-directed mutagenesis allow the identification of residues critical for catalysis, binding, folding, or structure, but it also made possible the first protein engineering experiments.[3] The bioorganic chemist, however, cannot be fully satisfied with this method because the changes allowed are very limited compared to the physical organic chemist's ability to subtly alter steric or electronic properties, or the synthetic chemist's ability to install useful functionalities or unique chemical handles for elaboration. Methods to alter proteins in more general ways have been developed over the last decade, inspiring the notion of "unnatural" life forms—living cells capable of using amino acids not accessible to life as we have observed it. Today's powerful in vitro methods of unnatural protein mutagenesis have become increasingly useful and accessible and have enhanced our understanding of protein function. In addition, the advent of living cells producing proteins with unnatural amino acids will allow a level of biophysical and cell-biological characterization that would have been difficult to imagine a decade ago.

Unnatural Amino Acid Mutagenesis in Vitro

A variety of chemical approaches have been employed to introduce novel functionalities into proteins, including post-translational chemical modification, total and partial chemical synthesis, and in vitro biosynthesis. Chemical modification of proteins is a field in itself and has been reviewed extensively elsewhere.[4] These methods typically involve the functionalization of exposed, reactive side-chains (cysteine, lysine or tyrosine) or of the N-terminal amino group.[5] For example, Gloss and Kirsch treated a Lys→Cys mutant of aspartate aminotransferase with bromoethylamine to generate a γ-thialysine at the active-site base with altered pK_a.[6] A related approach developed by Toney and Kirsch involved the treatment of an active-site Lys→Ala mutant with exogenous amines (e.g., methylamine) to recapitulate activity.[7,8] This noncovalent method allowed the examination of a wide range of related basic groups at the active site position, though not all residues of interest in proteins will tolerate reconstitution in this manner. Offord et al have used periodate oxidation on chemokines with N-terminal serine or threonine residues to generate a unique aldehyde functionality for conjugation with an oxime-linked

The Genetic Code and the Origin of Life, edited by Lluís Ribas de Pouplana.

fluorophore.[9] In general, these approaches are limited by the kinds of modifications that can be made and in the need for a uniquely reactive handle (e.g., a single surface-exposed cysteine).

Synthetic and Semi-Synthetic Routes

In principle, the simplest way to generate a protein containing unnatural amino acids at specific positions in vitro is to chemically synthesize the whole protein. Here, the chemist first synthesizes (or purchases) appropriately protected and activated amino acid monomers and then generates the full-length protein through iterative coupling of these monomers, typically on a solid support.[10] Modern methods of solid-phase peptide synthesis with Boc or Fmoc protection and uronium-, phosphonium-, or carbodiimide-mediated activation of amino acids typically result in excellent yields at each amide bond-forming step. One remarkable achievement was the total chemical synthesis of HIV-1 protease, a 99-residue protein.[11] Modern solid-phase peptide synthesis, however, can reliably produce peptides of only approximately fifty residues (about 6 kD), corresponding to the smallest proteins or even domains known.[12,13] Using the synthetic approach, unnatural amino acids can be incorporated into synthetic peptides with the same ease of incorporating proteinogenic building blocks. Protein synthesis has allowed the generation of proteins containing multiple site-specifically incorporated unnatural amino acids (such as γ-carboxyglutamate, which is found in the proteins of higher organisms but not bacteria or yeast).[12,14]

One way to overcome the length limitation for peptide synthesis is by the ligation of smaller, chemically synthesized peptides. An enzymic strategy for condensation of peptides involves the use of subtiligase, an engineered variant of the serine protease subtilisin BPN′ shown to ligate one peptide to a second peptide that contains a C-terminal ester bond.[15] Wells and coworkers defined the sequence specificity of subtiligase on both sides of the incipient amide bond (including the preferred types of C-terminal esters) and impressively demonstrated its use by synthesizing RNase A and by installing a binding ligand (biotin) or heavy atom (mercury) into proteins such as human growth hormone (hGH).[16,17] Modified subtiligases with improved stability and activity have since been developed as well.[18,19] This method remains limited, however, by the solubility of the peptide fragments and the sequence requirements of the enzyme.

A related chemical approach, dubbed native chemical ligation, was developed by Kent and colleagues (Fig. 1).[12] In the precursor to this method, a sulfur nucleophile from one peptide attacks an alkyl bromide in a second peptide to generate a full-length protein mimic with a single unnatural linkage (a thioester).[20] An important extension of this approach involved reaction of a peptide bearing an N-terminal cysteine with a second peptide bearing a C-terminal thioester. In this form, the cysteine's thiol group transthioesterifies onto the second peptide, and a spontaneous S-N acyl rearrangement follows to generate a native amide bond. Using native chemical ligation, Dawson et al produced full-length, 72-amino acid human interleukin 8 from two unprotected peptide fragments in high yield.[21] Two additional improvements in the method have been reported recently. By elaborating the N-terminus of a peptide with $HSCH_2CH_2O$- the rearrangement can be used to generate X-Gly and Gly-X bonds (in addition to X-Cys). The oxyethanethiol group can be removed afterwards by mild treatment with Zn and acid.[22] In order to improve the efficiency of multiple ligations needed to build large proteins, a solid-phase version of native chemical ligation was developed. Variations can be used to assemble unprotected peptides either N-to-C or C-to-N. Canne et al assembled human group V secretory phospholipase A2, a 118 amino acid protein, from four segments 25 to 33 amino acids each without purification between ligations.[23] The method is still limited by the need to synthesize all of the component peptides, but has allowed synthetic access to proteins (monomers) of approximately 15 kD. Because native chemical ligation uses synthetic peptides, unnatural amino acids may be incorporated facilely into the protein products.

A key insight that would extend this technology to much larger proteins was the observation that the natural process of protein splicing involves the excision of an intein with the intermediacy of a thioester. A mutant intein that traps the thioester intermediate has been exploited

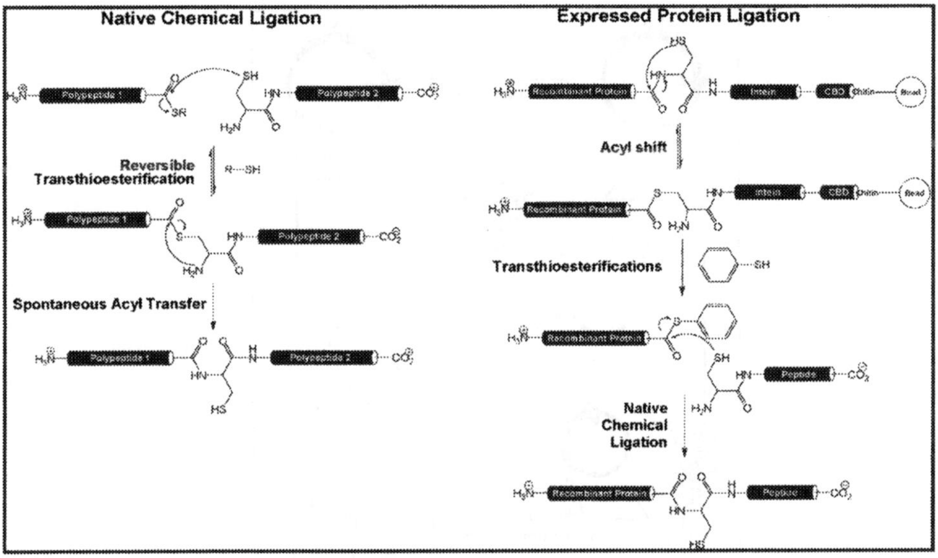

Figure 1. Native chemical ligation and expressed protein ligation. Schematic representation of techniques for assembling synthetic and semi-synthetic proteins from peptide fragments.

commercially as a means of protein purification, wherein the intein is linked to a chitin-binding domain and the recombinant protein is purified over chitin and released with DTT.[24,25] Muir and coworkers instead released the trapped thioester from the resin with a synthetic peptide bearing an N-terminal cysteine, which resulted in rearrangement to yield the native amide bond (Fig. 1). This process, called expressed protein ligation, is similar to the solid-phase ligation developed by Kent except that the immobilized N-terminal portion of the protein is produced recombinantly, which allows access to proteins of virtually any size.[26,27] The method is obviously most amenable to introduction of unnatural amino acids near the C-terminus of the protein, but it can be combined with synthetic approaches to label internal regions, as well.[28] Of course, this method, like all the in vitro methods, only allows examination of protein outside the context of living cells.

Expanding the Genetic Code in Vitro

Despite significant advances in synthetic methodology for peptides, there is still no comparison to the cell's ability to synthesize proteins. Even in a reconstituted in vitro format, transcription-translation is a robust and efficient means of producing folded soluble proteins of virtually any size. This fact inspired an in vitro biosynthetic approach for the site-selective insertion of unnatural amino acids into proteins (Fig. 2).[29,30] The method has three key requirements: a genetic signal for site-selective insertion (i.e., a codon); a translationally-competent tRNA to read the codon that is not charged by endogenous aminoacyl-tRNA synthetases; and a method of acylating that tRNA with an amino acid of choice. The redundancy of the genetic code provides the insertion signal: there are three stop codons (UAG, *amber*; UGA, *opal*; UAA, *ochre*), and only one is required to terminate any given protein.[31] Moreover, suppressor tRNAs that insert one of the natural amino acids in response to stop codons are known, and *amber* suppressors are especially robust, partially because the *amber* stop codon is the least used in *E. coli*.[32]

Schultz et al developed an *amber* suppressor tRNA for this purpose that meets the two key criteria: it is not acylated (or deacylated) by aminoacyl-tRNA synthetases (aaRSs) in the in vitro transcription-translation reaction (which is *Escherichia coli* derived), but it is accepted by

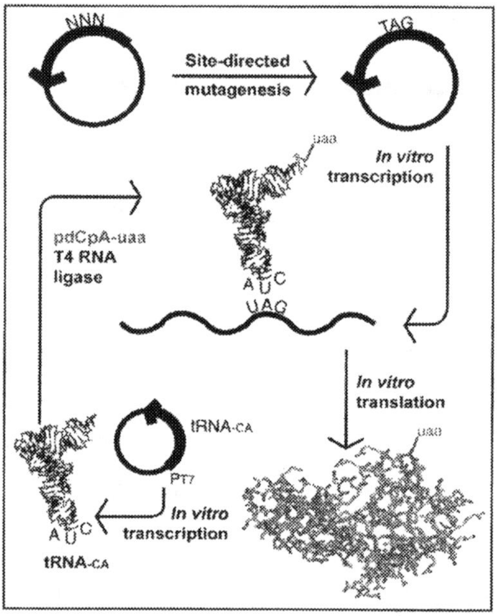

Figure 2. A biosynthetic approach to site-specific unnatural protein mutagenesis. Site-directed mutagenesis is first used to mutate the codon for the residue of choice to the *amber* stop codon. This is added to an in vitro transcription-translation mixture with *amber* suppressor tRNA chemically aminoacylated with unnatural amino acid to generate full-length protein bearing the unnatural amino acid. The *amber* suppressor tRNA is in vitro transcribed without its final pCpA-3′, and a synthetic pdCpA acylated with unnatural amino acid is ligated to the tRNA $_{-CA}$ with T4 RNA ligase.

the translational machinery, including EF-Tu and the ribosome. A yeast tRNA^{Phe} was chosen for its orthogonality to *E. coli* aaRSs, and its anticodon residues 34-37 were replaced with 5′-CUAA-3′ (the 3′-A in accord with Yarus's rule for efficient *amber* suppression).[33-35] Noren et al showed that this tRNA was not acylated in an *E. coli* suppression reaction, when pre-acylated enzymically by PheRS, does insert Phe into proteins in the same reaction.[30] Chamberlin et al took a different approach, generating an *E. coli* tRNA^{Gly}(CUA) by in vitro transcription and employing a rabbit reticulocyte suppression reaction.[29] Today, the state-of-the-art uses in vitro transcribed and chemically aminoacylated yeast tRNA^{Phe}(CUA) or *E. coli* tRNA^{Asn}(CUA) (the latter especially for small, polar amino acids) in an optimized *E. coli* transcription-translation mixture generated from an RF1-deficient strain of *E. coli* to increase suppression efficiency.[36-40]

Chemical misacylation of tRNAs has a long history, dating back to Raney nickel reduction of Cys-tRNA^{Cys} to Ala-tRNA^{Cys}, an experiment that proved Crick's adapter hypothesis that translational fidelity at the ribosome depends on codon-anticodon interaction rather than recognition of the amino acid acylated to the tRNA.[41,42] Hecht and coworkers largely pioneered this area, first acylating N-blocked amino acids onto a pCpA diribonucleotide and coupling this with T4 RNA ligase to tRNA lacking the ubiquitous pCpA-3′.[43,44] Although this method was eventually modified to allow addition of residues with free amino groups, it employed tRNAs with sense codons and resulted in insertion of unnatural amino acid at all such codons in the gene.[45,46] In addition to the use of the *amber* suppressor, innovations of Schultz and coworkers have included: the use of pdCpA, which facilitates chemical acylation of the tRNA by removing a nucleophilic 2′-OH with no ill effects on tRNA function; the use of cyanomethyl ester activation of the amino acid, which requires no additional protection of the pdCpA; and α-amino protection with nitroveratryloxy carbamate (NVOC), ester or ether, which protects the aminoacyl ester linkage from hydrolysis but can be easily removed photolytically prior to the suppression reaction.[47,48] (BPOC protection with removal in mild acid is useful when the amino acid is light-sensitive.[29,48])

Although the chemistry and suppression reactions are widely-applicable and robust (see II.E. below), this method has four major drawbacks: the protein yields are poor, on the scale of 50 μg per mL of suppression reaction; the method requires synthetic chemistry not easily performed by many biochemistry groups who would most benefit from it; the *amber* suppression method only permits site-selective insertion of a single amino acid; and the modified proteins can only be examined in vitro because of the chemical aminoacylation step.[49]

New Codons for Insertion of Unnatural Amino Acids

Two strategies have been examined to expand the number of possible ways to specify sites for biosynthetic insertion of unnatural amino acids: the use of natural codons such as the *amber* nonsense codon and expanded natural codons such as four-base codons, and development of unnatural codons containing unnatural bases. Sisido and coworkers attempted to use the sense codon AGG, *E. coli*'s rarest codon with low cognate tRNA abundance, as an insertion signal, but found that the background readthrough by the endogenous $tRNA^{Arg}$(CCU) was high.[50] Takaku, Sisido and coworkers have examined the use of sense codons for site-selection in in vitro protein synthesis with depletion of the endogenous sense tRNAs by antisense or RNase A treatment.[51-53] The *E. coli* S30 extract was treated with RNase A to destroy the endogenous tRNAs, and then a crude mixture of *E. coli* tRNAs was treated with RNase H and DNA oligomers antisense to specific tRNAs ($tRNA^{Asp}$ and $tRNA^{Phe}$). These tRNAs, chemically acylated with unnatural amino acids, were then added back into the mixture. Since, in principle, only a single codon is needed to specify each amino acid in a given gene, this is potentially a route to site-selective insertion of multiple unnatural amino acids. The biggest limitations to this method are the high levels of readthrough (presumably by tRNAs with near anticodons) and the need to heavily mutagenize the gene to contain only the appropriate array of codons.[54]

It has long been known from studies in yeast and *Salmonella* that naturally occurring frameshift suppressors are tRNAs with extended (usually eight nucleotide) anticodon loops that decode four-base codons.[55] Hardesty et al first used this as an approach to protein engineering, generating $tRNA^{Ala}$ with ACCU or CCUA anticodons to read AGGU or UAGG codons and insert alanine.[56] (AlaRS acylates many variants of its tRNA since its principal recognition element is in the tRNA's acceptor stem, far from the anticodon.) This is obviously preferable to use of sense codons, since misreading of the codon (i.e., readthrough by other endogenous tRNAs) produces an out-of-frame product that is typically truncated due to the relative abundance of stop codons out of frame. Hardesty has also used other tRNAs (Ser and fMet) that are acylated by their cognate synthetase enzymes even when their anticodon loops are expanded, and has found that four-base codons that are extensions of rare sense codons are generally most efficient.[57] Sisido's group used a $tRNA^{Phe}$(NCCU) to insert unnatural amino acids in response to AGGN codons, and found that efficient suppression could be achieved with virtually no background (i.e., full-length protein without unnatural amino acid).[58] An important extension of this was the use of two frameshift suppressor tRNAs reading the four-base codons AGGU and CGGG, chemically acylated with two different unnatural amino acids, to achieve the first simultaneous site-selective biosynthetic incorporation of multiple unnatural amino acids.[59]

Atkins and coworkers set out to examine the suppression in vivo of UAGN four-base codons by NCUA anticodons in all possible combinations in an effort to optimize tRNAs for genetic code expansion. Using $tRNA^{Leu}$ as a scaffold (LeuRS acylates $tRNA^{Leu}$ variants harboring mutations at the anticodon), they found $tRNA^{Leu}$(UCUA) suppressed UAGA efficiently with little decoding in the zero frame (i.e., non-frameshifted reading of UAG). Moreover, with partial inactivation of RF-1 (which acts at the *amber* stop codon), suppression efficiencies as high as 26 % could be achieved.[60] Two tandem UAGA codons could be suppressed with efficiencies as high at 10 % in this system.[61]

Magliery et al adopted a library approach to identify extended anticodon loop tRNAs capable of efficiently suppressing four-base codons. Here, $tRNA^{Ser}$ variants with randomized eight-nucleotide anticodon loops were crossed against a library of β-lactamase genes with all possible four-base codons at various sites. Survival on ampicillin identified four efficient tRNA/four-base codon pairs corresponding to AGGA, UAGA, CCCU and the previously unidentified CUAG.[62] This work was extended to examine the suppression of two through six base codons with tRNAs containing six to ten nucleotides in their anticodon loops. Several efficient tRNA/5-base codon pairs were found, including one for suppression of AGGAC. Moreover, these experiments demonstrated that the ribosome is capable of using 3-, 4- and 5-base codons, and that the corresponding tRNA suppressors have $N + 4$ nucleotides in their anticodon loops and exhibit Watson-Crick complementarity between codon and anticodon. The most efficiently suppressed codons of any size were based on rare codons, like AGG and UAG, the rarest sense and nonsense codons in the *E. coli* genome.[63]

Over a decade ago, Benner and colleagues suggested the use of a new base pair capable of being a substrate of DNA and RNA polymerases. The iso-C/iso-G pair (Fig. 3) was largely successful except that d-iso-G paired with, in addition to d-iso-C, dT, due to a minor tautomer with a hydrogen bonding pattern similar to dA.[64] Benner, Chamberlin and coworkers synthesized an RNA containing an (iso-C)AG codon and showed that in vitro suppression with tRNA(CU(iso-G)) was very efficient (90%) compared to UAG (*amber*) suppression (63%).[65] The translated peptide was only 17 residues long, however, due to the need to synthesize the mRNA chemically. Use of an unnatural pair that could be replicated in a plasmid and transcribed with high fidelity would eventually be needed to make this more useful. Although other pairs with differing H-bonding patterns were also developed in this effort (κ, X, π),[66] none of these was completely orthogonal to the natural bases, especially at the level of transcription.

In 1997, Kool and coworkers shed light on a different strategy for generating an orthogonal base pair using hydrophobic bases instead of bases with altered hydrogen bonding patterns. It was found that a hydrophobic near-isostere of thymine, difluorotoluene (dF), efficiently specified insertion of adenine by DNA polymerase even though the dF-dA "pair" destabilized the DNA duplex.[67] Matray and Kool went on to show that shape mimics of base pairs capable of stacking, even if nonpolar, can result in stable DNA duplexes, as in the case of pyrene paired against an abasic site.[68] Other pairs, such as dF and dimethylbenzimidazole (dZ, a nonpolar dA analog), were shown to be relatively stable and specify each other's insertion in DNA replication, although not with complete orthogonality (dF also pairs with dA efficiently and dZ weakly pairs with dT).[69]

Romesberg, Schultz and coworkers have also attempted to make nonpolar unnatural base pairs that form stable duplex DNA, have high specificity for each other and can be incorporated by polymerases. For example, the 7-propynyl isocarbostyril (PICS) self-pair is stable in duplex DNA but destabilizing when paired against the natural bases. Moreover, this nonpolar base specifies itself with good selectivity over the natural bases in DNA replication, but it acts as a chain terminator after insertion.[70] Other hydrophobic pairs were found to be stable, accepted and extended by DNA polymerase, notably with isocarbostyril (ICS) and 7-azaindole (7AI) bases.[71] The chief problem with this pair is that each base also specifies itself in DNA replication, which results in chain termination. Optimization led to a related pair, pyrrolopyrizine (PP) and methylisocarbostyril (MICS), which was inserted at rates within 20-fold of the rate of synthesis with the natural nucleotides and with some chain extension, although MICS self-pairing and orthogonality to the natural bases remain issues.[72] Another interesting pair designed by Meggers et al involves pyridine-2,6-dicarboxylate and pyridine bases that form a stable duplex only in the presence of copper, via a square planar complex. This pair shows good discrimination from the natural bases, thermodynamically.[73]

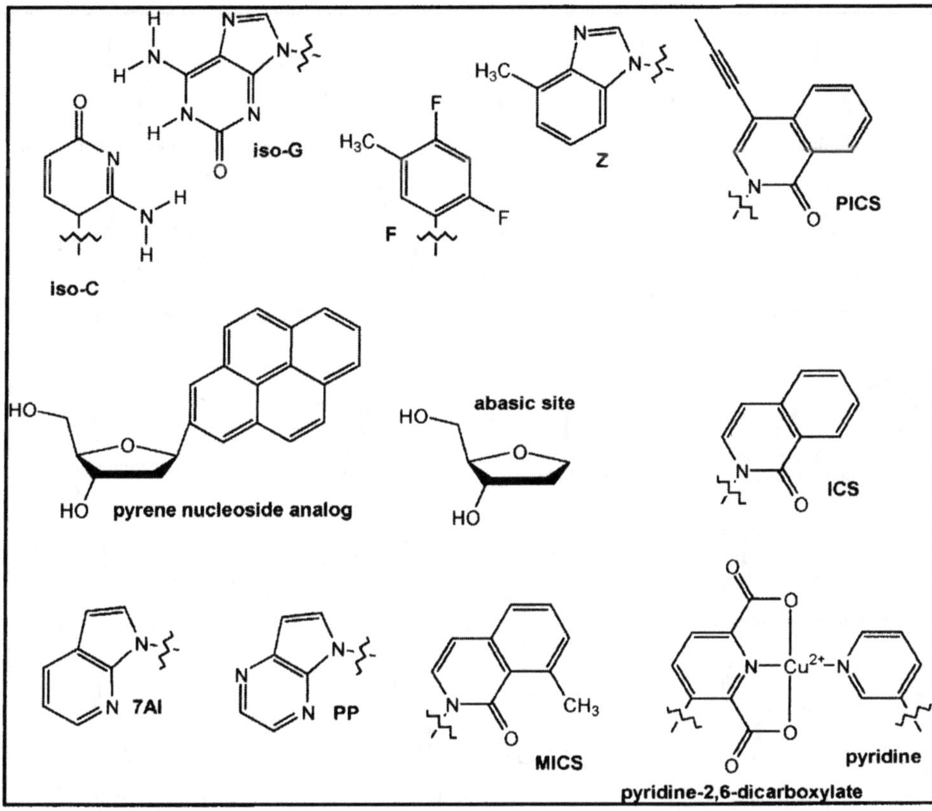

Figure 3. Structures of some unnatural bases for the generation of an orthogonal, unnatural base pair.

Though the development of unnatural base pairs is potentially very powerful, it remains to be seen if an unnatural base pair can be found that is stable, specific and orthogonal to the natural bases, efficiently incorporated and extended by DNA and RNA polymerases and, ultimately, accepted in vivo, including being uptaken from the growth media.[74]

Using in Vitro Acylation in Vivo: Oocytes

Dougherty and coworkers devised a way to make use of very sensitive voltage clamping techniques to examine single cells producing proteins containing unnatural amino acids through microinjection of chemically in vitro acylated tRNA. This technique allows the in vivo production (in *Xenopus* oocytes) of proteins with unnatural residues, and is especially useful for proteins in the membrane that could not be produced in vitro due to folding and expression problems.[75] It was found that the initial tRNA designed for this approach, while an improvement over $tRNA^{Phe}$ in protein yield, was not orthogonal to *Xenopus* aaRSs. A modified version of *Tetrahymena thermophila* $tRNA^{Gln}$(CUA), a tRNA that naturally inserts glutamine in response to UAG (which is not a stop codon in *Tetrahymena*) was both efficient and not a substrate for the endogenous aaRSs of the oocyte.[76]

Applications

Applications of the in vitro unnatural amino acid mutagenesis methodologies described above have been reviewed extensively, and so only a few notable and recent examples will be presented here.[49,77,78] Since these methods are all limited by the quantity of protein produced,

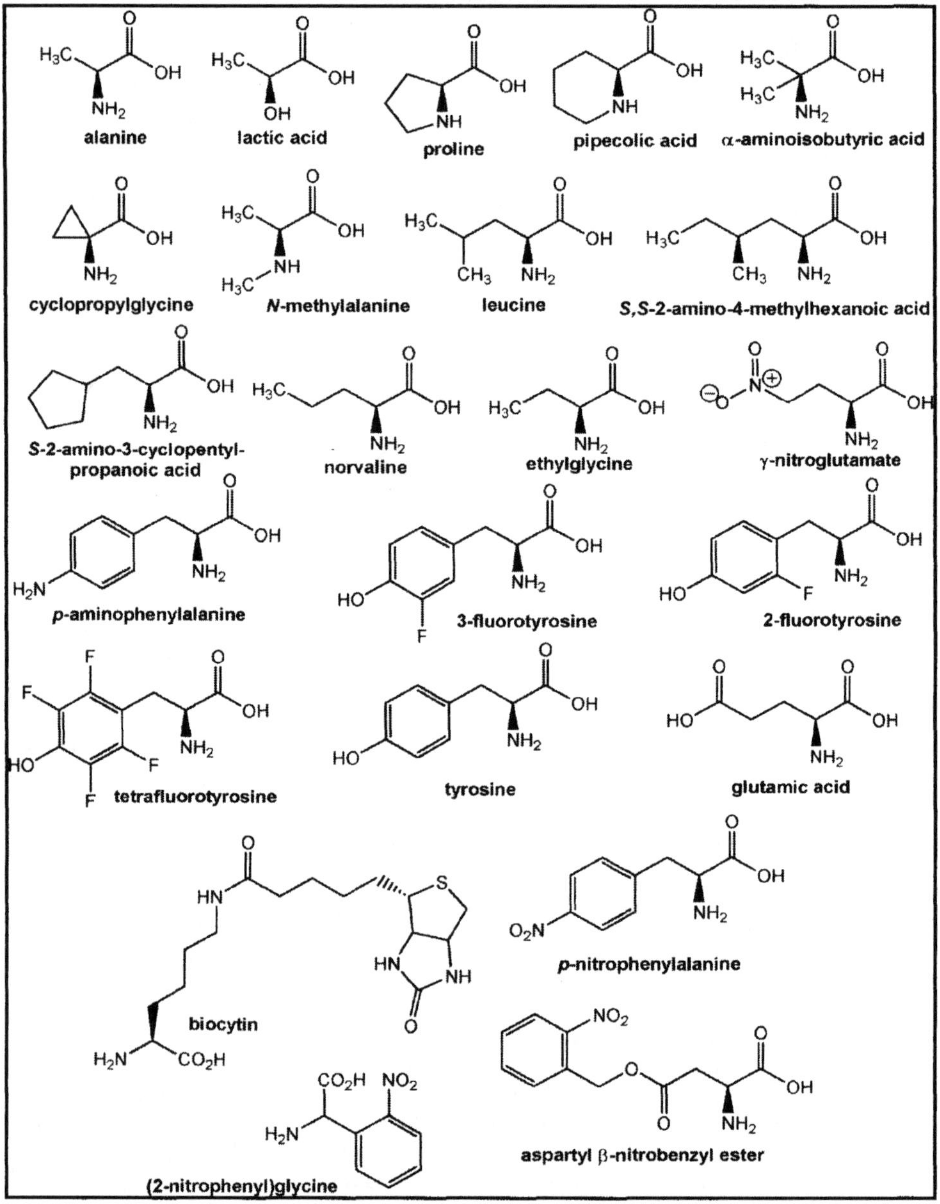

Figure 4. Structures of some amino acids inserted into proteins with in vitro methods.

applications that require only small amounts of protein have been most successful. Techniques of extraordinary sensitivity such as fluorescence, spectroscopic measurements of catalysis, and single-cell voltage clamping methods are of special note.

Schultz and coworkers have examined the effects of unnatural residues on the stability of proteins. Ala82 of T4 lysozyme (T4L), a surface residue between helices, was replaced with residues containing unnatural backbone structures including lactic acid, pipecolic acid, *N*-methylalanine, cyclopropylglycine and α-aminoisobutyric acid (Aib). (See Fig. 4 for structures of the amino acids discussed in this section.) It had already been shown that proline stabilized T4L at this position; interestingly, pipecolic acid slightly destabilized the protein while Aib stabilized it, presumably due to φ and ψ angles different from or similar to proline, respectively. The fact that lactic acid destabilized the protein much more than *N*-methylalanine or cyclopropylglycine suggests that electronic properties have a greater effect on stability than conformational restriction.[79] Studies such as this one demonstrated the remarkable ability of the protein biosynthetic machinery to incorporate bulky, conformationally restricted and β-amino acids.[80] At Leu133, a buried hydrophobic residue, larger amino acids (*S,S*-2-amino-4-methylhexanoic acid and *S*-2-amino-3-cyclopentylpropanoic acid) predicted by molecular modeling to fill a cavity in T4L better than the wild-type residue stabilized the protein slightly (0.6 and 1.2 kcal mol^{-1}), while successively smaller amino acids (norvaline and ethylglycine) destabilized the protein (1.1 and 3.3 kcal mol^{-1}).[81]

Thorson et al examined the role of hydrogen bonding in protein stability by altering the hydrogen bonding pair Tyr27-Glu10 in staphylococcal nuclease (SNase). Replacing either residue with an isosteric residue of weak hydrogen bonding ability (*p*-amino-L-phenylalanine or γ-nitro-L-glutamate) decreased protein stability 2.7 and 1.8 kcal mol^{-1}, respectively. Unnatural amino acid mutagenesis made possible in this case subtle changes that affect primarily hydrogen bond strength, since side-chain packing and solubility were very similar to the natural residues.[82] By substituting the 2-fluoro- (pK_a 9.3), 3-fluoro- (pK_a 8.8) and tetrafluoro- (pK_a 5.3) derivatives of Tyr (pK_a 10.0) at position 27, a linear free energy analysis was applied to hydrogen bond strength. These changes increased the stability of the protein about 0.5, 1 and 2 kcal mol^{-1}, respectively, as expected for hydrogen bonds of increasing strength. The derived value of $\alpha = 0.35$ for the linear free energy relationship $\log K_{app} = \alpha(pK_a) + C$ suggests a nearly equal sharing of the proton between donor and acceptor. Moreover, this study contributed to evidence that hydrogen bonds, in addition to specifying secondary and tertiary structure, are important for protein stability.[83] A similar analysis of the importance of cation-π interactions for protein stability in SNase placed the interaction strength at about 2.6 kcal mol^{-1}, similar to that of hydrogen bonds.[84] In a related study, Dougherty's group demonstrated the role of a cation-π interaction in the binding of acetylcholine to a tryptophan residue in the nicotinic acetylcholine receptor by measuring the effects of altered aromatic groups at the Trp site.[85]

A number of residues for biophysical studies have been inserted into proteins, including isotopically labeled residues for NMR;[86] spin-label amino acids, fluorophores and photoaffinity labels;[36,87] and uniquely modifiable residues bearing ketones.[88] Incorporation of biocytin, a biotin-containing amino acid, into various sites in ion channels expressed in *Xenopus* oocytes was used to determine the transmembrane topology of the channels.[89] Introduction of *p*-nitrophenylalanine into streptavidin, which bound to *N*-biotinyl-L-1-pyrenylalanine, allowed measurement of a distance decay constant (β) for photoinduced electron transfer in proteins.[90] Expressed protein ligation has been used to insert probes into proteins that can act as biosensors. Insertion of an environmentally sensitive 5-(dimethylamino)-napthylene-1-sulfonamide fluorophore into a peptide between Src homology domains SH2 and SH3 allowed sensing of interaction with a bidentate peptide ligand.[91] In addition, a Crk-II adapter protein was labeled with tetramethylrhodamine at the N-terminus and fluorescein at the C-terminus to generate a phosphorylation-sensitive protein by examination of FRET efficiency.[92] In vitro unnatural amino acid mutagenesis methodology may also aid single-molecule FRET experiments in the examination of protein folding.[93]

An especially useful kind of unnatural amino acid for biochemical studies has been the caged amino acid, which makes time-resolved studies possible via photoinduced activation. A light-activated form of T4 lysozyme was generated by replacing the active-site Asp20, which stabilizes the carbocation generated in degradation of the β-linked NAM-NAG cell wall, with a photocaged residue. The nitrobenzyl ester of the Asp20 side-chain carboxylate was efficiently removed with light to produce active enzyme.[94] The nitrobenzyl ether of tyrosine was used to cage residues in the nicotinic acetylcholine receptor (nAChR) in *Xenopus* oocytes, and pulse irradiation was used to confirm that at sites critical for ligand binding, two uncaged tyrosines

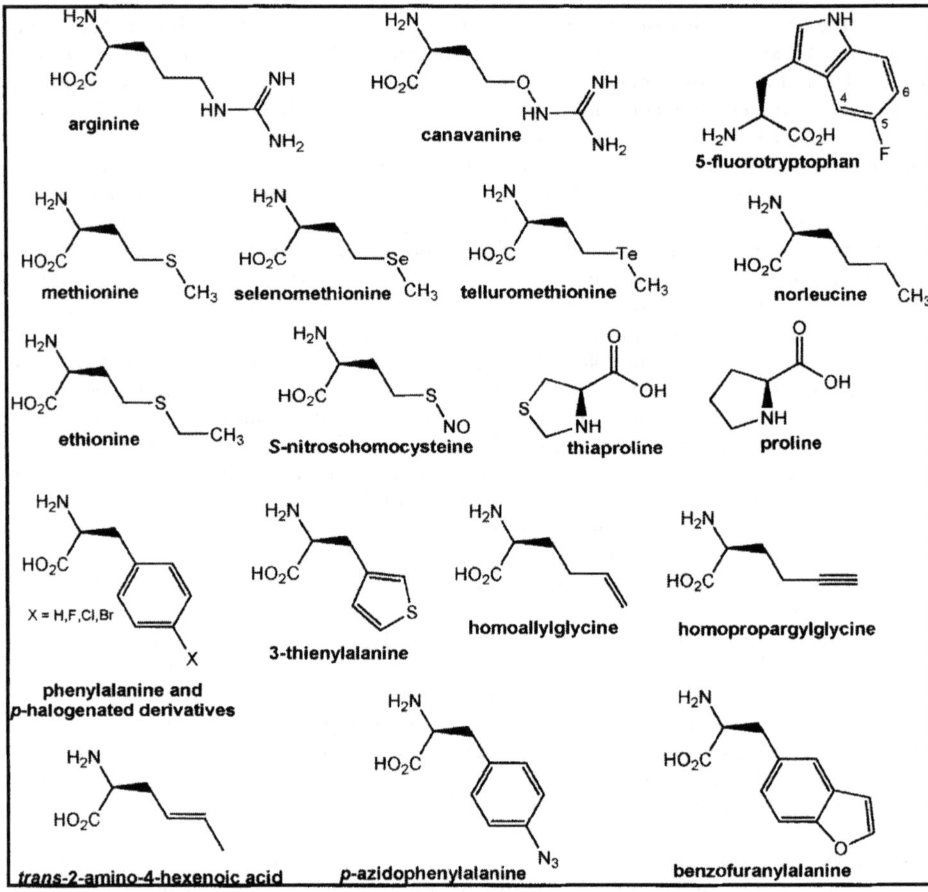

Figure 5. Structures of some unnatural amino acids inserted into proteins by *E. coli*. Some natural amino acids are included for comparison.

were required for receptor activation.[95] An interesting related method involves site-specific photochemical proteolysis upon irradiation of protein containing (2-nitrophenyl)glycine. The method was used to demonstrate the functional necessity of a loop formed as a result of a disulfide bond in nAChR.[96] Caged groups incorporated with in vitro methods have also been used to control protein-protein interactions, such as the dimerization of HIV-1 protease upon uncaging of Asp25 at the dimer interface,[97] or the interaction of ras with effector p120-GAP uncaging of Asp38 of ras.[98]

New Codes in Vivo

Naturally occurring non-proteinogenic amino acids are well known, and several act as growth inhibitors on different kinds of cells. For example, growth of *E. coli* is inhibited by L-canavanine (a very close analog of arginine, Fig. 5) because of massive mis-incorporation of canavanine into proteins. Cells cannot discriminate the natural amino acids from many close analogs, even through the many steps of protein biosynthesis. In the absence of these analogs, in fact, there is little reason for cells to have evolved mechanisms to distinguish natural substrates from unnatural ones. Presumably, this is why L-canavanine is erroneously uptaken and inserted into yeast and bacterial proteins, but ornithine and citrulline, non-proteinogenic amino acids that are intermediates in arginine biosynthesis, are discriminated against.[99] This has led to the development of strains and culture conditions for producing proteins with unnatural amino acids. Ideally, however, the protein biochemist would like to use this idea in a site-specific format, as with in vitro biosynthetic methods. Observations from in vitro site-specific methods and in vivo methods without site specificity have led to an approach for generating the first living cells with an expanded genetic code.

Alternate Substrates for Aminoacyl-tRNA Synthetases

Introduction of 5-fluorotryptophan (5-F-Trp) into proteins has been accomplished using strains auxotrophic for tryptophan, glyphosphate to block an enzymic step in the biosynthesis of tryptophan, and/or minimal media in which 5-F-Trp is added just before induction of protein expression (see Fig. 5 for structures of amino acids discussed in this section).[100-102] Incorporation of 5-F-Trp was used to assess the symmetry of the large (380 kD) F_1-ATPase from *E. coli*, using the fact that ^{19}F is NMR-active and detectable at unique NMR shifts. In the smaller *E. coli* glucose/galactose receptor (33 kD), all five of the individual 5-F-Trp residues could be distinguished in combination with site-directed mutagenesis.[103-106] Aided in part by the high yield of protein that can be produced in vivo, Armstrong and coworkers have produced a crystal structure of a 5-F-Trp-labeled glutathione transferase which has enhanced catalytic activity by virtue of its unnatural amino acids.[102] The 5-F-Trp is inserted at all the Trp positions in the protein and typically at 50-80 % for any individual site. More reliable insertion methods have recently been developed, and the rarity of Trp residues makes it possible in some cases to make single-site changes, as in replacement of the lone Trp187 in human annexin V with 4-fluoro-, 5-fluoro- and 6-fluoro-Trp to examine changes in enzyme stability and activity.[107]

The advent of an *E. coli* expression system that allowed high-level replacement of methionine with selenomethionine (SeMet) has made possible the widespread use of multiwavelength anomalous diffraction (MAD) for solving X-ray crystal structures of proteins. MAD provides an elegant solution to the phase problem in X-ray crystallography since, using a tunable radiation source like a synchrotron, crystals from a single heavy-atom derivative of an enzyme can be used to deduce the structure (as opposed to multiple crystals from imperfectly isomorphous protein derivatives with different heavy-atoms). SeMet is typically sufficiently well packed that the Se atom is well ordered, and the K-shell orbital excitation energy of Se is easily achievable with synchrotron radiation ($\lambda = 0.98$ Å). Importantly, some auxotrophic strains of *E. coli* can be used to express proteins with 100 % SeMet incorporation with careful Met-free culturing conditions.[108,109] The first X-ray crystal structure solved by this method was ribonuclease H, to 2 Å.[110] Many other Met analogs can be inserted this way, including 2-aminohexanoic acid (norleucine), ethionine, telluromethionine and *S*-nitrosohomocysteine.[111,112] Budisa and coworkers have used these analogs to examine subtle effects on protein folding from these "atomic" mutations.[113] Other replacements have been made for biophysical characterization, like ProRS-mediated insertion of thiaproline in *E. coli*.[114]

Tirrell, Fournier and colleagues have largely pioneered the field of homogeneous polymeric materials made by bacterial expression. In early work, T7 RNA polymerase-driven bacterial expression was used to generate polymers of a nonapeptide AGAGAGPEG with repeat lengths from 10 to 54 (i.e., 90 to 486 residues) to examine electrophoretic, structural and polymeric properties.[115] Using high-level induction from media depleted of a natural amino acid, a variety of unnatural amino acid analogs have been inserted into periodic proteins, including *p*-fluorophenylalanine, *p*-chlorophenylalanine, *p*-bromophenylalanine, 3-thienylalanine, homoallylglycine, homopropargylglycine, *trans*-2-amino-4-hexenoic acid and norleucine.[116-121] These studies have largely taken advantage of the natural permissivities of PheRS and MetRS, although *trans*-2-amino-hexenoic acid required overexpression of MetRS and *p*-Cl-Phe and *p*-Br-Phe required a mutant of PheRS with broadened substrate specificity.[122] Artificial proteins with unnatural amino acids have been investigated both for physical properties and as means of providing unique reactive handles, such as ruthenium-catalyzed olefin metathesis on homoallylglycine.[123]

Another approach to expanding the number of amino acids that can be inserted by the natural aminoacyl-tRNA synthetases is to disable the editing features inherent in some of the aaRSs. For example, a selection was carried out for a valyl-tRNA synthetase capable of misacylating valyl tRNA with cysteine, but with no subsequent hydrolytic removal of this cysteine. ValRS was known to acylate $tRNA^{Val}$ with threonine and then correct this misacylation with hydrolytic editing. As expected, then, the selected ValRS mutant conferred sensitivity to threonine, and moreover allowed the insertion of aminobutyric acid (Abu) by this pathway when Abu was supplied in the growth medium. However, only about 20% of the valine codons were replaced with Abu, and, even if this reached 100%, it should be noted that this would recode rather than expand the genetic code.[124]

Designing an in Vivo System for Site-Specific Unnatural Amino Acid Delivery

A method combining the best characteristics of the in vitro site selective methods and the in vivo non-site-selective methods would be of greatest utility to the protein biochemist. Protein engineering and our understanding of protein structure and function would benefit greatly from a method that allows all twenty of the natural amino acids to be inserted, in addition to the site-selective insertion of one or more additional unnatural amino acids. In vivo methods have the significant advantages of high yield of protein and easy scale-up, technical ease (simple addition of the unnatural amino acid to the medium), and the potential to observe the altered proteins in the living cell—that is, to do "unnatural cell biology" with caged proteins, affinity labeled proteins, or proteins bearing biophysical probes or other moieties to expand their functionality.

Many of the design considerations for in vitro biosynthetic site-selective incorporation of unnatural amino acids also apply to an analogous in vivo method. The common minimal requirements are a unique signal for site-selective insertion (i.e., a codon) and a translation-competent tRNA, capable of decoding the insertion signal, that is neither acylated nor deacylated by the endogenous aminoacyl-tRNA synthetases of the host organism. However, there are at least three additional design considerations for an in vivo method.

i. Since the "orthogonal" tRNA must be acylated in vivo, an aminoacyl-tRNA synthetase is required that uniquely acylates the orthogonal tRNA. This synthetase must itself be orthogonal, acylating the orthogonal tRNA and not endogenous tRNAs.
i. The synthetase must be capable of acylating the orthogonal tRNA with an unnatural amino acid but also incapable of using any other endogenous amino acid (proteinogenic or not) as a substrate.
iii. The unnatural amino acid must be uptaken by the cell or produced by it, and it cannot be accepted as a substrate by any endogenous synthetase or, more generally, be toxic to the cell.

Moreover, since a great deal of protein engineering will be required to achieve these goals, a well-understood organism capable of being transformed with large libraries (i.e., with high transformation efficiency) is desired. Likewise, since one will likely derive a synthetase capable of inserting an unnatural amino acid from one that inserts a natural amino acid, a well-characterized protein is desired as a starting point, especially one that has been structurally well-characterized.[125]

Schultz and coworkers have applied the following approach. For its genetic tractability, *E. coli* was chosen as the initial host organism. The ***amber*** stop codon was selected as the insertion signal due to the excellent suppression that is possible with known suppressor tRNAs isolated from *E. coli* strains. However, attempts to make orthogonal tRNA/synthetase pairs using four-base codons are underway, in light of our identification of easily suppressible extended codons (J.C. Anderson, T.J.M. and P.G. Schultz, unpublished work). Initially, we adopted an engineering approach to the generation of an orthogonal tRNA/synthetase pair, starting with the extremely well-characterized *E. coli* glutamine pair.[126] Since that time, several researchers have found it more advantageous to import *amber* suppressor tRNAs and aminoacyl-tRNA synthetases from other organisms (heterologous pairs, see Table 1). Of course, these orthogonal pairs insert a natural amino acid, and so robust screens and selections have been developed to find variants of these synthetases from carefully designed libraries of mutagenized enzymes capable of acylating tRNA with unnatural amino acids. Using these methods and other less general methods, the first successes in engineering living cells that are capable of site-selectively inserting unnatural amino acids into proteins have recently emerged.

Orthogonal tRNA/aaRS Pairs

The first orthogonal tRNA developed for the purpose of in vivo site selective delivery of unnatural amino acids was derived from *E. coli* $tRNA^{Gln}$. Glutaminyl-tRNA synthetase (GlnRS) was known to acylate its *amber*-suppressing derivative (anticodon 5′-CUA-3′) and biochemical and X-ray crystal structural information defined the nature of the interaction between tRNA and synthetase.[127-133] Three sites at which mutations were expected to modulate the ability of GlnRS to acylate the tRNA ("knobs") were selected, and tRNAs bearing mutations at each site (and in all possible combinations) were generated.[126,134] These mutations turned out to interact in complicated, non-additive ways both with respect to aminoacylation by GlnRS and performance as tRNAs for delivery of amino acids at the level of translation.[126]

For example, individual mutations in the acceptor stem (knob 1) or D-stem (knob 2) radically reduced the ability of the tRNA to be acylated by GlnRS, but together these mutations rescued this ability by about 6-fold over the knob 2 mutation alone. Overall, tRNAs with the knob 2 mutation and an additional mutation in the D-loop (knob 3), or with all three mutations, were not substrates for GlnRS (about 15,000-fold down in activity compared to wild-type $tRNA^{Gln}$). Likewise, when either of the *amber* suppressing tRNAs was added to an *E. coli* transcription-translation reaction, neither was acylated by the endogenous aaRSs, evidenced by the lack of full-length protein derived from a gene with an *amber* mutation. However, when these tRNAs were chemically acylated with valine, only the tRNA with mutations at *all three* knobs mediated insertion of valine to produce full-length protein.[126] Thus the tRNA with knob 2/knob 3 mutations was no longer a substrate for the *E. coli* translational apparatus (ribosome, EF-Tu, etc.), but the addition of knob 1 rescued the ability to act as a translation-competent *amber* suppressor. This *amber* suppressor tRNA with all three knob mutations, O-$tRNA^{Gln}$(CUA), therefore met the criteria for an orthogonal tRNA in *E. coli*: it was not a substrate for endogenous synthetases but was competent to act in translation.[126]

This tRNA was also characterized in vivo by transforming a plasmid causing the transcription of the O-$tRNA^{Gln}$(CUA) into an *E. coli* strain with an *amber* mutation in the gene for β-galactosidase (*lacZ*). The cells were incapable of surviving on lactose minimal media due to the fact that the O-$tRNA^{Gln}$(CUA) was not appreciably acylated to produce full-length LacZ and hydrolysis of lactose. This fact was used as the basis for a selection for a GlnRS mutant capable of aminoacylating the O-$tRNA^{Gln}$(CUA). The gene for GlnRS (*glnS*) was randomly mutagenized by the method of DNA shuffling[135,136] and co-transformed on a compatible plasmid into the *E. coli* strain bearing the O-$tRNA^{Gln}$(CUA) expression plasmid and the $lacZ_{am}$ mutation. Cells that survived on lactose minimal media contained library members with increased ability to acylate the O-$tRNA^{Gln}$(CUA) (presumably with glutamine); these were expressed, purified, examined in vitro for ability to acylate O-$tRNA^{Gln}$(CUA) and $tRNA^{Gln}$, pooled and resubmitted to mutagenesis and selection.[134]

After seven rounds of mutagenesis and selection, a mutant GlnRS was found that acylated the wild-type $tRNA^{Gln}$ substrate only 9-fold better than the O-$tRNA^{Gln}$(CUA) and was down only 250-fold with respect to acylation of $tRNA^{Gln}$ by wild-type GlnRS. This enzyme with overall 1,500-fold change in specificity was capable of acylating the O-$tRNA^{Gln}$(CUA) sufficiently to observe by Western blot full-length protein produced from an amber mutant of the gene for *E. coli* surface protein LamB. No full-length protein was observed with the

Table 1. Orthogonal tRNA/aaRS pairs for delivering unnatural amino acids

For Use In	tRNA	Synthetase	Notes	Ref.
E. coli	O-tRNA^{Gln}(CUA), with three "knob" mutations	selected mutant of *E. coli* GlnRS	tRNA is not as orthogonal as O-*Sc*tRNA^{Gln}(CUA); synthetase still acylates w.t. tRNA^{Gln}	134
E. coli	O-*Sc*tRNA^{Gln}(CUA)	yeast GlnRS	tRNA is highly orthogonal but *Sc*GlnRS activity is weak	140
E. coli	O-*Mj*tRNA^{Tyr}(CUA)	*M. jannaschii* TyrRS	tRNA is not as orthogonal as O-*Sc*tRNA^{Gln}(CUA) but *Mj*TyrRS is very active	143
E. coli	O-*Mj*tRNA^{Tyr}(CUA)*, mutant of O-*Mj*tRNA^{Tyr} (CUA)	*M. jannaschii* TyrRS	mutatations in tRNA reduce recognition by *E. coli* aaRSs while maintaining recognition by *Mj*TyrRS	146
E. coli	O-*Ec*tRNA^{fMet}, mutant of the *E. coli* initiator tRNA	Selected mutant of yeast TyrRS	Mutant *Sc*TyrRS selected not to acylate *Ec*tRNA^{Pro}	148
E. coli	O-*Sc*tRNA^{Asp}(CUA)	D188K mutant of yeast AspRS	AspRS(D188K) has very weak activity	153
yeast	O-*Hs*tRNA^{fMet}(CUA), mutant of human initiator tRNA	*E. coli* GlnRS	First pair for use in eukaryotic cells	148

O-tRNA^{Gln}(CUA) only. It is also of interest that this enzyme, with 55-fold greater activity toward the O-tRNA^{Gln}(CUA), also exhibited 28-fold less activity toward wild-type tRNA^{Gln}, even though the selection only demanded higher activity toward acylation of the orthogonal suppressor tRNA substrate.[134] Analysis of mutations in GlnRS mutants with desired activities revealed changes both near the sites that make contact with the knob mutations in the tRNA and, also changes throughout the protein, often far from the tRNA binding site. Despite the remarkable change in activity, this mutant GlnRS was still not ideal because it acylated the wild-type tRNA^{Gln} about as well as the orthogonal substrate. This would presumably cause toxicity via misincorporation of an unnatural amino acid throughout the *E. coli* proteome if this enzyme could be mutated to deliver such a substrate.[134,137]

Recently, Schimmel et al showed that *E. coli* GlnRS (*Ec*GlnRS) does not acylate *Saccharomyces cerevisiae* tRNA^{Gln} (*Sc*tRNA^{Gln}) due to the lack of an N-terminal RNA-binding domain that *S. cerevisiae* GlnRS (*Sc*GlnRS) possesses.[138,139] Liu and Schultz showed that the *amber* suppressing derivative of *Sc*tRNA^{Gln} (O-*Sc*tRNA^{Gln}(CUA)) and *Sc*GlnRS constitute an orthogonal tRNA/synthetase pair in *E. coli*.[140] The O-*Sc*tRNA^{Gln}(CUA) was neither acylated by purified *Ec*GlnRS, nor did it mediate suppression of an *amber* mutation in vitro. However, when chemically acylated with valine, O-*Sc*tRNA^{Gln}(CUA) caused efficient *amber* suppression, indicating that it is orthogonal to *E. coli* aaRSs and translationally competent. *Sc*GlnRS does not appreciably acylate *E. coli* tRNAs in vitro, but it does acylate O-*Sc*tRNA^{Gln}(CUA). This pair was also characterized in vivo by co-transforming *E. coli* with a plasmid driving the transcription of O-*Sc*tRNA^{Gln}(CUA) and a compatible plasmid with an *amber* mutant of the gene for β-lactamase (*amp*) and the gene for *Sc*GlnRS or a mutant thereof. This *amber* mutation occurs at a permissive site (Ala184), so that insertion of virtually any amino acid confers resistance to ampicillin.[141] With an inactive mutant of *Sc*GlnRS, these cells exhibited an IC_{50} of about 20 μg ml^{-1} ampicillin, indicating virtually no acylation by endogenous synthetases. With an active *Sc*GlnRS, cells exhibit an IC_{50} of about 500 μg ml^{-1} ampicillin, indicating that the *Sc*GlnRS acylates the O-*Sc*tRNA^{Gln}(CUA) in *E. coli.*[140]

Ohno et al found conditions under which *S. cerevisiae* TyrRS could be expressed in the presence of the *amber* suppressing derivative of *S. cerevisiae* tRNA^{Tyr} in *E. coli* resulting in tyrosylation of the tRNA that was absent without expression of *Sc*TyrRS.[142] However, using the very sensitive β-lactamase *amber* suppression system, Wang et al discovered that this *amber* suppressing *Sc*tRNA^{Tyr} was in fact acylated in *E. coli* (but presumably not by *Ec*TyrRS; see below). Inspired by this and the work of Schimmel's group, the *amber* suppressing derivative of *Methanococcus jannaschii* tRNA^{Tyr} and *M. jannaschii* TyrRS were shown to be an orthogonal pair in *E. coli.*[143] *Mj*TyrRS recognizes a C1:G72 pair in the tRNA^{Tyr} acceptor stem, while *Ec*TyrRS strongly favors G1:C72. In fact, *Mj*TyrRS was shown to acylate crude tRNA from yeast (whose tRNA^{Tyr} has C1:G72) but not crude *E. coli* tRNA in vitro.[144,145] Using the ampicillin resistance test for *amber* suppression, O-*Mj*tRNA^{Tyr}(CUA) expression alone confers an IC_{50} of about 55 μg ml^{-1}, but co-expression with the *Mj*TyrRS confers resistance to an IC_{50} of about 1,200 μg ml^{-1}. This indicates both that the O-*Mj*tRNA^{Tyr}(CUA) is slightly less orthogonal to endogenous *E. coli* synthetases than O-*Sc*tRNA^{Gln}(CUA) but that the *Mj*TyrRS is more active than the *Sc*GlnRS under the expression conditions examined. The additional fact that the TyrRS active site accommodates a relatively large, hydrophobic amino acid makes it a suitable starting point for attempts to acylate with interesting unnatural hydrophobic amino acids such as fluorophores or affinity labels.

Wang and colleagues set out to improve the orthogonality of this *Mj*tRNA^{Tyr}(CUA) with a selection strategy. Here, eleven nucleotides were identified in the tRNA that were thought not to interact directly with the *Mj*TyrRS. These nucleotides were randomized and the resulting library was first passed through a negative selection, wherein acylation of the *amber* suppressor tRNA resulted in translation of a toxic gene product, barnase. This step removed tRNA variants that can be acylated by endogenous *E. coli* tRNAs. The products of this selection were then passed through a positive selection step in the presence of the *Mj*TyrRS; here, survival of the cells, grown in the presence of antibiotic, required that the *Mj*TyrRS acylate the tRNA variant to support translation of an antibiotic resistance gene. (This is a variation of the general, double sieve selection introduced by Liu et al[140]; see III.D. below for discussion.) The resulting O-*Mj*tRNA^{Tyr}(CUA)* supported survival on ampicillin in the β-lactamase suppression assay at an IC_{50} of only 12.4 μg mL^{-1}, making it about four-fold more orthogonal than the unmodified *Mj*tRNA^{Tyr}(CUA). Nevertheless, the modified tRNA was still acylated sufficiently by *Mj*TyrRS to support survival at an IC_{50} of 436 μg mL^{-1} ampicillin, down about three-fold from the unmodified suppressor.[146,147]

RajBhandhary and coworkers also found that *Sc*TyrRS aminoacylated some *E. coli* tRNA, and went on to find that the substrate was *E. coli* tRNA^{Pro}. Since this misacylation is lethal to *E. coli*, this was used as the basis for a negative selection for a mutant *Sc*TyrRS incapable of acylating *Ec*tRNA^{Pro}. It was also shown that an *amber* suppressing mutant of *E. coli* initiator tRNA^{fMet} was not acylated in *E. coli* but was acylated by *Sc*TyrRS due to the C1:G72 recognition element in this O-*Ec*tRNA^{fMet}(CUA). Thus, by co-expressing this tRNA and a library of mutant *Sc*TyrRSs in an *E. coli* strain with an *amber* mutation in the gene for chloramphenicol acetyltransferase (CAT), survival on chloramphenicol demanded a synthetase both capable of acylating the O-*Ec*tRNA^{fMet}(CUA) and incapable of acylating *Ec*tRNA^{Pro}. Some such mutant *Sc*TyrRSs were isolated, one with a specificity factor for O-*Ec*tRNA^{fMet}(CUA) 15-fold greater than that for *Ec*tRNA^{Pro}. It was also demonstrated that *E. coli* GlnRS and an *amber* suppressing derivative of human initiator tRNA^{fMet} constitute an orthogonal pair in yeast cells, the first such pair demonstrated in cells other than *E. coli.*[148] Interestingly, RajBhandary and coworkers had previously demonstrated that a variant of the *E. coli* tRNA^{Gln}(CUA) is expressed and orthogonal in COS-1 and CV-1 cells, but that co-expression of *E. coli* GlnRS in these cells results in *amber* suppression.[149] This may constitute an orthogonal pair in mammalian cells, although it is not known to what degree *Ec*GlnRS acylates mammalian tRNAs. Although protein engineering using library approaches in eukaryotic cells is difficult due to poor transformation efficiencies, it is possible that an unnatural amino acid-inserting active site from a bacterial selection system could be "transplanted" into a synthetase to be used in mammalian cells due to the high homology of synthetase active sites between species.

Due to the fact that the anticodon of $tRNA^{Asp}$ is a critical recognition element of AspRS and that yeast $tRNA^{Asp}$ is not acylated by *E. coli* synthetases, Pastrnak et al speculated that an *amber* suppressing derivative of *Sc*$tRNA^{Asp}$ might be orthogonal in *E. coli*.[35,150,151] Moreover, it was known that the Asp93→Lys mutant of *E. coli* AspRS was able to acylate the *amber* suppressing derivative of *Ec*$tRNA^{Asp}$.[152] With the related Asp188→Lys mutation in *Sc*AspRS, this enzyme and O-*Sc*$tRNA^{Asp}$(CUA) constituted an orthogonal pair in *E. coli*, albeit with weak activity. (In fact, the activity is similar to the amount of background acylation of the O-*Mj*$tRNA^{Tyr}$(CUA) by endogenous *E. coli* synthetases.) However, to make this pair useful, expression levels of the synthetase and tRNA were increased and an RF-1 deficient strain of *E. coli* was employed, substantially increasing the amount of *amber*-suppression mediated ampicillin or chloramphenicol resistance upon co-expression of the O-*Sc*AspRS with the O-*Sc*$tRNA^{Asp}$(CUA) (554 $\mu g\ mL^{-1}$ ampicillin) versus the tRNA alone (135 $\mu g\ mL^{-1}$ ampicillin).[153]

It is interesting to speculate on the limits of orthogonality for tRNAs and aaRSs. For example, it is known that even weak misacylation of some tRNAs (as with yeast TyrRS and *E. coli* $tRNA^{Pro}$ above and engineered GlxRSs with $tRNA^{Gln}$, next section) causes toxicity to host cells. Therefore, when alterations are made to synthetases to change amino acid specificity (see next section), there is already a negative selection against broadened tRNA specificity. However, it is not clear that site-specific delivery of unnatural amino acids actually requires a tRNA that cannot be acylated by any *E. coli* aaRSs, so long as the orthogonal synthetase acylates the suppressor tRNA more rapidly than any of the endogenous synthetases. Yarus pointed out some time ago that tRNAs from *E. coli* can be misacylated in vitro, but he demonstrated mathematically that competition for tRNA substrates among the twenty aaRSs adds to specificity.[154] By altering the in vivo abundance of tRNAs, Söll and coworkers showed that competition is indeed a mechanism of regulating aaRS specificity in cells.[155] Conversely, impaired GlnRS mutants lose some tRNA specificity as a result of lowered affinity for $tRNA^{Gln}$ rather than because of greater affinity toward other tRNAs.[156] Due to the fact that negative selections (see next section) require that there be virtually no charging of the orthogonal tRNA by endogenous synthetases, it is likely that these tRNAs are *more orthogonal* to *E. coli* aaRSs than *E. coli* tRNAs.

Making Mutations to Alter Amino Acid Specificity of aaRSs

There have been limited attempts to alter the amino acid specificity of aminoacyl-tRNA synthetases. This has proven to be a more difficult problem than tRNA specificity, probably because of the very large interaction surface with the tRNA (25 kD) in contrast to that of the amino acid (120 Daltons). One simple modification that has been useful is the Ala294→Gly mutant of *E. coli* PheRS, which allows insertion of *p*-Cl-Phe and *p*-Br-Phe in addition to *p*-F-Phe and Phe, both of which are accepted by the wild-type PheRS. The mutation was discovered as a result of the knowledge that Ala294→Ser conferred resistance to *p*-F-Phe, suggesting that this residue bounds the amino acid binding cavity at the *para* ring position.[122,157] Recently, Ibba and colleagues have shown that *p*-azidophenylalanine and benzofuranylalanine (Bfa) act as inhibitors of phenylalanylation by G294PheRS, and moreover that only the L enantiomorph of Bfa is effective (see Fig. 5 for structures). This finding suggests that these two photoreactive amino acids are also substrates for G294PheRS.[158]

Two groups have demonstrated that mutants of GlnRS can be made to weakly accept glutamate. Mirande and coworkers analyzed the consensus sequences for the GlxRSs from different organisms to identify two mutations, Cys456→Arg and Gln481→Ile or Ala, to make the human GlnRS functionally more GluRS-like. (See Fig. 6 for the structure of GlnRS and its active site.) The GlnRS(C456R/Q481I) mutant displayed a specificity constant (k_{cat}/K_M) for Gln only 1.8-fold that of Glu (compared to 105,000-fold for wild-type GlnRS).[159] Hong et al

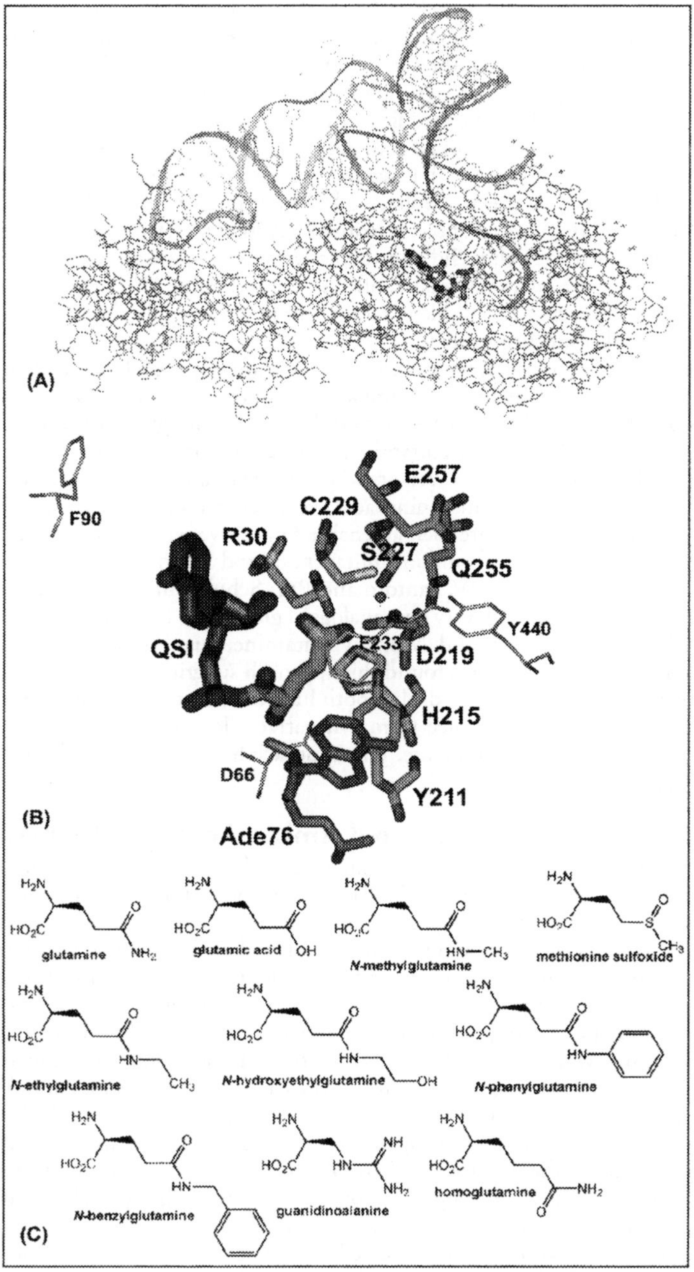

Figure 6. Altering the amino acid specificity of glutaminyl-tRNA synthetase. A) The structure of *E. coli* GlnRS co-crystallized with $tRNA^{Gln}$ (light ribbon) and an analog of Gln-AMP (marked by arrow). B) The active site of GlnRS. The ligand Gln-AMP analog (QSI) is in the center. Asp66, Phe90, Tyr211, Cys 229, Phe233, Gln255 and Tyr 440 have been shown to affect amino acid specificity. Arg30, His215, Asp219, Cys229, Gln255, Glu 257 and Ser227 (in the context of Tyr211 or Phe211) are randomized in libraries for altering GlnRS specificity toward the amino acids in (C). Ade76 of the tRNA and water 1050, which hydrogen bonds to the ligand glutamine, are shown for reference. A and B were generated using InsightII from the PDB file 1qtq.

randomly mutagenized a portion of *E. coli* GlnRS and found that mutations at Phe90 and Tyr240 improved recognition of glutamate. Neither of these amino acids make direct contacts to the substrate glutamine, but rather are nearby the active site.[160] Other experiments have shown that amino acid specificity is dependent upon tRNA recognition; for example, $tRNA^{Gln}$ mutants with changes at U35 (the middle of the anticodon) cause approximately 15-fold increases in the K_M for glutamine (and 20-fold decreases in k_{cat}). For this to occur, recognition of elements far from the active site of the enzyme (which binds the acceptor end of the tRNA) must be transduced back to the active site.[161,162] Mutation of Asp66, which binds to the substrate glutamine's α-NH_3^+ group and is proximal to A76 of the tRNA, and of Tyr211 and Phe 233, which form a hydrophobic "lid" on the active site by stacking with A76 of the tRNA, resulted in enzymes with markedly poorer affinity for glutamine.[163] These studies suggest that altering amino acid specificity will likely require more than simple alteration of residues that bind the natural substrate amino acid.

Due to this fact, Schultz et al have taken a semi-rational approach to the design of libraries for altering amino acid specificity. For example, one library of *Sc*GlnRS variants was created by randomization of seven residues that either bind the substrate glutamine, bind a water molecule that is hydrogen bonded to glutamine, or directly position these residues (Fig. 6).[132] Since virtually all of these *Sc*GlnRS variants will differ dramatically from wild-type at multiple residues proximal to the substrate amino acid, very few are even weakly active (about 1 in 5,000). These types of libraries are then amenable to positive selection or screening (see below III.E.), followed by iteration of random mutagenesis and recombination by DNA shuffling and further selection (T.J.M., S.W. Santoro and P.G. Schultz, unpublished results). These libraries were designed for use with very near analogs of glutamine, including *N*-methyl-, *N*-ethyl-, *N*-hydroxyethylene-, *N*-benzyl- and *N*-phenylglutamine, guanidinoalanine, homoglutamine and methionine sulfoxide. We have found this approach superior to relying purely on random mutagenesis, which requires a great deal of both luck (i.e., that suitable enzymes are common in sequence space) and performance from a negative selection to remove the many active, near-variants of the wild-type synthetase. This approach has recently been applied to *Mj*TyrRS, as well (see III.F. below).

Selections and Screens for Altered Amino Acid Specificity of aaRSs

Developing a selection for the insertion of an unnatural amino acid requires one to devise a way to tie the survival of a cell to something it fundamentally does not need (i.e., is "unnatural"). A way around this conundrum is the use of a general, double-sieve scheme that in two steps demands both active variants of the aaRS and rejection of natural amino acids as substrates[140] (Fig. 7). One formulation of this method involves, first, selection from a pool of variants of an orthogonal aaRS in *E. coli* bearing an orthogonal *amber* suppressor tRNA and an antibiotic resistance gene with an *amber* mutation corresponding to a permissive position in the enzyme. In the presence of unnatural amino acids and antibiotic, survivors of the selection must contain synthetases capable of acylating the orthogonal tRNA(CUA) with some amino acid, natural or unnatural. The selected aaRSs are then transformed into a second strain of *E. coli* with the O-tRNA(CUA) and the gene for a toxic protein bearing an *amber* mutation at a permissive site. Survivors of this selection, grown in the *absence* of unnatural amino acids, must therefore contain an active synthetase that is capable of rejecting all endogenous amino acids—overall, an enzyme capable of uniquely acylating with an unnatural amino acid.[140]

Our original formulation of this involved the yeast glutamine orthogonal pair, positive selection with ampicillin in the context of β-lactamase bearing an Ala184→*amber* mutation, and negative selection with the toxic enzyme barnase bearing two or three *amber* mutations (Gln2, Asp44 and, optionally, Gly65). For the positive selection step, enrichment factors as high as 200,000 could be achieved for cells containing wild-type *Sc*GlnRS in a high dilution of cells containing an inactive form of the synthetase, *Sc*GlnRSΔ500 (1:10^7), using 500 μg mL^{-1} ampicillin. For the negative selection step, enrichment factors as high as 3 million for the three-*amber* barnase and 3 x 10^7 for the two-*amber* barnase could be achieved for selection of inactive *Sc*GlnRSΔ500 diluted 1:10^7 into *Sc*GlnRS.[140]

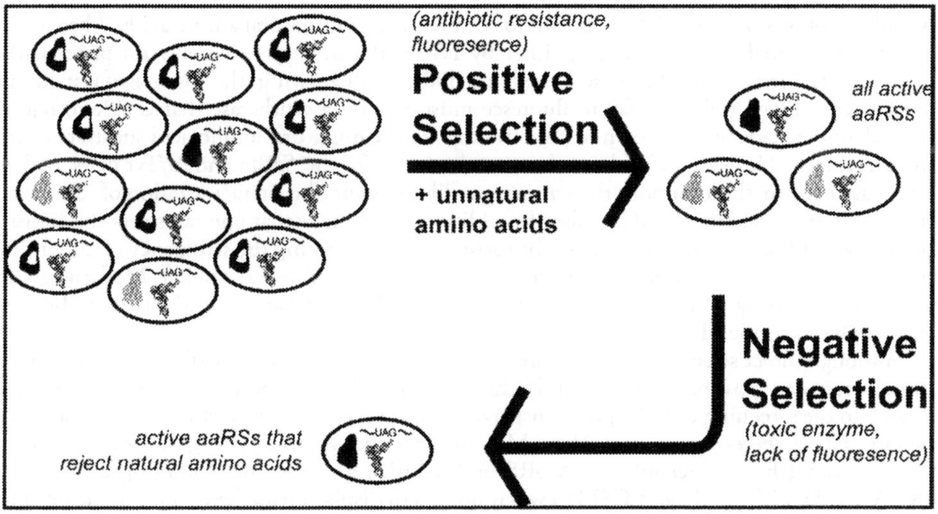

Figure 7. A general, double-sieve scheme for selecting aminoacyl-tRNA synthetases with unnatural amino acid specificity. A library of orthogonal aaRS variants is co-transformed into bacteria with an orthogonal amber-suppressing tRNA. First, a selection or screen for aaRS activity in the presence of unnatural amino acids is applied, where amber suppression confers survival or some marker that can be scored. Cells with aaRSs that acylate with natural or unnatural amino acids survive, but those with very weak activity are not amplified. Second, a selection or screen against aaRS activity in the absence of amino acids is applied, where undesired amber suppression leads to expression of a toxic gene product or a marker that can be scored. Survivors of both selections must, overall, contain synthetases capable of acylating the orthogonal tRNA but also capable of rejecting natural amino acids as substrates.

Among the factors that allowed for substantially higher enrichment from the negative selection than from the positive, ampicillin-based selection is the fact that β-lactamase acts from the periplasm to hydrolyze the ampicillin in the media. While this clearly confers a growth advantage upon the cells that possess active β-lactamase, it also allows rescue in *trans*, since nearby cells are protected (on plates) and all cells are rescued once a sufficient amount of ampicillin has been hydrolyzed. A second problem associated with ampicillin selection is that ampicillin is bacteriocidal, and it is difficult to know what dosage is applicable when one is selecting for potentially weak synthetases from a pool of virtually inactive synthetases. It is even possible to prevent the growth of cells bearing the wild-type *Sc*GlnRS if they are sufficiently dilute and the ampicillin concentration is sufficiently high. For an improved positive selection, an *amber* mutant of chloramphenicol acetyltransferase (CAT) was generated at a site known to be fairly permissive (Ser27).[164,165] In contrast to ampicillin, chloramphenicol inhibits growth instead of killing cells, and therefore the amplification of cells is simply tied to the amount of *amber* suppression in the given cells, without loss of cells containing weakly active synthetases. Chloramphenicol acts at the ribosome, and CAT is expressed cytosolically, reducing the action in *trans* of this mechanism of resistance. Moreover, increasing the concentration of chloramphenicol increases the selective advantage for cells bearing wild-type synthetase versus cells containing an inactive synthetase, so a nearly arbitrary concentration of chloramphenicol can be used for selection from a library (T.J.M. and P.G. Schultz, unpublished results). Other sites in CAT have been tested for permissivity, including Thr10 and Asp112.[153]

A variation on the general, double sieve selection has been introduced that uses fluorescence-activated cell sorting (FACS) and a variant of green fluorescent protein (GFP) as a reporter. In a first step, cells containing the gene for T7 RNA polymerase with multiple *amber* mutations, GFP under the control of the T7 promoter, orthogonal tRNA(CUA) and a library of variants of orthogonal aaRS are grown in the presence of unnatural amino acids. These cells are then examined for fluorescence, either by FACS, fluorimetry or visually on plates with long-wave UV irradiation. Fluorescent cells are diluted and grown in the absence of unnatural amino acids. Here, cells that fail to fluoresce must contain a synthetase that is able to reject natural amino acids but is known to be active toward the unnatural amino acid substrate from the first screen. This system has been shown to be useful for the *Sc*Gln and *Mj*Tyr pairs, and is being used as a method for both the screening of libraries and the characterization of selectants from antibiotic selections. FACS offers the additional advantage that one can select the appropriate level of fluorescence as a cut-off for sorting, thereby altering the stringency of either the negative or positive step (S.W. Santoro and P.G. Schultz, unpublished results). Modern FACS is capable of sorting a billion bacterial cells a day, which is comparable to the largest libraries that can be conveniently generated in *E. coli*.

Although direct selections for unnatural amino acid insertion are very difficult to conceive, direct screens are possible. Pastrnak and Schultz developed an antibody recognition-based approach to the screening of M13 phage displaying a surface epitope into which amino acids can be inserted via *amber* suppression. Here, M13 phage harbor genes directing the expression of a variant aaRS (the engineered O-*Sc*AspRS or *Sc*GlnRS) and the corresponding orthogonal tRNA(CUA). Helper-phage VCSM13 was modified to display an *amber* mutant of the immunogenic C3 epitope derived from poliovirus; also, since the *amber* mutation corresponds to a residue near the N-terminus of the C3-pIII fusion, phage production requires *amber* suppression. Therefore, this is both a selection for *amber* suppression (i.e., active aaRS) and a means of displaying an amino acid on the surface of M13 phage for screening. Antibodies were elicited from synthetic peptides of the C3 epitope containing the amino acid of choice at the position corresponding to the *amber* mutation in the C3-pIII fusion. To use this system with the relatively weak O-*Sc*AspRS, an RF-1 deficient version of the excellent cloning strain DH10B was constructed. In a model selection in which M13 phage with C3-pIII fusions with an Asp residue in the C3 epitope were screened from a large excess of C3-pIII-containing phage with Asn at the same residue, 900-fold enrichments were possible with a monoclonal antibody elicited against C3(Asp) peptide.[166] Since antibodies against C3 epitopes with natural or unnatural amino acids can be generated, both positive and negative screening can in theory be carried out.

The First "Unnatural" Organisms

Taking advantage of the fact that mutants of PheRS are known that reject *p*-F-Phe, Furter and coworkers were partially successful in engineering a bacterium capable of inserting *p*-F-Phe in a site-selective manner. The G37A mutant of yeast $tRNA^{Phe}$(CUA) was found to be nearly orthogonal to *E. coli* synthetases, although it was a poor substrate of LysRS in *E. coli*. When yeast PheRS was co-expressed with this tRNA in *E. coli*, 95 % of the amino acid inserted into an *amber* mutant of dihydrofolate reductase (DHFR) via O-$tRNA^{Phe}$(CUA) was phenylalanine. Expression of yeast PheRS in a *p*-F-Phe-resistant, Phe-auxotrophic strain of *E. coli*, growth in the presence of high levels of *p*-F-Phe resulted in largely site-specific *p*-F-Phe insertion at *amber* mutations. Under optimal conditions, about 75 % of the *amber*-encoded site in DHFR was occupied by *p*-F-Phe, while 20 % was Phe and 5 % was Lys. This indicates both that the O-$tRNA^{Phe}$(CUA) is being promiscuously acylated by *Ec*LysRS and that the *Sc*PheRS inserts Phe in addition to *p*-F-Phe. Also, when the same site in DHFR was replaced with a Phe codon, 93 % of this site was occupied by Phe, but 7 % was *p*-F-Phe, indicating that the endogenous *Ec*PheRS incorporates a small amount of *p*-F-Phe in addition to Phe. The yield of DHFR was high, however (about 10 mg L^{-1} culture), and since the natural amino acids are silent in ^{19}F NMR, this system may be useful for this application despite its lack of specificity. Ultimately, engineering this pair for better orthogonality and amino acid specificity would be beneficial, as would engineering the *E. coli* synthetase to better reject the unnatural amino acid. The most significant limitation to this approach, however, is that it requires synthetases already known to use an unnatural amino acid as a substrate.[167]

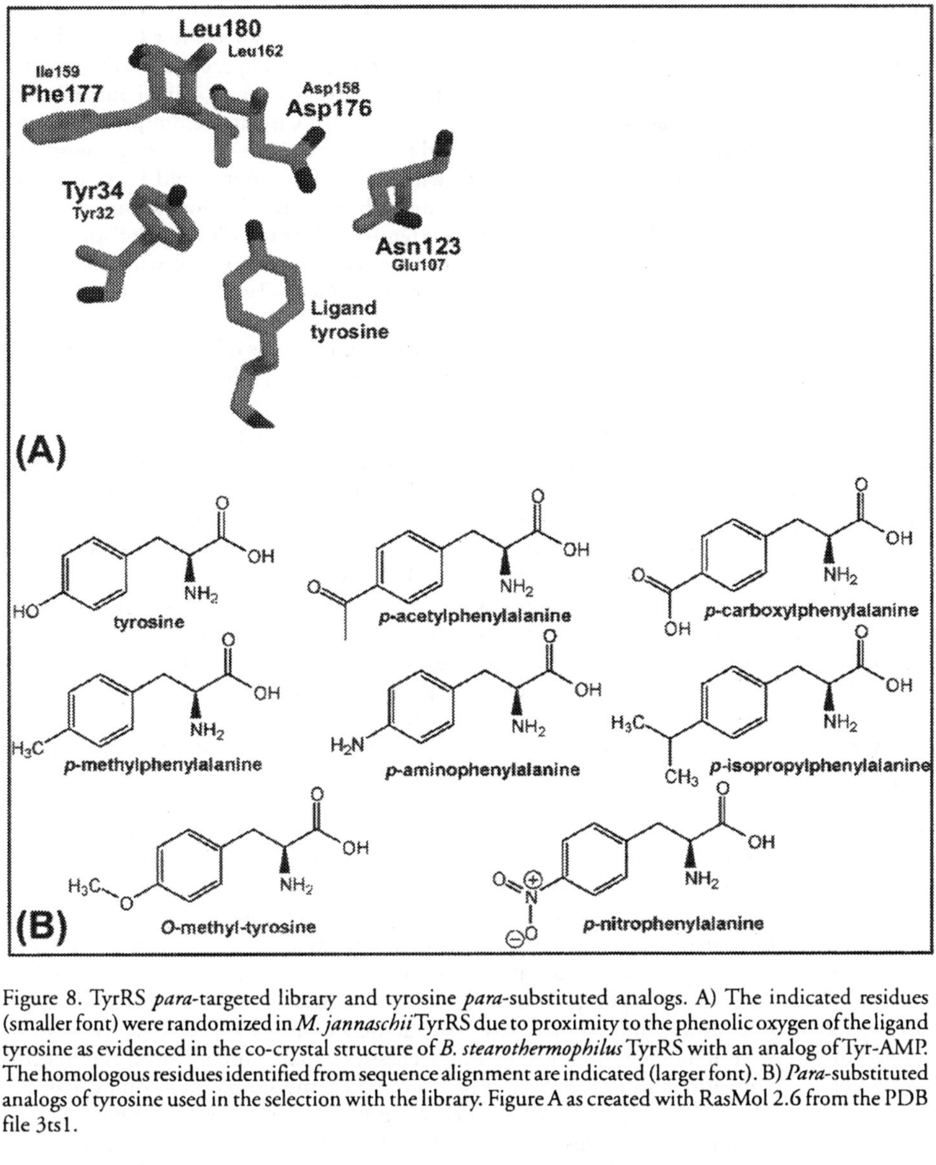

Figure 8. TyrRS *para*-targeted library and tyrosine *para*-substituted analogs. A) The indicated residues (smaller font) were randomized in *M. jannaschii* TyrRS due to proximity to the phenolic oxygen of the ligand tyrosine as evidenced in the co-crystal structure of *B. stearothermophilus* TyrRS with an analog of Tyr-AMP. The homologous residues identified from sequence alignment are indicated (larger font). B) *Para*-substituted analogs of tyrosine used in the selection with the library. Figure A as created with RasMol 2.6 from the PDB file 3ts1.

Recently, Schultz and coworkers applied a general, directed library strategy combined with the improved, double-sieve selection based on chloramphenicol resistance to find a mutant of *Mj*TyrRS capable of acylating the selected O-*Mj*tRNA^{Tyr}(CUA)* specifically with the methyl ether of tyrosine (*O*-Me-Tyr). A small library was constructed by randomization of five of the residues of *Mj*TyrRS proximal to the phenolic oxygen of the ligand tyrosine (Fig. 8). This library was subjected to positive, chloramphenicol-based selection in the presence of *para*-substituted analogs of tyrosine (Fig. 8) and negative, barnase-based selection in the absence of the unnatural amino acids. Selectants were pooled, subjected to DNA shuffling and reselected twice. However, the final selectant with activity toward *O*-Me-Tyr had mutations only at the originally-randomized positions from the first library, suggesting that the shuffling steps were unnecessary here. The relative importance of the four amino acid changes in the protein (Tyr32→Gln, Glu107→Thr, Asp158→Ala, Leu162→Pro) is not yet known. However, the specificity of this mutant *Mj*OMeTyrRS is remarkable. When the tRNA and synthetase are present with an *amber* mutant of the gene coding for dihydrofolate reductase, no DHFR can be detected unless *O*-Me-Tyr is added to the media. Moreover, using Fourier Transform Ion Cyclotron Resonance Mass Spectrometry, the DHFR isolated from growth in the presence of the unnatural amino acid was confirmed to carry the *O*-Me-Tyr at a single position. Not only is this mutant capable of adenylating with *O*-Me-Tyr about 8-fold faster than tyrosine at saturation (as measured by pyrophosphate exchange), the K_M for tyrosine is about 13-fold higher than for *O*-Me-Tyr.[146] Efforts are currently underway to "transplant" this active site into TyrRSs from other organisms in order to transfect mammalian cells with a heterologous aaRS/tRNA(CUA) pair capable of specifically inserting *O*-Me-Tyr (S.W. Santoro, J. Chin, T.J.M. and P.G. Schultz, unpublished).

Perhaps one of the most important factors that led to the success of this effort is the relatively high activity of the wild-type synthetase; here, the mutant synthetase is still capable of acylating a modified tRNA with an unnatural substrate with a k_{cat} of 0.84 min^{-1}, while the wild-type synthetase acylates its wild-type tRNA with tyrosine with a k_{cat} of 9.0 min^{-1}.[144,146] Recent work by Tirrell and coworkers suggests that, assuming amino acid is present in the cell at a concentration near or above the K_M, a k_{cat} above about 0.1 min^{-1} is required for the synthetase to support moderate to high levels of protein synthesis in *E. coli*.[121] In contrast, *Ec*GlnRS has a k_{cat} of about 156 min^{-1}, and *Sc*GlnRS has a k_{cat} of only about 12 min^{-1} (for wild-type substrates).[128,133] While this seems comparable to *Mj*TyrRS at first, it is important that the GlnRSs recognize their *amber* suppressing tRNAs much more poorly than wild-type tRNAs (1,700-fold for *E. coli*), while *Mj*TyrRS is less sensitive to changes in the anticodon (100-fold for G34C).[168] It is quite possible that for orthogonal pairs in the activity range of *Sc*GlnRS, one will have to find a mutant capable of inserting an unnatural amino acid at a rate comparable to the rate of insertion of wild-type substrate (here, glutamine) by wild-type synthetase. Further success of this strategy may depend upon the improvement of existing orthogonal pairs ($k_{cat}/K_{M(tRNA)}$) or the development of new, more active pairs.

Conclusion

State-of-the-art methods for protein synthesis, semi-synthesis with native chemical ligation and expressed protein ligation, and in vitro biosynthesis provide powerful ways of studying protein structure and function by introducing amino acids not represented by the standard genetic code into proteins of interest. Recent successes in engineering new genetic codes to introduce site-specifically unnatural amino acids into proteins in living cells expands even further our ability to control the composition of proteins. The general methods discussed above for expanding and editing the standard genetic code may eventually lead to a new field of study: "unnatural" cell biology. Site-directed mutagenesis, introduction of plasmids into cells, and addition of unnatural amino acids to media are sufficiently simple procedures that any biochemist or cell biologist will have access to cells expressing proteins with amino acids not found in the natural repertoire. This may allow the direct observation of the effects of subtle perturbations to protein structure on cellular function, the insertion of biophysical probes including fluorophores or affinity labels into proteins of interest, or the insertion of photoactivated switches to turn protein function on or off in a time-resolved manner.

The challenges that lie ahead in this field are daunting, but significant inroads have been made to suggest that they are surmountable. Application of selections for novel unnatural amino acid specificity toward side chains significantly different from the natural set will be needed to make useful tools for cell biology. As these new aaRSs are generated, additional novel ways of inserting these unnatural amino acids site-specifically will have to be demonstrated, with, for example, four-base codons or unnatural codons. Finally, transitioning these systems into eukaryotes like yeast and mammalian cells will make unnatural cell biology possible in a milieu in which human disease and drug targets can be studied in more relevant context.

Note Added in Proof

A number of unnatural amino acid analogs of tyrosine have now been successfully inserted in vivo into both *E. coli* and yeast proteins.[169-175]

Acknowledgements

The authors thank Dr. Stephen W. Santoro, J. Christopher Anderson, Dr. Jason Chin and Professor Peter G. Schultz (The Scripps Research Institute, La Jolla, CA) for unpublished results cited in this manuscript. Additionally, we thank Dr. Santoro and Dr. Stuart Licht (SUNY Buffalo) for critical reading of portions of this manuscript.

References

1. Zoller MJ, Smith M. Oligonucleotide-directed mutagenesis of DNA fragments cloned into M13 vectors. Methods Enzymol 1983; 100:468-500.
2. In vitro mutagenesis protocols; Trower MK, ed. Humana Press: Totowa, NJ, 1996; Vol. 57, 408.
3. Knowles JR. Tinkering with enzymes: What are we learning? Science 1987; 236:1252-8.
4. Hermanson GT. Bioconjugate Techniques; Academic Press: 1996, 785.
5. Kaiser ET, Lawrence DS, Rokita SE. The chemical modification of enzymatic specificity. Ann Rev Biochem 1985; 54:565-95.
6. Gloss LM, Kirsch JF. Decreasing the basicity of the active site base, Lys-258, of Escherichia coli aspartate aminotransferase by replacement with gamma-thialysine. Biochemistry 1995; 34:3990-8.
7. Toney MD, Kirsch JF. Direct Brønsted analysis of the restoration of activity to a mutant enzyme by exogenous amines. Science 1989; 243:1485-8.
8. Toney MD, Kirsch JF. Bronsted Analysis of Aspartate Aminotransferase Via Exogenous Catalysis of Reactions of an Inactive Mutant. Protein Science 1992; 1:107-119.
9. Offord RE, Gaertner HF, Wells TN et al. Proudfoot. Synthesis and evaluation of fluorescent chemokines labeled at the amino terminal. Methods Enzymol 1997; 287:348-69.
10. Merrifield B. Solid phase synthesis. Science 1986; 232:341-7.
11. Schneider J, Kent SB. Enzymatic activity of a synthetic 99 residue protein corresponding to the putative HIV-1 protease. Cell 1988; 54:363-8.
12. Dawson PE, Kent SBH. Synthesis of native proteins by chemical ligation. Ann Rev Biochem 2000; 69:923-960.
13. Kent SB. Chemical synthesis of peptides and proteins. Ann Rev Biochem 1988; 57:957-89.
14. Humphrey JM, Chamberlin AR. Chemical synthesis of natural product peptides: Coupling methods for the incorporation of noncoded amino acids into peptides. Chemical Reviews 1997; 97:2243-2266.
15. Abrahmsen L, Tom J, Burnier J et al. Engineering Subtilisin and Its Substrates For Efficient Ligation of Peptide Bonds in Aqueous Solution. Biochemistry 1991; 30:4151-4159.
16. Chang TK, Jackson DY, Burnier JP et al. Subtiligase—a Tool For Semisynthesis of Proteins. Proc Natl Acad Sci USA 1994; 91:12544-12548.
17. Jackson DY, Burnier J, Quan C et al. A Designed Peptide Ligase For Total Synthesis of Ribonuclease a With Unnatural Catalytic Residues. Science 1994; 266:243-247.
18. Braisted AC, Judice JK, Wells JA. Synthesis of proteins by subtiligase. Methods Enzymol 1997; 289:298-313.
19. Atwell S and Wells JA. Selection for improved subtiligases by phage display. Proc Natl Acad Sci USA 1999; 96:9497-502.
20. Schnolzer M and Kent SBH. Constructing Proteins By Dovetailing Unprotected Synthetic Peptides—Backbone-Engineered Hiv Protease. Science 1992; 256:221-225.
21. Dawson PE, Muir TW, Clarklewis I et al. Synthesis of Proteins By Native Chemical Ligation. Science 1994; 266:776-779.
22. Canne LE, Bark SJ, Kent SBH. Extending the Applicability of Native Chemical Ligation. Amer Chem Soc 1996; 118:5891-5896.
23. Canne LE, Botti P, Simon RJ et al. Chemical protein synthesis by solid phase ligation of unprotected peptide segments. Amer Chem 1999; 121:8720-8727.
24. Xu MQ, Perler FB. The Mechanism of Protein Splicing and Its Modulation By Mutation. EMBO J 1996; 15:5146-5153.
25. Chong SR, Mersha FB, Comb DG et al. Single-column purification of free recombinant proteins using a self-cleavable affinity tag derived from a protein splicing element. Gene 1997; 192:271-281.
26. Severinov K, Muir TW. Expressed protein ligation, a novel method for studying protein-protein interactions in transcription. Bio Chem 1998; 273:16205-16209.

27. Muir TW, Sondhi D, Cole PA. Expressed protein ligation: A general method for protein engineering. Proc Natl Acad Sci USA 1998; 95:6705-6710.
28. Blaschke UK, Cotton GJ, Muir TW. Synthesis of multi-domain proteins using expressed protein ligation: Strategies for segmental isotopic labeling of internal regions. Tetrahedron 2000; 56:9461-9470.
29. Bain JD, Glabe CG, Dix TA et al. Biosynthetic site-specific incorporation of a non-natural amino acid into a polypeptide. Amer Chem Soc 1989; 111:8013-8014.
30. Noren CJ, Anthony-Cahill SJ, Griffith MC et al. A general method for site-specific incorporation of unnatural amino acids into proteins. Science 1989; 244:182-8.
31. Crick FHC, Barret L, Brenner S et al. General nature of the genetic code for proteins. Nature (London) 1961; 192:1227-1232.
32. Bossi L, Roth JR. The influence of codon context on genetic code translation. Nature (London) 1980; 286:123-127.
33. Ayer D. Yarus M. The context effect does not require a fourth base pair. Science 1986; 231:393-5.
34. Bruce AG, Atkins JF, Wills N et al. Replacement of anticodon loop nucleotides to produce functional tRNAs: amber suppressors derived from yeast tRNAPhe. Proc Natl Acad Sci USA 1982; 79:7127-31.
35. Kwok Y, Wong JT. Evolutionary relationship between Halobacterium cutirubrum and eukaryotes determined by use of aminoacyl-tRNA synthetases as phylogenetic probes. Canadian J Biochem 1980; 58:213-8.
36. Steward LE, Collins CS, Gilmore MA et al. In vitro site-specific incorporation of fluorescent probes into beta-galactosidase. Amer Chem Soc 1997: 119:6-11.
37. Cload ST, Liu DR, Froland WA et al. Development of improved tRNAs for in vitro biosynthesis of proteins containing unnatural amino acids. Chem Biol 1996; 3:1033-1038.
38. Thorson JS, Cornish VW, Barrett JE et al. In Protein Synthesis: Methods and Protocols; R. Martin, Ed.; Humana Press: Totowa, NJ, 1998; 77:43-73.
39. Short GF, Golovine SY, Hecht SM. Effects of release factor 1 on in vitro protein translation and the elaboration of proteins containing unnatural amino acids. Biochemistry 1999; 38:8808-8819.
40. Kim DM, Kigawa F, Choi CY et al. A Highly Efficient Cell-Free Protein Synthesis System From Escherichia Coli. Eur J Biochem 1996; 239:881-886.
41. Crick FHC. On Protein Synthesis. Symposium of the Society for Experimental Biology 1958; 12:138-163.
42. Chapeville F, Lipmann F, von Ehrenstein G et al. On the Role of Soluble Ribonucleic Acid Encoding for Amino Acid. Proc Natl Acad Sci USA 1962; 48:1086-1092.
43. Pezzuto JM, Hecht SM. Amino acid substitutions in protein biosynthesis. Poly(A)-directed polyphenylalanine synthesis. Bio Chem 1980; 255:865-9.
44. Heckler TG, Chang LH, Zama Y et al. T4 RNA ligase mediated preparation of novel "chemically misacylated" tRNAPheS. Biochem 1984; 23:1468-73.
45. Payne CH, Nichols BP, Hecht SM. Escherichia coli tryptophan synthase: synthesis of catalytically competent alpha subunit in a cell-free system containing preacylated tRNAs. Biochemistry 1987; 26:3197-205.
46. Baldini G, Martoglio B, Schachenmann A et al. Mischarging Escherichia coli tRNAPhe with L-4'-[3-(trifluoromethyl)-3H-diazirin-3-yl]phenylalanine, a photoactivatable analogue of phenylalanine. Biochemistry 1988; 27:7951-9.
47. Robertson SA, Noren CJ, Anthonycahill SJ et al. The Use of 5'-Phospho-2 Deoxyribocytidylylriboadenosine As a Facile Route to Chemical Aminoacylation of Transfer Rna. Nucleic Acids Res 1989; 17:9649-9660.
48. Robertson SA, Ellman JA, Schultz PG. A General and Efficient Route For Chemical Aminoacylation of Transfer Rnas. Amer Chem Soc 1991; 113:2722-2729.
49. Cornish VW, Mendel D, Schultz PG. Probing Protein Structure and Function With an Expanded Genetic Code. Angewandte Chemie-International Edition in English 1995; 34:621-633.
50. Hohsaka T, Sato K, Sisido M et al. Site-Specific Incorporation of Photofunctional Nonnatural Amino Acids Into a Polypeptide Through in Vitro Protein Biosynthesis. Febs Letters 1994; 344:171-174.
51. Kanda T, Takai K, Yokoyama S et al. Knocking out a specific tRNA species within unfractionated Escherichia coli tRNA by using antisense (complementary) oligodeoxyribonucleotides. Febs Letters 1998; 440:273-276.
52. Kanda T, Takai K, Yokoyama S et al. Specific inactivation of Escherichia coli tRNA(Phe) by antisense DNA-treatment under Mg2+-deficient conditions. Bioorg Med Chem 2000; 8:675-679.
53. Kanda T, Takai K, Yokoyama S et al. An easy cell-free protein synthesis system dependent on the addition of crude Escherichia coli tRNA. J Biochem 2000; 127:37-41.

54. Kanda T, Takai K, Hohsaka T et al. Sense codon-dependent introduction of unnatural amino acids into multiple sites of a protein. Biochem Biophys Res Commun 2000; 270:1136-1139.
55. Atkins JF, Weiss RB, Thompson S et al. Towards a genetic dissection of the basis of triplet decoding, and its natural subversion: programmed reading frame shifts and hops. Annu Rev Genet 1991; 25:201-28.
56. Ma CH, Kudlicki W, Odom OW et al. In vitro Protein Engineering Using Synthetic Transfer RNA(Ala) With Different Anticodons. Biochem 1993; 32:7939-7945.
57. Kramer G, Kudlicki W, Hardesty B. In Protein Synthesis: Methods and Protocols; R. Martin, Ed.; Humana Press: Totowa, NJ, 1998:105-116.
58. Hohsaka T, Ashizuka Y, Murakami H et al. Incorporation of nonnatural amino acids into streptavidin through in vitro frame-shift suppression. Amer Chem Soc 1996; 118:9778-9779.
59. Hohsaka T, Ashizuka Y, Sasaki H et al. Incorporation of two different nonnatural amino acids independently into a single protein through extension of the genetic code. Amer Chem Soc 1999; 121:12194-12195.
60. Moore B, Persson BC, Nelson CC et al. Quadruplet codons: Implications for code expansion and the specification of translation step size. J Mol Biol 2000; 298:195-209.
61. Moore B, Nelson CC, Persson BC et al. Decoding of tandem quadruplets by adjacent tRNAs with eight-base anticodon loops. Nucleic Acids Res 2000; 28:3615-3624.
62. Magliery TJ, Anderson JC, Schultz PG. Expanding the genetic code: selection of efficient suppressors of four-base codons and indentification of "shifty" four-base codons with a library approach in Escherichia coli. J Mol Biol 2001; 307: 755-769.
63. Anderson JC, Magliery TJ, Schultz PG. Exploring the limits of codon and anticodon size. Chem Biol 2001, submitted.
64. Switzer C, Moroney SE, Benner SA. Enzymatic incorporation of a new base pair into DNA and RNA. Amer Chem Soc 1989; 111:8322-8323.
65. Bain JD, Switzer C, Chamberlin AR et al. Ribosome-mediated incorporation of a non-standard amino acid into a peptide through expansion of the genetic code. Nature 1992; 356:537-539.
66. Piccirilli JA, Krauch T, Moroney SE et al. Enzymatic incorporation of a new base pair into DNA and RNA extends the genetic alphabet. Nature 1990; 343:33-37.
67. Moran S, Ren RXF, Rumney S et al. Difluorotoluene, a nonpolar isostere for thymine, codes specifically and efficiently for adenine in DNA replication. Amer Chem Soc 1997; 119:2056-2057.
68. Matray TJ, Koo ET. Selective and stable DNA base pairing without hydrogen bonds. Amer Chem Soc 1998; 120:6191-6192.
69. Morales JC, Kool ET. Efficient replication between non-hydrogen-bonded nucleoside shape analogs. Nat Struct Biol 1998; 5: 950-954.
70. McMinn DL, Ogawa AK, Wu YQ et al. Efforts toward expansion of the genetic alphabet: DNA polymerase recognition of a highly stable, self-fairing hydrophobic base. Amer Chem Soc 1999; 121:11585-11586.
71. Ogawa AK, Wu YQ, McMinn DL et al. Efforts toward the expansion of the genetic alphabet: Information storage and replication with unnatural hydrophobic base pairs. Amer Chem Soc 2000; 122:3274-3287.
72. Wu YQ, Ogawa AK, Berger M et al. Efforts toward expansion of the genetic alphabet: Optimization of interbase hydrophobic interactions. Amer Chem Soc 2000; 122:7621-7632.
73. Meggers E, Holland PL, Tolman WB et al. A novel copper-mediated DNA base pair. Amer Chem Soc 2000; 122:10714-10715.
74. Kool ET. Synthetically modified DNAs as substrates for polymerases. Curr Opin Chem Biol 2000; 4:602-608.
75. Nowak MW, Kearney PC, Sampson JR et al. Nicotinic receptor binding site probed with unnatural amino acid incorporation in intact cells. Science 1995; 268:439-442.
76. Saks ME, Sampson JR, Nowak MW et al. An engineered tetrahymena Trna(Gln) for in vivo Incorporation of unnatural amino acids into proteins by nonsense suppression. Bio Chem 1996; 271:23169-23175.
77. Cotton GJ, Muir TW. Peptide ligation and its application to protein engineering. Chem Biol 1999; 6:R247-R256.
78. Dougherty DA. Unnatural amino acids as probes of protein structure and function. Curr Opin Chem Biol 2000; 4:645-652.
79. Ellman JA, Mendel D, Schultz PG. Site-Specific Incorporation of Novel Backbone Structures Into Proteins. Science 1992; 255:197-200.
80. Mendel D, Ellman J, Schultz PG. Protein Biosynthesis With Conformationally Restricted Amino Acids. Amer Chem Soc 1993; 115:4359-4360.

81. Mendel D, Ellman JA, Chang ZY et al. Probing protein stability with unnatural amino acids. Science 1992; 256:1798-1802.
82. Thorson JS, Chapman E, Schultz PG. Analysis of hydrogen bonding strengths in proteins using unnatural amino acids. Amer Chem Soc 1995; 117:9361-9362.
83. Thorson JS, Chapman E, Murphy EC et al. Linear free energy analysis of hydrogen bonding in proteins. Amer Chem Soc 1995; 117:1157-1158.
84. Ting AY, Shin I, Lucero C et al. Energetic analysis of an engineered cation-pi interaction in staphylococcal nuclease. Amer Chem Soc 1998; 120:7135-7136.
85. Zhong WG, Gallivan JP, Zhang YN et al. Dougherty. From ab initio quantum mechanics to molecular neurobiology: A cation-pi binding site in the nicotinic receptor. Proc Natl Acad Sci USA 1998; 95:12088-12093.
86. Ellman JA, Volkman BF, Mendel D et al. Site-Specific Isotopic Labeling of Proteins For Nmr Studies. Amer Chem Soc 1992; 114:7959-7961.
87. Cornish VW, Benson DR, Altenbach CA et al. Site-Specific Incorporation of Biophysical Probes Into Proteins. Proc Natl Acad Sci USA 1994; 91:2910-2914.
88. Cornish VW, Hahn KM, Schultz PG. Site-Specific Protein Modification Using a Ketone Handle. Amer Chem Soc 1996; 118:8150-8151.
89. Gallivan pg, Lester HA, Dougherty DA. Site-specific incorporation of biotinylated amino acids to identify surface-exposed residues in integral membrane proteins. Chem Biol 1997; 4:739-749.
90. Murakami H, Hohsaka T, Ashizuka T et al. Site-directed incorporation of p-nitrophenylalanine into streptavidin and site-to-site photoinduced electron transfer from a pyrenyl group to a nitrophenyl group on the protein framework. Amer Chem Soc 1998; 120:7520-7529.
91. Cotton GJ, Ayers B, Xu R et al. Insertion of a synthetic peptide into a recombinant protein framework: A protein biosensor. Amer Chem Soc 1999; 121:1100-1101.
92. Cotton GJ, Muir TW. Generation of a dual-labeled fluorescence biosensor for Crk-II phosphorylation using solid-phase expressed protein ligation. Chem Biol 2000; 7:253-261.
93. Deniz AA, Laurence TA, Beligere GS et al. Single-molecule protein folding: Diffusion fluorescence resonance energy transfer studies of the denaturation of chymotrypsin inhibitor 2. Proc Natl Acad Sci USA 2000; 97:5179-5184.
94. Mendel D, Ellman JA Schultz PG. Construction of a light-activated protein by unnatural amino acid mutagenesis. Amer Chem Soc 1991; 113:2758-2760.
95. Miller JC, Silverman SK, England PM et al. Flash decaging of tyrosine sidechains in an ion channel. Neuron 1998; 20:619-624.
96. England PM, Lester HA, Davidson N et al. Site-specific, photochemical proteolysis applied to ion channels in vivo. Proc Natl Acad Sci USA 1997; 94:11025-11030.
97. Short GF, Lodder M, Laikhter AL et al. Caged HIV-1 protease: Dimerization is independent of the ionization state of the active site aspartates. Amer Chem Soc 1999; 121:478-479.
98. Pollitt SK, Schultz PG. A photochemical switch for controlling protein-protein interactions. Angewandte Chemie-International Edition 1998, 37, 2104-2107.
99. Budisa N, Minks C, Alefelder S et al. Toward the experimental codon reassignment in vivo: Protein building with an expanded amino acid repertoire. Faseb Journal 1999; 13:41-51.
100. Browne DR, Kenyon GL, Hegeman GD. Incorporation of monoflurotryptophans into protein during the growth of Escherichia coli. Biochem Biophys Res Commun 1970; 39:13-9.
101. Kim HW, Perez JA, Ferguson SJ et al. The Specific Incorporation of Labelled Aromatic Amino Acids Into Proteins Through Growth of Bacteria in the Presence of Glyphosate—Application to Fluorotryptophan Labelling to the H+-Atpase of Escherichia-Coli and Nmr Studies. Febs Letters 1990; 272:34-36.
102. Parsons JF, Xiao GY, Gilliland GL et al. Enzymes harboring unnatural amino acids: Mechanistic and structural analysis of the enhanced catalytic activity of a glutathione transferase containing 5-fluorotryptophan. Biochem 1998; 37:6286-6294.
103. Luck LA, Falke JJ. Open Conformation of a Substrate-Binding Cleft—F-19 Nmr Studies of Cleft Angle in the D-Galactose Chemosensory Receptor. Biochem 1991; 30:6484-6490.
104. Luck LA, Falke JJ. F-19 Nmr Studies of the D-Galactose Chemosensory Receptor .1. Sugar Binding Yields a Global Structural Change. Biochem 1991; 30:4248-4256.
105. Luck LA, Falke JJ. F-19 Nmr Studies of the D-Galactose Chemosensory Receptor .2. Ca(Ii) Binding Yields a Local Structural Change. Biochem 1991; 30:4257-4261.
106. Luck LA, Vance JE, Oconnell TM et al. F-19 Nmr Relaxation Studies On 5-Fluorotryptophan- and Tetradeutero-5-Fluorotryptophan-Labeled E-Coli Glucose/Galactose Receptor. J Biomolecular Nmr 1996;7:261-272.
107. Minks C, Huber R, Moroder L et al. Atomic mutations at the single tryptophan residue of human recombinant annexin V: Effects on structure, stability, and activity. Biochem 1999; 38:10649-10659.

108. Geisow M. New attempts to solve protein structures lead to madness. Trends Biotechnol 1991; 9:4-5.
109. Hendrickson WA, Horton JR, Lemaster DM. Selenomethionyl proteins produced for analysis by multiwavelength anomalous diffraction (Mad)—a vehicle for direct determination of 3-dimensional structure. EMBOJ 1990; 9:1665-1672.
110. Yang W, Hendrickson WA, Crouch RJ et al. Structure of ribonuclease-H phased at 2-a resolution by Mad analysis of the selenomethionyl protein. Science 1990; 249:1398-1405.
111. Budisa N, Steipe B, Demange P et al. High-level biosynthetic substitution of methionine in proteins by its analogs 2-aminohexanoic acid, selenomethionine, telluromethionine and ethionine in Escherichia Coli. Eur J Biochem 1995; 230:788-796.
112. Jakubowski H. Accelerated publication—Translational incorporation of S-nitrosohomocysteine into protein. Bio Chem 2000; 275:21813-21816.
113. Budisa H, Huber R, Golbik R et al. Moroder. Atomic mutations in annexin V—Thermodynamic studies of isomorphous protein variants. Eur J Biochem 1998; 253:1-9.
114. Budisa N, Minks C, Medrano FJ et al. Residue-specific bioincorporation of non-natural, biologically active amino acids into proteins as possible drug carriers: Structure and stability of the per-thiaproline mutant of annexin V. Proc Natl Acad Sci USA 1998; 95:455-459.
115. McGrath KP, Fournier MJ, Mason TL et al. Genetically directed syntheses of new polymeric materials—expression of artificial genes encoding proteins with repeating (Alagly)3proglugly elements. Amer Chem Soc 1992; 114:727-733.
116. Kothakota S, Mason TL, Tirrell DA et al. Biosynthesis of a periodic protein containing 3-Thienylalanine—a step toward genetically engineered conducting polymers. Amer Chem Soc 1995; 117:536-537.
117. Yoshikawa E, Fournier MJ, Mason TL et al. Genetically engineered fluoropolymers—synthesis of repetitive polypeptides containing P-fluorophenylalalanine residues. Macromolecules 1994; 27:5471-5475.
118. vanHest JCM, Tirrell DA. Efficient introduction of alkene functionality into proteins in vivo. Febs Letters 1998; 428:68-70.
119. van Hest JCM, Kiick KL, Tirrell DA. Efficient incorporation of unsaturated methionine analogues into proteins in vivo. Amer Chem Soc 2000; 122:1282-1288.
120. Sharma N, Furter R, Kast P et al. Efficient introduction of aryl bromide functionality into proteins in vivo. Febs Letters 2000; 467:37-40.
121. Kiick KL, Tirrell DA. Protein engineering by in vivo incorporation of non-natural amino acids: Control of incorporation of methionine analogues by methionyl-tRNA synthetase. Tetrahedron 2000; 56:9487-9493.
122. Ibba M, Kast P, Hennecke H. Substrate specificity is determined by amino acid binding pocket size in Escherichia coli phenylalanyl-Trna synthetase. Biochem 1994; 33:7107-7112.
123. Clark TD, Ghadiri MR. Supramolecular design by covalent capture—design of a peptide cylinder via hydrogen-bond-promoted intermolecular olefin metathesis. Amer Chem Soc 1995; 117:12364-12365.
124. Doring V, Mootz HD, Nangle LA et al. Enlarging the amino acid set of Escherichia coli by infiltration of the valine coding pathway. Science 2001; 292:501-504.
125. Ibba M, Soll D. Aminoacyl-tRNA synthesis. Ann Rev Biochem 2000; 69:617-650.
126. Liu DR, MaglieryTJ, Schultz PG. Characterization of an 'orthogonal' suppressor tRNA derived from E-coli tRNA(2)(Gln). Chem Biol 1997; 4:685-691.
127. Bradley D, Park JV, Soll L. TRNA2Gln Su+2 mutants that increase amber suppression. J Bacteriology 1981; 145:704-12.
128. Jahn M, Rogers JM, Soll D. Anticodon and acceptor stem nucleotides in transfer Rnagln are major recognition elements for E-coli glutaminyl-transfer RNA synthetase. Nature 1991; 352:258-260.
129. Hayase Y, Jahn M, Rogers MJ et al. Recognition of bases in Escherichia-coli transfer Rna(Gln) by glutaminyl-transfer RNA synthetase—a complete identity set. EMBO J 1992; 11:4159-4165.
130. Englisch-Peters S, Conley J, Plumbridge J et al. Mutant enzymes and transfer RNAs as probes of the glutaminyl-transfer RNA synthetase—transfer RNA Gln interaction. Biochimie 1991; 73:1501-1508.
131. Ibba M, Hong KW, Soll D. Glutaminyl-tRNA synthetase: From genetics to molecular recognition. Genes Cells 1996; 1:421-427.
132. Rath VL, Silvian LF, Beijer B et al. How glutaminyl-tRNA synthetase selects glutamine. Structure 1998; 6:439-449.
133. Freist D, Gauss DH, Ibba M et al. Glutaminyl-tRNA synthetase. Biological Chemistry 1997; 378:1103-1117.

134. Liu DR, Magliery TJ, Pasternak M et al. Engineering a tRNA and aminoacyl-tRNA synthetase for the site-specific incorporation of unnatural amino acids into proteins in vivo. Proc Natl Acad Sci USA 1997; 94:10092-10097.
135. Stemmer WPC. DNA shuffling by random fragmentation and reassembly—in vitro recombination for molecular evolution. Proc Natl Acad Sci USA 1994; 91:10747-10751.
136. Patten PA, Howard RJ, Stemmer WPC. Applications of DNA shuffling to pharmaceuticals and vaccines. Curr Opin Biotech 1997; 8:724-733.
137. Schimmel P, Soll D. When protein engineering confronts the tRNA world. Proc Natl Acad Sci USA 1997; 94:10007-10009.
138. Whelihan EF, Schimmel P. Rescuing an essential enzyme RNA complex with a non-essential appended domain. EMBOJ 1997; 16:2968-2974.
139. Wang CC, Schimmel P. Species barrier to RNA recognition overcome with nonspecific RNA binding domains. Bio Chem 1999; 274:16508-16512.
140. Liu DR, Schultz PG. Progress toward the evolution of an organism with an expanded genetic code. Proc Natl Acad Sci USA 1999; 96:4780-4785.
141. Huang WZ, Petrosino J, Hirsch M et al. Amino Acid Sequence Determinants of Beta-Lactamase Structure and Activity. J Mol Biol 1996; 258:688-703.
142. Ohno S, Yokogawa T, Fujii I et al. Co-expression of yeast amber suppressor tRNA(Tyr) and tyrosyl-tRNA synthetase in Escherichia coli: Possibility to expand the genetic code. J Biochem 1998; 124:1065-1068.
143. Wang L, Magliery TJ, Liu DR et al. A new functional suppressor tRNA/aminoacyl-tRNA synthetase pair for the in vivo incorporation of unnatural amino acids into proteins. Amer Chem Soc 2000; 122:5010-5011.
144. Steer BA, Schimmel P. Major anticodon-binding region missing from an archaebacterial tRNA synthetase. Bio Chem 1999; 274:35601-35606.
145. Steer BA, Schimmel P. Domain-domain communication in a miniature archaebacterial tRNA synthetase. Proc Natl Acad Sci USA 1999; 96:13644-13649.
146. Wang L, Brock A, Herberich B et al. Expanding the genetic code of Escherichia coli. Science 2001; 292:498-500.
147. Wang L, Schultz PG. 2001, manuscript in preparation.
148. Kowal AK, Kohrer C, RajBhandary UL. Twenty-first aminoacyl-tRNA synthetase-suppressor tRNA pairs for possible use in site-specific incorporation of amino acid analogues into proteins in eukaryotes and in eubacteria. Proc Natl Acad Sci USA 2001; 98:2268-2273.
149. Drabkin HJ, Park HJ, Rajbhandary UL. Amber Suppression in Mammalian Cells Dependent Upon Expression of an Escherichia Coli Aminoacyl-Trna Synthetase Gene. Mol Cell Biol 1996; 16:907-913.
150. Doctor BP, Mudd JA. Species specificity of amino acid accepter ribonucleic acid and aminoacyl soluble ribonucleic acid synthetases. Bio Chem 1963; 238:3677-3681.
151. Giege R, Florentz C, Kern D et al. Aspartate Identity of Transfer Rnas. Biochimie 1996; 78:605-623.
152. Martin F. In Universite Louis Pasteur: Strasbourg, France, 1995;pp 186.
153. Pastrnak M, Magliery TJ, Schultz PG. A new orthogonal suppressor tRNA/aminoacyl-tRNA synthetase pair for evolving an organism with an expanded genetic code. Helvetica Chimica Acta 2000; 83:2277-2286.
154. Yarus M. Intrinsic precision of aminoacyl-tRNA synthesis enhanced through parallel systems of ligands. Nature. New Biology 1972; 239:106-8.
155. Sherman JM, Rogers MJ, Soll D. Competition of aminoacyl-transfer RNA synthetases for transfer RNA ensures the accuracy of aminoacylation. Nucleic Acids Res 1992; 20:2847-2852.
156. Sherman JM, Soll D. Aminoacyl-tRNA synthetases optimize both cognate tRNA recognition and discrimination against noncognate tRNAs. Biochem 1996; 35:601-607.
157. Ibba M, Hennecke H. Relaxing the substrate specificity of an aminoacyl-tRNA synthetase allows in vitro and in vivo synthesis of proteins containing unnatural amino acids. Febs Letters 1995; 364:272-275.
158. Behrens C, Nielsen JN, Fan XJ et al. Development of strategies for the site-specific in vivo incorporation of photoreactive amino acids: p-azidophenylalanine, p-acetylphenylalanine and benzofuranylalanine. Tetrahedron 2000; 56:9443-9449.
159. Agou F, Quevillon S, Kerjan P et al. Switching the amino acid specificity of an aminoacyl-tRNA synthetase. Biochem 1998; 37:11309-11314.
160. Hong KW, Ibba M, Soll D. Retracing the evolution of amino acid specificity in glutaminyl-tRNA synthetase. Febs Letters 1998; 434:149-154.

161. Ibba M, Hong KW, Sherman JM et al. Interactions between tRNA identity nucleotides and their recognition sites in glutaminyl-tRNA synthetase determine the cognate amino acid affinity of the enzyme. Proc Natl Acad Sci USA 1996; 93:6953-6958.
162. Hong KW, Ibba M, Weygandduraseviс I et al. Transfer RNA-dependent cognate amino acid recognition by an aminoacyl-tRNA synthetase. EMBO J 1996; 15:1983-1991.
163. Liu JH, Ibba M, Hong KW et al. The terminal adenosine of tRNA(Gln) mediates tRNA-dependent amino acid recognition by glutaminyl-tRNA synthetase. Biochem 1998; 37:9836-9842.
164. Shaw WV, Leslie AGW. Chloramphenicol acetyltransferase. Ann Rev Biophys Chem 1991; 20:363-386.
165. Capone JP, Sedivy JM, Sharp PA et al. Introduction of UAG, UAA, and UGA nonsense mutations at a specific site in the Escherichia coli chloramphenicol acetyltransferase gene: use in measurement of amber, ochre, and opal suppression in mammalian cells. Mol Cell Biol 1986; 6:3059-67.
166. Pastrnak M, Schultz PG. Phage selection for site-specific incorporation of unnatural amino acids into proteins in vivo. Bioorganic Medicinal Chem 2001, in press.
167. Furter R. Expansion of the genetic code: Site-directed p-fluoro-phenylalanine incorporation in Escherichia coli. Protein Science 1998; 7:419-426.
168. Fechter P, Rudinger-Thirion J, Tukalo M et al. Major tyrosine identity determinants in Methanococcus jannaschii and Saccharomyces c erevisiae tRNA(Tyr) conserved but expressed differently. Eur J Biochem 2001; 268:761-767.
169. Chin JW, Martin AB, King DS et al. Addition of a photocrosslinking amino acid to the genetic code of Escherichia coli. Proc Natl Acad Sci USA 2002; 99:11020-11024.
170. Chin JW, Santoro SW, Martin AB et al. Addition of p-azido-L-phenylaianine to the genetic code of Escherichia coli. J Am Chem Soc 2002; 124:9026-9027.
171. Santoro SW, Wang L, Herberich B et al. An efficient system for the evolution of aminoacyl-tRNA synthetase specificity. Nat Biotech 2002; 20:1044-1048.
172. Wang L, Brock A, Schultz PG. Adding L-3-(2-naphthyl)alanine to the genetic code of E. coli. J Am Chem Soc 2002; 124:1836-1837.
173. Zhang ZW, Wang L, Brock A et al. The selective incorporation of alkenes into proteins in Escherichia coli. Angew Chem Int Ed 2002; 41:2840-2842.
174. Chin JW, Cropp TA, Anderson JC et al. An expanded eukaryotic genetic code. Science 2003; 301:964-967.
175. Deiters A, Cropp TA, Mukherji M et al. Adding amino acids with novel reactivity to the genetic code of Saccharomyces cerevisiae. J Am Chem Soc 2003; 125:11782-11783.

INDEX

I

L

M

N

O

P

R

S

T

U

V

MIX
Papier aus verantwortungsvollen Quellen
Paper from responsible sources
FSC® C105338

If you have any concerns about our products,
you can contact us on
ProductSafety@springernature.com

In case Publisher is established outside the EU,
the EU authorized representative is:
Springer Nature Customer Service Center GmbH
Europaplatz 3, 69115 Heidelberg, Germany

Printed by Libri Plureos GmbH
in Hamburg, Germany